T0231097

Topical Drug Delivery Formulations

DRUGS AND THE PHARMACEUTICAL SCIENCES

A Series of Textbooks and Monographs

Edited by

James Swarbrick

School of Pharmacy
University of North Carolina
Chapel Hill, North Carolina

Topical Drug Delivery Formulations

edited by

David W. Osborne
Anton H. Amann
The Upjohn Company
Kalamazoo, Michigan

informa
healthcare

New York   London

First published in 2008 by Informa Healthcare, Telephone House, 69-77 Paul Street, London EC2A 4LQ, UK.

Simultaneously published in the USA by Informa Healthcare, 52 Vanderbilt Avenue, 7th Floor, New York, NY 10017, USA.

Informa Healthcare is a trading division of Informa UK Ltd. Registered Office: 37–41 Mortimer Street, London W1T 3JH, UK. Registered in England and Wales number 1072954.

A CIP record for this book is available from the British Library.

Library of Congress Cataloging-in-Publication Data available on application

ISBN-13: 9780824781835

Orders may be sent to: Informa Healthcare, Sheepen Place, Colchester, Essex CO3 3LP, UK
Telephone: +44 (0)20 7017 5540
Email: CSDhealthcarebooks@informa.com
Website: http://informahealthcarebooks.com/

For corporate sales please contact: CorporateBooksIHC@informa.com
For foreign rights please contact: RightsIHC@informa.com
For reprint permissions please contact: PermissionsIHC@informa.com

Preface

This book is designed to provide the pharmaceutical formulator with the fundamental understanding necessary to prepare efficacious topical drug delivery formulations. In the past the ability to formulate topical creams and ointments has been described as both "art" and "magic." The demanding expectations of topicals include (1) formulations that have both chemical and physical stability for at least two years, (2) formulations that have components both alone and in combination that are nonirritating, nonsensitizing, and nonallergenic, (3) formulations that are at least cosmetically acceptable and preferably cosmetically elegant, and (4) formulations that are efficacious because of their ability to release and/or deliver therapeutic levels of a drug. Note that these expectations are for an inherently multiple-phase system (usually an emulsion or suspension) that contains components having a wide range of polarities and physical properties.

With such demanding expectations, it is of little surprise that most scientists, when challenged with the assignment of formulating a new topical drug, select a traditional formulation taken from one of the many formulatories available. This tactic will ultimately provide a suitable formulation (usually after appropriate, often frustrating, adaptations or "fine-tuning"). However, the realization that drugs can be delivered transdermally to systemically therapeutic levels (as evidenced by commercially available nitroglycerin, and scopolamine patch systems) and that changes in vehicles can significantly change efficacy has permanently altered the way pharmaceutical scientists view topicals. No longer can we consider creams and ointments as inert drug carriers. Rather, we as formulators are required to design vehicles that target follicles or other specific skin regions, vehicles that optimize drug—skin and prodrug—skin interactions, or vehicles that release drugs for sustained periods of time. Traditional formulations can seldom meet these additional demands.

Therefore, a symposium was organized as part of the 61st Colloid and Surface Science Symposium held in Ann Arbor, Michigan, June 24, 1987. The symposium, entitled Topical Drug Delivery Formulations, was designed to bring together academic and industrial scientists who had (1) developed techniques that aid in selection of an optimized formulation, (2) futhered current understanding of how vehicle components and drugs interact with the skin, or (3) investigated non-traditional drug delivery formulations. By using this symposium as a core, additional material was included that would provide the formulator with state-of-the-art understanding in three areas: (1) how to design stability and in vitro testing to obtain optimized formulations before conducting clinical investigations; (2) the role of the skin as an immune response organ and barrier to percutaneous absorption, and how this barrier can be compromised using enhancers; and (3) how nontraditional topical drug delivery formulations can be utilized as pharmaceuticals. By combining the latest insights with established principles, we sincerely hope that this book will be a useful tool in formulating the next generation of highly effective nontraditional topical drug delivery formulations.

David W. Osborne
Anton H. Amann

Contents

Part II: DRUG FORMULATION SELECTION AND TESTING FOR TOPICAL USE

Contributors

William J. Addicks* University of Michigan College of Pharmacy, Ann Arbor, Michigan

Eugene R. Cooper Sterling Drug, Inc., Malverne, Pennsylvania

Rane L. Curl University of Michigan College of Engineering, Ann Arbor, Michigan

Gary R. Dukes The Upjohn Company, Kalamazoo, Michigan

Peter M. Elias Veterans Administration Medical Center, University of California School of Medicine, San Francisco, California

Gordon L. Flynn Cygnus Research Corporation, Redwood City, California

Stig E. Friberg Clarkson University, Potsdam, New York

Douglas A. Hatzenbuhler The Upjohn Company, Kalamazoo, Michigan

Martin Katz Advanced Polymer Systems, Inc., Redwood City, California

Ibrahim H. Kayali Clarkson University, Potsdam, New York

David H. Lynch Immunex Corporation, Seattle, Washington

Howard I. Maibach University of California School of Medicine, San Francisco, California

---

Current affiliation: E. I. du Pont de Nemours and Company, Wilmington, Delaware

Marion Margosiak   Clarkson University, Potsdam, New York

Sergio Nacht   Advanced Polymer Systems, Inc., Redwood City, California

Kilian J. O'Neill   University College Dublin, Dublin, Ireland

David W. Osborne   The Upjohn Company, Kalamazoo, Michigan

Dinesh C. Patel   Theratech, Inc., Salt Lake City, Utah

Lorraine E. Pena   The Upjohn Company, Kalamazoo, Michigan

Lee K. Roberts   University of Utah School of Medicine, Salt Lake City, Utah

Kenneth B. Sloan   University of Florida, Gainesville, Florida

Michael S. Starch   Dow Corning Corporation, Midland, Michigan

Regina Tallon   University College Dublin, Dublin, Ireland

Paul S. Uster   Liposome Technology, Inc., Menlo Park, California

F. L. Vaughan   University of Michigan School of Public Health, Ann Arbor, Michigan

Anthony J. I. Ward*   University College Dublin, Dublin, Ireland

Norman D. Weiner   University of Michigan College of Pharmacy, Ann Arbor, Michigan

Ronald C. Wester   University of California School of Medicine, San Francisco, California

---

Current affiliation: Clarkson University, Potsdam, New York

Topical Drug Delivery Formulations

I

PRINCIPLES OF SKIN
AND FORMULATION INTERACTIONS

1

Practical Considerations for Topical Drug Formulations With and Without Enhancers

EUGENE R. COOPER   *Sterling Drug, Inc., Malverne, Pennsylvania*

DINESH C. PATEL   *Theratech, Inc., Salt Lake City, Utah*

I. INTRODUCTION

The purpose of topical dosage forms is to conveniently deliver drugs across a localized area of the skin. To develop an ideal dosage form one must take into account the flux of drug across skin, retention of the dosage form on the skin's surface, the reservoir capacity of the dosage form, and the patients' acceptability of the formulation. The problem of formulating a drug is complex because of the wide diversity of drug solubility in vehicle components and the vast range in cutaneous fluxes (six orders of magnitude).

The objective of this chapter is to simplify and make more efficient the development of optimum topical dosage forms. There is always a certain amount of empiricism to this process, but a strategy based upon fundamental principles of thermodynamics and diffusion can greatly reduce the time required to develop good topical dosage forms. Another topic to be addressed in this chapter is how to compare two different topical dosage forms. In some organizations, physical and chemical stabilities, plus the formulator's opinion of aesthetics, are all that is required of a formulation. The science and technology available today make this approach obsolete, and one must have a quantitative measure of delivery to develop a competitive topical dosage form.

II. VEHICLE DESIGN

A. General Considerations

The purpose of dissolving a drug in a solvent or mixture of solvents is to facilitate the transport of drug to the surface of the

skin. Most drugs are crystalline, and the mere rubbing of small
crystals on the surface of skin will result in very poor transfer to
the skin. Delivery from volatile solvents, such as ethanol and acetone, will transfer small portions of drugs into the outer layers of
the skin, and the flux will quickly drop off. In fact, any transport
study conducted with such solvents, will more than likely measure
the dissolution rate of the drug, rather than the transport properties of skin.

B. Thermodynamic Factors

The driving force for transport is the chemical potential gradient,
and within a single phase, this reduces to the concentration gradient. To create this gradient within the skin, one normally dissolves
the drug in a solvent or vehicle and establishes a certain concentration of drug in the outer surface of the skin. Different vehicles
can provide different concentrations of drug at this interface and,
thus, the driving force and, hence, the flux will be a function of
the vehicle. In general, one does not know the concentration of
drug in the outer surface of the skin, but there is a simple thermodynamic relationship to compare the concentration of drug in the
skin for different vehicles.

The *chemical potential* or *activity* is the continuous variable
across interfaces (1) and, thus, one has the following equation at
the skin−vehicle boundary:

$$a_V = a_S \tag{1}$$

where a_V is the activity of the drug in the vehicle, and a_S is the
activity of the drug in the skin (assumed to be a lipid medium for
most situations). Thus, to produce equivalent activity in the skin
with different vehicles, one only has to ensure equal activities in
the vehicles. The activities are usually written as

$$a = \gamma C \tag{2}$$

where γ is the activity coefficient and C is the concentration. Thus,
for equivalent activity for vehicle 1 and 2, one has

$$\gamma_1 C_1 = \gamma_2 C_2$$

To estimate the concentration in vehicle 1 to give equal activity to
vehicle 2, one has

$$C_1 = \frac{\gamma_2}{\gamma_1} C_2 \qquad [3]$$

The simplest way to estimate γ_2 and γ_1 is to measure the solubility, where S is the solubility.

$$\frac{\gamma_2}{\gamma_1} = \frac{S_1}{S_2} \qquad [4]$$

Thus, Eq. 3 becomes

$$C_1 = \frac{S_1}{S_2} C_2 \qquad [5]$$

Another way to state Eq. 5 is that equal fractions of the solubility will provide for equal activities in various vehicles. In most applications one wishes to optimize the flux in a formulation, and, the simplest way to do this is to work with saturated vehicles. An example of equal flux at saturation for markedly different solubilities is shown in Figure 1, and further use of this concept will be dealt with later by way of a practical example. Despite that the solubility of salicylic acid is 100 times greater in propylene glycol than in water, the fluxes across human epidermis are nearly the same.

C. Finite Doses and Vehicle Evaporation

Except for enclosed systems, such as transdermal patches, one is limited to a very thin film. Product application of a cream is usually limited to a few milligrams per square centimeter, because application of more material will result in a sticky film that will easily be removed by clothing and such. The consequences of thin films can be easily seen by considering the solution of the appropriate diffusion equations. If a finite dose of volume V is applied to surface of area A and it is assumed that the vehicle does not penetrate or evaporate, the equations for the flux J, are

$$V \frac{dC_V}{dt} = AD \left(\frac{\partial C}{\partial x} \right)_{x = 0} \qquad [6]$$

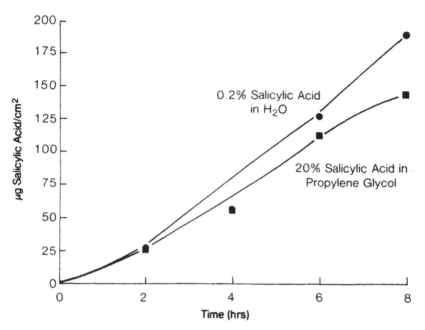

Figure 1.1 Penetration of salicylic acid across human epidermis in vitro from saturated solutions.

$$\frac{\partial C}{\partial t} = D \frac{\partial^2 C}{\partial x^2} \tag{7}$$

$$C(x, 0) = 0, \; C(h, t) = 0 \tag{8}$$

$$C(0, t) = R \, C_V(t) \tag{9}$$

where h is the thickness of the stratum corneum, D is the diffusion coefficient, C_V is the concentration in the vehicle, R is the partition coefficient between skin and the vehicle, C is the concentration in the skin. These equations reflect the assumptions that diffusion in the vehicle is much faster than in the skin, and that sink conditions exist below the skin. The solution of Eqs. 6 through 9 can be obtained from the existing solution for heat conduction (2) and is

$$J = -D \left(\frac{\partial C}{\partial x} \right)_{x=h} = 2J_s \sum_{n=1}^{\infty} \frac{\alpha_n^2 \exp[-\alpha_n^2 \tau/6]}{\cos \alpha_n [\beta^2 + \beta + \alpha_n^2]} \tag{10}$$

where

$$\beta = \frac{Rh}{h_v}$$ [11]

α_n is a root of $\alpha \tan \alpha = \beta$ [12]

$$\tau = \frac{6D}{h^2} t$$ [13]

$$J_s = \frac{D\,C(0,\,0)}{h}$$

and h_v is the thickness of the vehicle. The dimensionless parameter, τ, is used for a general solution in terms of number of lag
times. Thus, a graph of flux versus τ can be used for any diffusion coefficient and membrane thickness.

A plot of the ratio J/J_s versus τ is given in Figure 2 for several values of β. For $\beta < 0.1$ the flux is near the steady-state
value for many lag times. For $\beta \simeq 1$, however, the maximum flux
is only 60% of the steady-state value, and it decays rapidly after
two lag times. For example, consider the application of 5 mg/cm^2
of cream containing 20% oil and 1% salicylic acid. After evaporation
one is left with an oil film about 10 μm thick (corresponding to 1
mg/cm^2). Here, $R \simeq 1$, and β is unity if we use $h \simeq 10$ μm. For
a molecule like salicylic acid that penetrates well (short lag time),
one cannot reach steady-state flux, nor maintain it, without a thicker film or a solvent in which salicylic is more soluble (e.g., polyethylene glycol).

This type of behavior is observed in vivo (monitoring urinary
excretion) for benzoic acid penetration from a lipophilic matrix (3).
The mathematical description of the in vivo situation is similar to
that described here, but it is more involved because of the addition
of whole-body pharmacokinetics (4).

A simple in vitro test to determine the effect of film thickness is
useful before conducting an in vivo study so that one can know the
film thickness required to obtain steady state. Evaporation and
penetration of the vehicle components that are not extremely volatile
will complicate the analysis but will serve to keep the concentration
higher and, thus, maintain the flux more in a plateau region, although the decay rate will be faster when the reservoir is depleted.

D. Penetration Enhancers

The performance of a topical dosage form is linked to the flux of
drug across the skin, unless one has reached beyond the plateau on

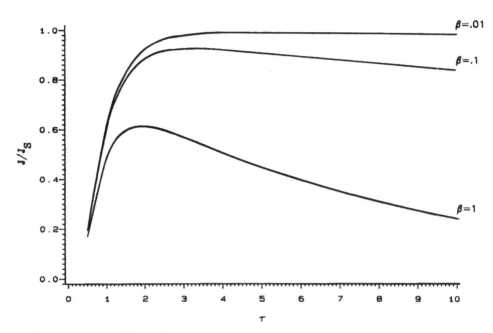

Figure 1.2 Effect of thickness on cutaneous flux.

the dose—response curve. Such response is easily observed in the
effect of topically applied nonsteroidal anti-inflammatory agents on
guinea pigs irradiated with ultraviolet light. The blanching of the
erythema can reach a maximum, after which higher concentrations
provide no further benefit. Human skin is a much greater barrier
and, thus, enhanced penetration is of greater value for human skin
applications.
    Penetration enhancers can be of great value, and formulations
including enhancers should generally be prepared as an option to be
evaluated in the in vivo system. Enhancers can be categorized conveniently in terms of the type of drug to be delivered. For polar
molecules, surface active agents (5,6) are very effective enhancers.
For lipophilic molecules, dimethyl sulfoxide (7,8) is the classic enhancer, but other agents such as laurocapram (Azone; 9) and polar
lipids (10) are very effective. A simple model for viewing the effects of enhancers on skin is to regard the barrier as consisting of
two parallel pathways. The polar pathway is thought to be hydrated
protein that is quite sensitive to conformational changes induced by
surfactants, heat, and the like. A marked contrast between the
effects of surfactants on polar versus nonpolar molecules serves to
illustrate this point (6). Enhancers for the nonpolar pathway are
thought to fluidize the lipids (11). This concept is quite reasonable

when considering diffusion processes in a continuous medium. The formulation of these enhancers is not simple because they often interact with emollients and such, that are put into a cream and can be rendered ineffective. Because of potential irritation problems with enhancers, they are most practical for short-term use. For prolonged application, more attention will have to be placed on regulating cutaneous irritation.

III. EVALUATION OF VEHICLE PERFORMANCE

A. In Vitro Evaluation

As described in earlier sections one can use thermodynamics and elementary diffusion theory as guides for optimizing the flux. In the final analysis, however, one should measure the penetration of the drug from the actual vehicles. A variety of experimental methods (described elsewhere in this book) are available for determining cutaneous flux in vitro, and they are quite simple to use. These in vitro techniques are an invaluable guide to the formulation of topical dosage forms and should be a standard laboratory tool. Without a measurement of penetration, there is virtually no way to compare different dosage forms. When employing enhancers, it is absolutely necessary to make these measurements.

We prefer to use human skin, when possible, although animal skin is acceptable, to compare different formulations. Our experience has been that for comparing vehicles there is good correlation between human and animal skin, but sometimes animal skin will give a false-positive result when studying skin penetration enhancers. Although there are a number of diffusion apparatus available, we prefer the one shown in Figure 3 (12) because it is relatively inexpensive and occupies very little laboratory space.

Typical data from in vitro experiments are depicted in Figure 4 for penetration from large reservoir systems in which the effect of an enhancer is shown. Here, the effect of a lipid fluidizing agent is shown for the enhancement of lipophilic drugs (10). The lag time is an important factor to minimize for treatment of acute symptoms such as pruritis and burns. For treatment of less acute symptoms, this factor is of less importance. Although for transport across homogeneous membranes the lag time is only a function of the diffusion coefficient and not the partition coefficient, it is a function of partition coefficient for heterogeneous membranes (13). The lag time may not be well defined in the presence of enhancers, especially if the membrane barrier properties change over a reasonable time period. The enhancer, however, should shorten the onset time for pharmacologic response.

For drugs that penetrate rapidly, one can use a vehicle with a lower partition coefficient to sustain the delivery, rather than have

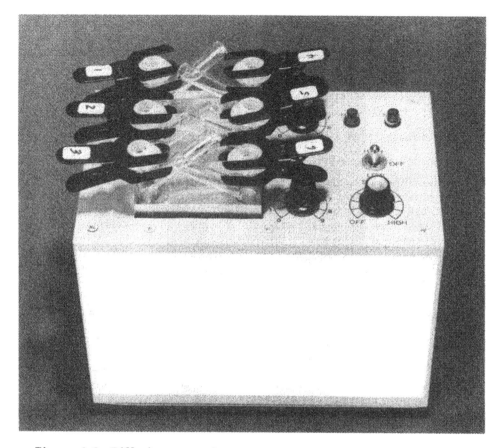

Figure 1.3 Diffusion apparatus.

pulsed delivery. Again the in vitro method is required to determine the actual fluxes, although the theory can be used as a guide. Transport across diseased skin can potentially be quite different (14), and one must consider this possibility, particularly if the skin is thickened or broken.

B. In Vivo Evaluation

The most quantitative in vivo evaluations of topically applied drugs have been done with skin grading, in which blanching of normal skin by steroids or the blanching of erythema by nonsteroidal anti-inflammatory agents is measured. In these models, the onset of re-

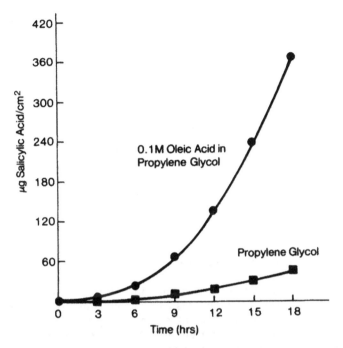

Figure 1.4 Effect of oleic acid on salicylic acid (1%) in vitro penetration across human epidermis.

sponse is related to lag time, and shorter lag times are viewed as good. For disease states, it is not clear whether a large, pulsed delivery is better, or a slow, prolonged delivery is preferred. Accurate knowledge of vehicle delivery rates correlated with clinical results could lead to new vehicles with very different delivery profiles. These systems, in turn, could be evaluated clinically to determine the best delivery rate. The methods and skill certainly exist to answer some of these key questions, which undoubtedly will be answered as the dosage forms become more controllable for delivery rate.

Topical treatment of skin cancer, warts, baldness, and so on, may be possible, not only because of new drugs, but also because better vehicles can be designed to enhance delivery. The skin is an accessible tissue of large surface area and should be a viable port of entry for drugs. Vehicles with superior delivery will undoubtedly play a substantial role in increasing the usefulness of this port of entry.

Table 1.1 Vehicle Composition

Ingredient	Vehicle 1 (wt%)	Vehicle 2 (wt%)
Sorbitan stearate	1.2	1.2
Polysorbate 60	3.6	3.6
Propylene glycol	11.9	0
Caprylic/capric triglyceride	0	11.9
Water	83.3	83.3

IV. FORMULATION EXAMPLE: DESONIDE

It is often desirable to change key ingredients in a formulation, and this substitution can occur without any loss in vehicle performance. As an example consider the substitution of caprylic/capric triglyceride for propylene glycol in a formulation for the steroid desonide. The reason for such a replacement is to move from a hydrophilic base, such as propylene glycol, to a lipophilic base. It is also expected that, under occlusion, the triglyceride system (15,16) would be less irritating than the propylene glycol vehicle.

The experimental vehicles saturated with desonide are listed in Table 1. According to thermodynamics, the flux should be the same from both systems, unless there is an effect on the barrier properties of the skin. The fluxes across hairless mouse skin and the vehicle solubilities (17) are given in Table 2, from which it is observed that the fluxes are indeed the same (within experimental variations), and the solubilities are only slightly different. When

Table 1.2 Desonide Solubility and Flux Across Hairless Mouse Skin

Vehicle		Solubility (mg/ml)	Flux (mg/hr $\times 10^4$)
1	Propylene glycol	0.40 ± 0.02	8.3 ± 4
2	Caprylic/capric triglyceride	0.53 ± 0.03	13 ± 8

these two systems were compared in the vasoconstrictor assay (18) there was no difference in biological response between them. Under closed-patch irritation studies, however, the propylene glycol-based vehicle was more irritating, as expected (17).

It may not always be possible to carry each different vehicle through all the steps just described. One part that can always be conducted is the saturation step. If the solubilities are tenfold or greater in difference, one might encounter reservoir limitations, as discussed earlier, for their films. Unless there are toxicity concerns, one should seek to find a very soluble vehicle to provide a long-lasting reservoir. The in vitro penetration step may have to be omitted, but it is highly recommended that it be included because, for example, solvents such as polyethylene glycol (10,19—21) or glycerol have a tendency to retard penetration compared with propylene glycol.

REFERENCES

1. A. Katchalsky, P. F. Curran, in *Nonequilibrium Thermodynamics in Biophysics*, Harvard University Press, Cambridge, Mass., p. 113, 1965.
2. H. S. Carslaw, and J. C. Jaeger, *Conduction of Heat in Solids*. Oxford University Press, Oxford, Eng., p. 128, 1959.
3. T. J. Franz, personal communication.
4. E. R. Cooper, and B. Berner, *J. Pharm. Sci.*, 74:1100, 1985.
5. F. R. Bettley, and E. Donoghue, *Nature*, 17:185, 1960.
6. E. R. Cooper, in *Solution Behavior of Surfactants: Theoretical and Applied Aspects* (K. L. Mittal, and E. J. Fendler, eds.). Plenum Press, New York, p. 1505, 1982.
7. S. K. Chandrasekaren, P. S. Campbell, and A. S. Michaels, *AICHE J.*, 23:810, 1977.
8. S. K. Chandrasekaren, *Drug Dev. Ind. Pharm.*, 9:627, 1983.
9. R. B. Stoughton, *Arch. Dermatol.*, 118:474, 1982.
10. E. R. Cooper, *J. Pharm. Sci.*, 73:1153, 1984.
11. E. R. Cooper, in *Percutaneous Absorption: Principles and Practices* (R. Bronaugh, and H. I. Maibach, eds.), Marcel Dekker, New York, 1985.
12. E. R. Cooper, and E. W. Merrit, *J. Controlled Release*, 1:161, 1984.
13. R. M. Barrer, in *Diffusion in Polymers* (J. Crank and G. S. Park, eds.). Academic Press, New York, p. 165, 1968.
14. B. W. Barry, *Dermatological Formulations Percutaneous Absorption*. Marcel Dekker, New York, p. 132, 1983.
15. Technical Bulletin, Kay Fries, Inc., Rockleigh, New Jersey.

16. Technical Bulletin, Henkel, Hoboken, New Jersey.
17. D. Patel, D. Welsh, and M. Baker, *J. Soc. Cosmet. Chem.* (in press).
18. A. W. McKenzie, R. B. Stoughton, *Arch. Dermatol.*, *86*:608, 1962.
19. J. A. Faucher, E. D. Goddard, and R. Kulkarni, *J. Am. Oil Chem. Soc.*, *56*:776, 1979.
20. A. A. Belmonte, and W. Tsai, *J. Pharm. Sci.*, *67*:517, 1978.
21. J. L. Zatz, *CTFA Cosmet. J.*, *15*:6, 1983.

2

The Importance of Epidermal Lipids for the Stratum Corneum Barrier

PETER M. ELIAS   *Veterans Administration Medical Center, and University of California School of Medicine, San Francisco, California*

I. EVOLUTION OF CURRENT CONCEPTS OF STRATUM CORNEUM STRUCTURE

The old image of mammalian stratum corneum, based on routine histological preparations, is one of a loosely bound layer in various stages of disorganization and shedding (Table 1). This misleading image, refuted by physicochemical (1), frozen-section (2–4), and freeze-fracture (5–7) studies, led to the long-held view that the barrier resides at the stratum granulosum (SG)–stratum corneum (SC) interface (8). New appreciation for the integrity of the SC initially led to the concept that the full thickness of this layer is functionally competent (1,9), a view that analogized the stratum corneum to a homogeneous film ("plastic wrap" hypothesis; 9), a view supported by x-ray diffraction studies that demonstrated highly ordered lipid structures in the SC (10). The homogeneous film analogy views percutaneous transport as occurring transcellularly, without regard to membrane, intercellular, or intracellular compartments (9,11), and assumes that the SC is devitalized (i.e., metabolically inert, with permeability phenomena regulated solely as a passive membrane; 9).

A. The Two-Compartment Model

Studies outlining the evidence supporting a two-compartment model are outlined in Table 2. In 1968, Middleton first reported certain straightforward, yet elegant, studies that tended to refute the homogeneous film concept (11). Whereas Irvin Blank had shown years

Table 2.1   Evolution in Understanding of the Stratum Corneum

1. Disorganized, nonfunctional layer in various stages of shedding

2. Homogeneous film—the "plastic wrap" model

3. Lipid—protein compartmentalization—the "two-compartment" model

4. Metabolically active tissue with ongoing modulations in structure
    and composition

earlier that organic solvent extraction destroyed the water-holding
capacity of SC (12), Middleton showed that pulverization was as effective as solvent extraction, thereby providing evidence that SC
lipids might be organized as osmotically active membranes within the
SC.   Shortly thereafter, the existence of separate lipophilic versus
hydrophilic pathways of percutaneous absorption was suggested from
physicochemical studies (13).

The destruction of stratum corneum during processing for routine histological and ultrastructural preparations obscured further
advances until the early 1970s, when the cells of the SC were shown
by frozen sectioning to comprise tightly arrayed, polyhedral structures in vertical, interlocking columns (2—4).   The initial awareness
of lipid—protein segregation to specific tissue compartments came
with freeze-fracture replication, which showed, for the first time,

Table 2.2   Lines of Evidence for Lipid—Protein Compartmentalization
in the Stratum Corneum

Pulverization destroys the water-holding capacity of SC (11).

Hydrophilic versus lipophilic substances cross separate SC pathways (13).

Freeze-fracture reveals lipid lamellae in SC interstices (5—7).

Frozen sections display neutral lipids in SC interstices (6).

SC can be dispersed into individual cells with organic solvents (14,
15).

Isolated SC membrane sandwiches account for most SC lipids (16).

X-ray diffraction shows ordered lipids in isolated SC membranes (17).

Catabolic enzymes colocalize with lipids in SC interstices (18—20).

the presence of multiple, broad lamellations in the interstices of several types of mammalian keratinizing epithelia (5—7). Several other lines of indirect evidence also supported such structural heterogeneity, including lipid histochemistry of frozen sections (6), as well as the ability of SC to be dispersed by certain organic solvents (14,15). Definitive evidence for the compartmentalization of lipids came with the isolation of SC membrane "sandwiches," containing trapped intercellular lipids. These preparations (16):(1) comprised about 50% lipid by weight, and accounted for over 80% of SC lipid; (2) displayed the same lipid profile as whole SC; (3) contained the same broad lamellae on freeze-fracture as are present in the interstices of whole SC; and (4) generated the same ordered x-ray diffraction pattern previously ascribed to the "interfilamentous lipid matrix" (17). More recently, the colocalization of lipid catabolic enzymes to SC membrane domains, both by ultrastructural cytochemistry and by enzyme biochemistry, can be considered further evidence for the structural heterogeneity of mammalian SC (18—20).

B. Localization of the Barrier

The localization of the barrier continued to be debated: Is the entire SC functionally competent, or does the principal barrier reside in the lower layers? Tracer perfusion studies, as early as 1969, showed that water-soluble molecules, injected into the dermis, do not reach the SC (21,22). Outward percolation halted in the outer SG, at intercellular sites engorged with discharged lamellar body contents (5). These studies, although admittedly employing tracers considerably larger than the water molecule itself, pointed to the presence of a barrier in the outer SG. Direct evidence of the barrier capabilities of different layers of the SC came with the recent isolation of intact sheets of porcine stratum compactum, after prior enzymatic stripping of the stratum disjunctum (23). Whereas these studies did not address the capacity of the stratum disjunctum to contribute to barrier function, they did demonstrate the functional integrity of the stratum compactum.

II. EPIDERMAL DIFFERENTIATION

The ultimate goal of epidermal differentiation is the production of the stratum corneum, a process more correctly called cornification rather than keratinization.

A. Major Structural Components

Modern research in epidermal cellular and molecular biology now recognizes four distinct cellular events that occur during the process of cornification, including *keratinization*, the synthesis of the principal fibrous proteins of the keratinocyte; *keratohyalin* deposition, associated with synthesis of "histidine-rich protein," stratum corneum basic protein, or filaggrin; formation of a highly cross-linked, *insoluble peripheral envelope* of the corneocyte, composed of two or more precursor proteins, including involucrin and keratolinin; and the generation of neutral lipid-enriched intercellular domains, resulting from the secretion of distinctive structures termed *lamellar bodies* (membrane-coating granules, Odland bodies, keratinosomes, cementsomes; 24).

Although the precise role of these components in protective function remains to be discovered, the insoluble stratum corneum envelope appears to provide a rigid structural ectoskeleton for the cornified cells, a scaffold for insertion of keratin filaments, and a highly resistant barrier to external chemical assault. Additionally, in conjunction with extracellular lipids, the peripheral envelope may selectively regulate the permeability of the cornified cell to water, hydrophobic electrolytes, and nonelectrolytes (11). Although both keratin filaments and keratohyalin-derived proteins are major epidermal differentiation products, their function remains conjectural. Clearly, the filament—matrix complex in the cornified cells imparts structural and chemical integrity, acts as a filter to incident ultraviolet radiation, and acts as an absorptive "sponge" for water and other small hydrophobic molecules. Whether the cornified cell interior also represents a potential pathway or reservoir for substances in transit across the stratum corneum is less clear.

B. Lamellar Bodies

The lamellar body, a 0.2 to 0.3 μm diameter, ovoid, secretory organelle, which is considered the central actor in the formation of the intercellular "mortar," is synthesized primarily within the spinous cell and then displaced to the apex and periphery of the granular cell (24). In response to an unknown signal, it fuses with the plasma membrane, secreting its contents into the intercellular spaces, thereby generating an expanded intercellular compartment that constitutes from 10% to 40% of the total volume of this tissue (25). Thus, secretion is one of two cellular events associated with lamellar body exocytosis; fusion also occurs, and the "splicing" of additional organelle membrane into the plasma membrane may contribute a large reservoir of surface area that could explain the stratum corneum's remarkable water-holding capacity (25).

Lamellar bodies appear to contain three types of materials: first, sugars, in the form of glycosphingolipids and possibly glycoproteins; second, free sterols and phospholipids; and third, a selective array of hydrolytic enzymes possibly charged with degrading

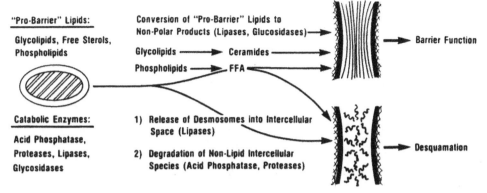

Figure 2.1 Model that summarizes available data about dynamic changes in the stratum corneum interstices that result from the secretion of lamellar body lipids and hydrolytic enzymes.

intercellular materials (26,27; Fig. 1). After outward migration into the stratum corneum, dramatic changes occur in both the morphologic appearance and histochemical reactions of the secreted material: in thin sections and in freeze-fracture replicas, large planar membranes gradually replace the short disks of the initially secreted material (5—7,20). A family of lipid catabolic enzymes, including steroid sulfatase, sphingomyelinase, phospholipase A, acid lipase, and possibly, glycosidases contribute to the ultimate degradation of residual polar lipids (glycosphingolipids, phospholipids, and cholesterol sulfate), leading to the formation of these broad membrane bilayers (18—20; see Fig. 1).

III. COMPOSITION OF STRATUM CORNEUM LIPIDS

Despite the 4-decade-old observation that stratum corneum lipids are generally nonpolar and enriched in cholesterol, the full panoply of mammalian stratum corneum lipids has only recently been unraveled. The stratum corneum is virtually devoid of phospholipids and is selectively enriched in ceramides, free sterols, and free fatty acids, with smaller quantities of glycolipids, sterol esters, triglycerides, cholesterol sulfate, and hydrocarbons present as well (28—30; Table 3). Yet, there is a gradient within the stratum corneum itself: whereas the stratum compactum still contains demonstrable levels of phospholipids, glycosphingolipids, and cholesterol sulfate, only the last persists into the stratum disjunctum (20; Fig. 2). This change

Table 2.3 Composition of Mammalian Epidermal Lipids

Lipid type	Living layers (%)	Stratum corneum (%)
Phospholipids	40	Trace
Sphingolipids	10	35
Cholesterol	15	20
Triglycerides	25	Trace
Fatty acids	5	25
Other	5	10
Totals:	100	100

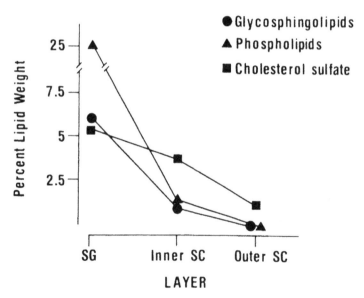

Figure 2.2 Changes in lipid composition that occur during stratum corneum transit. Note that small amounts of phospholipids (diamonds), glycolipids (circles), and cholesterol sulfate (squares) remain in the inner stratum corneum (*SC*) (stratum compactum), whereas only cholesterol sulfate persists in the outer SC. *SG*, stratum granulosum.

in composition presumably reflects ongoing metabolic activity, further dispelling the notion of the stratum corneum as an inert tissue (20; see Table 1). Despite the paucity of phospholipids, glycosphinogolipids, and cholesterol sulfate, these constituents apparently can arrange themselves into membrane bilayers, possibly by exploiting the amphipathic properties of sphingolipids (25,29—31), which apparently are capable of extensive hydrogen bonding. Yet, the hydrophobic, long-chain bases (31,32), the long-chain, fully saturated fatty acids (28,29,31,31; Table 4), and the enrichment of sphingolipids in linoleic acid (31,33—35), all make sphingolipids particularly good candidates for epidermal water-proofing (Table 5). Hence, sphingolipids are bipolar, possessing both a relatively hydrophilic terminus, capable of intermolecular bonding that can form ordered membrane structures, and highly hydrophobic moieties that could be highly water-repellent. Still unresolved is the fate of these lamellations during the first stages of shedding: Is membrane bilayer break-up a prerequisite for shedding? Do further changes in composition or the physicochemical properties of these bilayers lead to desquamation?

IV.   LIPID BIOSYNTHETIC GRADIENTS
IN THE EPIDERMIS

Certain striking features of the modulations that occur in lipid composition during epidermal differentiation suggest that the nonpolar mixture that ultimately resides in the stratum corneum is important for barrier function. Yet, a lipid analytic approach provides only indirect evidence about the function of stratum corneum lipids; furthermore, it can not define the role, if any, of neutral lipids, particularly free fatty acids, free sterols, and the smaller quantities of alkanes, sterol esters, triglycerides, and cholesterol sulfate in barrier function.

In the early 1980s, we noted that the skin of both rodents and primates synthesized abundant cholesterol and other nonsaponifiable lipids (36,37). In fact, the synthetic activity of the skin rivalled that of the liver and gastrointestinal tract, the two major putative sites of sterologenesis (38). Moreover, in contrast with hepatic and gastrointestinal synthesis, cutaneous sterologenesis was not influenced by circulating sterol levels (38), suggesting an unusual degree of independence from systemic regulation. To determine whether or not barrier function can influence epidermal lipid synthesis, we have employed a functional, rather than a purely analytical, approach. The basic strategy is first, to perturb epidermal barrier function, typically with an organic solvent such as acetone

Table 2.4 Fatty Acid Composition of Lipids from Human Abdomen Stratum Corneum (mol%)

C no.	Free fatty acids	Sterol/wax esters	Ceramides
14:0	3.8		
16:0	36.8	20.0	7.7
16:1	3.6	15.9	
18:0	9.9	5.8	4.8
18:1	33.1	49.4	6.3
18:2	12.5	6.6	14.0
20:0	0.3		
20:1	Trace		
20:2			
20:3			
20:4			
22:0	Trace		
22:1			
22:0		0.9	50.8
24:1		1.6	
26:0			10.5
Total	100.0	100.2	100.0

Source: Modified from Ref. 28.

Table 2.5 Evidence for Role of Sphingolipids in Barrier

Carriers of extremely long-chain, saturated fatty acids

Replacement of long-chain by short-chain fatty acids in marine mammals

Principal repositories of linoleic acid (essential fatty acid deficiency)

Removal required to completely break the barrier

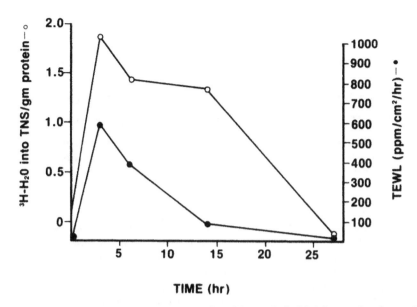

Figure 2.3 The time course of epidermal lipid biosynthesis and barrier function after treatment with acetone. Note that transepidermal water loss (*TEWL*), total nonsaponifiable lipid biosynthesis (*TNS*), and cholesterol (C) synthesis exhibit a parallel return toward normal over 24 hr (from Ref. 39, reprinted with permission).

and, then, to correlate barrier status with both an assessment of lipid replenishment in oil red O- and nile-red-stained frozen sections, and with rates of lipid biosynthesis (from tritiated H_2O) in the same samples. These results have shown (1) that the epidermis is a major site of sterol synthesis, accounting for about 30% of total cutaneous sterologenesis (38); and (2) that both epidermal sterol and fatty acid synthesis are stimulated by perturbation of the permeability barrier (39,40). Moreover, such stimulation is localized to treated sites, is limited to the epidermis (39—41), corrects as barrier function returns to normal (39; Fig. 3), and correlates, first, with removal and, subsequently, with repletion of lipids in the stratum corneum (39,41). (3) Epidermal sterologenesis is stimulated in essential fatty acid deficiency in relation to the defect in barrier function, (i.e., lipogenesis rates normalize with occlusion despite persistence of the underlying deficiency state; 42). (4) Synthesis is normalized when an impermeable membrane is applied to the perturbed skin (39—41). (5) Finally, the relationship of epidermal lipogenesis to barrier function is underscored further by the lack of

Table 2.6   Evidence That Barrier Function (Water Loss) Regulates Epidermal Lipogenesis

---

Dietary or solvent-induced barrier disruption stimulates epidermal lipogenesis.

Extent of lipid biosynthetic rates parallels severity of barrier defect.

Normalization of lipid biosynthetic rates parallels barrier recovery.

Occlusion with impermeable, but not vapor-permeable membranes after barrier disruption blocks acceleration of lipid biosynthesis.

---

modulations in cutaneous sterologenesis in response to either vitamin D deficiency or excess, despite the known importance of the cutaneous free sterol, 7-dehydrocholesterol, for vitamin D synthesis (43).

In more recent studies, the role of the water molecule itself as the regulatory signal has been explored in greater detail (41; Table 6). Whereas occlusive membranes blunted the expected burst in lipid biosynthesis that follows barrier disruption, application of semipermeable membranes did not block synthesis, retard the rate of return of normal barrier function, nor impede the return of stainable lipid in the stratum corneum. These studies lend strong support to the concept that the rate of water loss itself may be the regulatory signal for epidermal lipogenesis.

Although the foregoing body of evidence demonstrates the capacity of the epidermis to synthesize lipids in response to barrier requirements, they leave unanswered the sites of synthesis in the epidermis and the relative importance of specific SC lipids for barrier function. The rapid rate of return of barrier function to normal [i.e., by 15−20 hr, even after exhaustive delipidization accompanied by transepidermal water loss (TEWL) rates > 1000 ppm/(cm^2/hr (39,40)] suggests that lipid synthesis may not be limited to the basal layer but may extend to, or even be accelerated in, the outer layers of the viable epidermis. Indeed, in both in vivo and in organ culture, the rates of epidermal lipogenesis in the SG of neonatal mouse epidermis approach those in the subjacent basal/spinous layers, even under basal conditions (44; Fig. 4), that is, in the absence of barrier perturbation. The importance of this observation is underscored further by the simultaneous, precipitous decrease in protein, DNA, and CO_2 generation in the same layer (44). Whether the SG is the site of accelerated synthesis after barrier perturbation is currently under investigation.

Certainly, the foregoing studies establish that nonsaponifiable lipids (i.e., sterols and possibly hydrocarbons) are important for

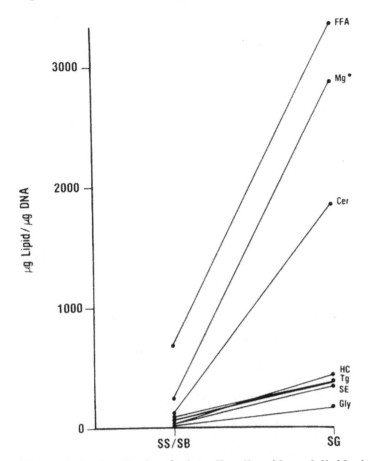

Figure 2.4 Synthesis of virtually all epidermal lipids is accelerated in the stratum granulosum in comparison with the basal layer.

barrier function. Although the role of sphingolipids has not been assessed by the metabolic approach, the observation that fatty acid synthesis is also regulated by barrier requirements (40) is consistent with a role for acylated lipids, including sphingolipids, in barrier function. Recently, we assessed the importance of relatively polar SC lipid species (i.e., sphingolipids and free sterols) compared with nonpolar species (i.e., free fatty acids, sterol esters, and hydrocarbons) for barrier function (45). With use of a highly nonpolar organic solvent, petroleum ether, instead of the more bipolar solvent, acetone, we found that removal of highly nonpolar species alone produced a significant break in the barrier, but TEWL rates never exceeded 150 mg/(cm^2/hr). Whereas petroleum ether removed large quantities of nonpolar lipids, it left the polar species in place. In contrast, acetone treatment caused profound barrier defects, as more and more lipid was removed. Moreover, whereas petroleum ether removed only nonpolar lipids, acetone removed a much larger proportion of polar species. These experiments show (1) a linear relationship between lipid content and barrier function; (2) that nonpolar lipids, in the absence of sphingolipids, provide a "first-line" of barrier function; and (3) that relatively polar lipids appear to provide a more profound level of barrier integrity and cohesion.

V. CLINICAL AND PATHOPHYSIOLOGIC IMPLICATIONS OF THE TWO-COMPARTMENT MODEL

The two-compartment model of the stratum corneum has stood the test of several predictions. First, in situations for which barrier function is defective, such as in essential fatty acid deficiency, there is a demonstrable abnormality in the lamellar body secretory system that leads to inadequate intercellular lipid deposition in the stratum corneum (46; Table 7). Moreover, lipophilic substances appear to traverse intercellular, rather than transcellular, routes in passing across the stratum corneum (47), consistent with sequestration of lipids to intercellular domains. There is also evidence that quantitative differences in lipid content are more accurate predictors of regional variations in skin permeability than either stratum corneum thickness or cell number (Table 8), that is, neither the thickness nor the number of cell layers in the stratum corneum correlate with permeability (48). These findings explain both the observation that lipophilic agents, such as topical steroids, readily traverse facial stratum corneum (10%–20% lipid by weight) and the poor barrier properties of the palm and sole stratum corneum, (1%–2% lipid by weight). This model also may explain why *eczema* occurs most read-

Table 2.7 Importance of Stratum Corneum Lipids for Barrier

Solvents and detergents destroy barrier while extracting lipids.

Barrier properties of different skin sites (face, leg, abdomen, palms) is related directly to lipid content.

In pathological conditions in which the barrier is defective, there is a decrease in lipid content (e.g., essential fatty acid deficiency).

Table 2.8 Regional Variations in Lipid-Weight Percent and Distribution of Major Lipid Species

	Site			
	Abdomen (*n*=4)	Leg (*n*=4)	Face (*n*=3)	Plantar (*n*=3)
Lipid weight (%)	6.5 ± 0.5	4.3 ± 0.8	7.2 ± 0.4	2.0 ± 0.6
Polar lipids	4.9 ± 1.6	5.2 ± 1.1	3.3 ± 0.3	3.2 ± 0.89
Cholesterol sulfate	1.5 ± 1.6	6.0 ± 0.9	2.7 ± 0.3	3.4 ± 1.2
Neutral lipids	77.7 ± 5.6	65.7 ± 1.8	66.4 ± 1.4	60.4 ± 0.9
Sphingolipids	18.2 ± 2.8	25.9 ± 1.3	26.5 ± 0.9	34.8 ± 2.1

Source: Ref. 28.

Table 2.9 Potential Practical Applications Based upon Current Knowledge

Therapy of dry skin conditions

   severe-to-mild

   localized-to-generalized

   seasonal

Therapy of aging skin

   prevention of further damage

   reversal of preexisting damage

Therapy of defective skin barrier

   palms and soles

   burn wounds, blisters, stasis ulcers, bedsores

Manipulation of skin barrier

   enhancement of percutaneous drug delivery

   reinforcement of barrier (occupational and warfare considerations)

ily on the palms and soles, (i.e., these sites would be most suscep-
tible to further depletion from exposure to hot water, detergents,
or solvents). Furthermore, the two-compartment model also possesses important implications for desquamation. That lamellar body-
derived lipids regulate desquamation is supported by numerous studies that link abnormal desquamation to inherited, lipid metabolic disorders and to drug-induced ichthyoses. Because of the limited
scope of this review, the reader who desires further information
about the role of lipids in desquamation is referred to several recent
reviews (49,50). Finally, Table 9 points to several areas for potential exploration based upon the foregoing information.

ACKNOWLEDGMENTS

This work was supported by NIH grant AM 19098 and the Medical
Research Service, Veterans Administration. We appreciate the typ-
ing assistance of Mr. Bil Chapman and Ms. Sally Michael.

REFERENCES

1. A. M. Kligman, and E. Christophers, *Arch. Dermatol.*, *88*:702,
    1964.
2. E. Christophers, *J. Invest. Dermatol.*, *56*:165, 1971.
3. I. C. MacKenzie, *Nature*, 222:881, 1969.
4. D. N. Menton, and A. Z. Eisen, *J. Ultrastructure Res.*, *35*:
    247, 1971.
5. P. M. Elias, and D. S. Friend, *J. Cell Biol.*, *65*:180, 1975.
6. P. M. Elias, J. Goerke, and D. Friend, *J. Invest. Dermatol.*,
    69:535, 1977.
7. P. M. Elias, N. S. McNutt, and D. Friend, *Anat. Rec. 189*:
    577, 1977.
8. H. Stupel, and A. Szakall, *Die Wirkung von Waschmitteln auf
    die Haut.* Heidelberg, Huthig, 1957.
9. R. J. Scheuplein, and I. H. Blank, *Physiol. Rev.*, *51*:702,
    1971.
10. G. Swanbeck, *Acta Dermato-Venereol.*, *39*(suppl. 43):00, 1959.
11. J. D. Middleton, *Br. J. Dermatol.*, *80*:437, 1968.
12. I. H. Blank, *J. Invest. Dermatol. 21*:259, 1953.
13. A. S. Michaels, S. K. Chandrasekaran, and J. E. Shaw, *J.
    Am. Inst. Chem. Eng.*, *21*:985, 1975.
14. P. M. Elias, *Int. J. Dermatol.*, *20*:1, 1981.
15. W. P. Smith, M. S. Christensen, S. Nacht, and E. H. Gans,
    J. Invest. Dermatol., *78*:7, 1982.

16. S. Grayson, and P. M. Elias, *J. Invest. Dermatol.*, *78*:128, 1982.
17. P. M. Elias, L. Bonar, S. Grayson, and H. P. Baden, *J. Invest. Dermatol.*, *80*:213, 1983.
18. P. M. Elias, M. L. Williams, M. E. Maloney, J. A. Bonifas, B. E. Brown, S. Grayson, and E. H. Epstein, Jr., *J. Clin. Invest.*, *74*:1414, 1984.
19. G. K. Menon, S. Grayson, and P. M. Elias, *J. Invest. Dermatol.*, *86*:591, 1986.
20. P. M. Elias, G. K. Menon, S. Grayson, and B. E. Brown, *J. Invest. Dermatol.* (in press), 1988.
21. E. Schreiner, and K. Wolff, *Arch. Klin. Exp. Dermatol. 235*: 78, 1969.
22. C. A. Squier, *J. Ultrastructure Res.*, *43*:160, 1973.
23. D. A. Bowser, and R. J. White, *Br. J. Dermatol. 112*:1, 1985.
24. G. P. Odland, and K. Holbrook, *Curr. Probl. Dermatol.*, *9*:29, 1987.
25. P. M. Elias, *J. Invest. Dermatol.*, *80*:44, 1983.
26. R. K. Freinkel, and T. N. Traczyk, *J. Invest. Dermatol.*, *85*: 295, 1985.
27. S. Grayson, A. G. Johnson-Winegar, B. U. Wintroub, E. H. Epstein Jr., and P. M. Elias, *J. Invest. Dermatol.*, *85*:289, 1985.
28. M. A. Lampe, A. L. Burlingame, J. Whitney, M. L. Williams, B. E. Brown, E. Roitman, and P. M. Elias, *J. Lipid Res.*, *24*: 120, 1983.
29. P. M. Elias, B. E. Brown, P. O. Fritsch, R. J. Goerke, G. M. Gray, and R. J. White, *J. Invest. Dermatol.*, *73*:339, 1979.
30. G. M. Gray, and H. J. Yardley, *J. Lipid Res.*, *16*:441, 1975.
31. G. M. Gray, and R. J. White, *J. Invest. Dermatol.*, *70*:336, 1978.
32. P. W. Wertz, and D. T. Downing, *J. Lipid Res.*, *24*:759, 1983.
33. G. M. Gray, R. J. White, and J. R. Majer, *Biochim. Biophys. Acta 528*:122, 1978.
34. P. W. Wertz, and D. T. Downing, *J. Lipid Res.*, *24*:753, 1982.
35. P. A. Bowser, D. H. Nguteren, R. J. White, U. M. T. Houtsmuller, and C. Prottey, *Biochim. Biophys. Acta, 834*:419, 1985.
36. K. R. Feingold, M. H. Wiley, G. MacRae, S. R. Lear, A. H. Moser, G. Zsigmond, and M. D. Siperstein, *Metabolism, 32*:75, 1983.
37. K. R. Feingold, M. H. Wiley, A. H. Moser, S. R. Lear, and M. D. Siperstein, *J. Lab. Clin. Med.*, *100*:405, 1982.
38. K. R. Feingold, B. E. Brown, S. R. Lear, A. H. Moser, and P. M. Elias, *J. Invest. Dermatol.*, *81*:365, 1983.
39. G. K. Menon, K. R. Feingold, A. H. Moser, B. E. Brown, and P. M. Elias, *J. Lipid Res.*, *26*:418, 1985.

40. G. Grubauer, K. R. Feingold, and P. M. Elias, *J. Lipid Res.*, 28:746, 1987.
41. G. Grubauer, P. M. Elias, and K. R. Feingold, *Clin. Res.* (In Press) 1988.
42. K. R. Feingold, B. E. Brown, S. R. Lear, A. H. Moser, and P. M. Elias, *J. Invest. Dermatol.*, 87:588, 1986.
43. K. R. Feingold, M. L. Williams, S. Pillai, G. K. Menon, B. P. Halloran, D. D. Bikle, and P. M. Elias, *Biochim. Biophys. Acta*, 930:193, 1987.
44. D. J. Monger, M. L. Williams, K. R. Feingold, B. E. Brown, and P. M. Elias, *J. Lipid Res.* 29:603, 1988.
45. G. Grubauer, K. R. Feingold, and P. M. Elias, *J. Invest. Dermatol.*, 88:492, 1987 (abstract).
46. P. M. Elias, and B. E. Brown, *Lab. Invest.*, 39:574, 1978.
47. M. K. Nemanic, and P. M. Elias, *J. Histochem. Cytochem.*, 28:573, 1980.
48. P. M. Elias, E. R. Cooper, A. Korc, and B. E. Brown, *J. Invest. Dermatol.*, 76:297, 1981.

3

Stratum Corneum Structure and Transport Properties

STIG E. FRIBERG, IBRAHIM H. KAYALI, and MARION MARGOSIAK
Clarkson University, Potsdam, New York

DAVID W. OSBORNE *The Upjohn Company, Kalamazoo, Michigan*

ANTHONY J. I. WARD* *University College Dublin, Dublin, Ireland*

I. INTRODUCTION

The outermost layer of the skin, the stratum corneum, has an essential role as a barrier against the transport of water and of chemical and biological agents (1,2). This barrier is functionally the same, for short periods, in both living and dead skin (3), with lipids being responsible for the essential function (4,5). The biological and medical importance of this cannot be overstated, and the investigations into the individual structure and overall organization of the stratum corneum lipids has been extensive (6—12).

These investigations and the scientific examinations remain an ongoing process (13), but recent investigations allow a limited discussion of the relationship between structure and transport of water through the stratum corneum. This chapter will describe the layered structure and review its importance for transdermal transport.

II. LIPID ORGANIZATION

The basis for the discussion is the layered structure proposed by Elias (13) and analyzed by Downing (14) and by us (15). The essential experimental information was suggested by electron micrographs (13) that disclosed the layered structure. Built on this information Elias (13) proposed the layered structure shown in Figure 3.1. The polar head groups of the lipids (Table 3.1) are gathered in layers with the nonpolar chains pointed in opposite directions forming layers of methyl groups in the plane where the hydrocarbon ends

---

Current affiliation: Clarkson University, Potsdam, New York

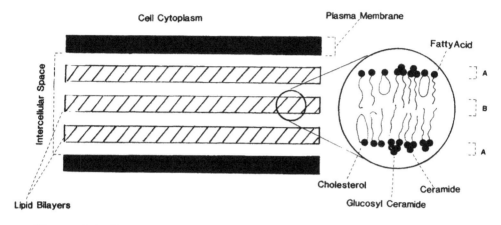

Figure 3.1   Layered structure of epidermal lipids as proposed by
Elias (13).

meet.   White (16) has recently made a very extensive x-ray investigation into the stratum corneum lipid structure, finding a layered
structure only at high temperatures.   At physiological temperatures
solid phases were also present and the structure was complex.

Our discussion will be limited to the features of the layered
structure, which are more complex than indicated by Figure 3.1.
Low-angle x-ray diffractometry shows that in a model such as Elias'
(13) not all of the lipids are positioned with their polar groups localized in the polar layer (see Fig. 3.1).   Some of them are actually
located in the region between the methyl groups.   In addition, the
low-angle x-ray diffraction patterns provide information with which
to estimate penetration by water into the space between the lipid
molecules.   Order parameters of lipid groups by nuclear magnetic
resonance (NMR) (17) are used to confirm the x-ray results and to
provide information on the mobility of groups within the lipid molecules.

The results of low-angle x-ray diffraction and NMR, as well as
their interpretation, are comparatively new to the dermatological
constituency; therefore, the methods used and the interpretation of
results will be briefly discussed.

A.   Low-Angle X-Ray Diffraction

The low-angle x-ray diffraction pattern from a lamellar structure is
unique and, at the same time, directly and unambiguously interpreted.   A structure with alternating layers of electron-rich (polar

Table 3.1 Composition of Model Epidermal Lipid

Component	Source	Purity (%)	Wt% in mixture
PE	Avanti Polar Lipids	99	5
Cholesteryl Sulfate	Research Plus	98	2
Cholesterol	Fisher		14
Triolein	Sigma	99	25
Free fatty acids	Sigma		19
myristic	Sigma	99	3.8
linoleic	Sigma	99	12.5
oleic	Sigma	99	33.1
palmitic	Sigma	99	36.8
palmitoleic	Sigma	99	3.6
stearic	Sigma	99	9.9
Oleic acid Palmityl ester	Sigma	98	6
Squalene	Aldrich	98	7
Pristane	Aldrich	96	4
Ceramides	Sigma	99	18

groups) and electron-poor material (nonpolar parts) will reflect x-rays according to Bragg's law and give a series of reflections as shown in Figure 3.2. Hence, the appearance of diffraction lines in the ratio 1:2:3:4... means that the structure is layered.

The distinction between a crystalline and a liquid crystalline structure is made on the basis of the x-ray diffraction at higher angles. A crystalline structure gives rise to a series of sharp reflections in the range 3.5 to 5 Å because of the crystalline packing of the hydrocarbon chains (16). Conversely, a liquid crystal has no short-range ordering of the hydrocarbon chains and, hence, displays a single diffuse reflection at about 4.5 Å.

The information given by low-angle x-ray diffraction on the location of individual molecules within the bilayer is well illustrated

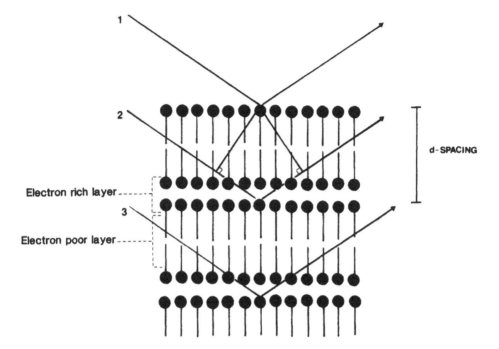

Figure 3.2 X-rays are reflected off of the parallel planes of the layered structure. These reflections may then be converted to the interlayer spacing, d, by using Bragg's equation.

by recent results from an analysis of stratum corneum lipids (18). The interlayer spacings versus water content for different combinations of lipids from the stratum corneum spectrum (18) are shown in Figure 3.3.

The basic feature of the x-ray diffraction evaluation is the location of an added substance, as directly reflected in the change of interlayer spacing, as a function of the amount of substance added (Fig. 3.4). Location of the added substance between the lipid *chains* (see Fig. 3.4(a)) leads to no interlayer spacing change, whereas localization to the space between the methyl group layers (Fig. 3.4(b)) (or in the aqueous layer) leads to a strong increase in spacing.

With knowledge of this relationship, a cursory glance at the results in Figure 3.3 may be given a straightforward (but erroneous) interpretation. It would predict that the layered structure of the stratum corneum free fatty acids plus water will accommodate added phosphatidylethanolamine, mostly between the methyl groups, and that a subsequent addition of cholesterol will be localized to the same site.

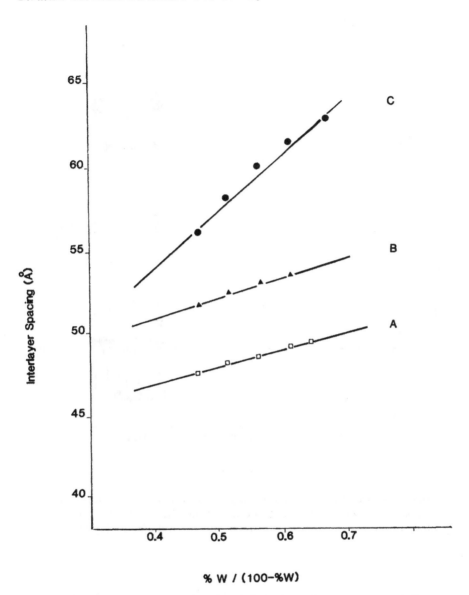

Figure 3.3 Interlayer spacing as a function of water content for the systems: (A) partially saponified free fatty acids; (B) phosphatidylethanolamine + A; (C) cholesterol + B.

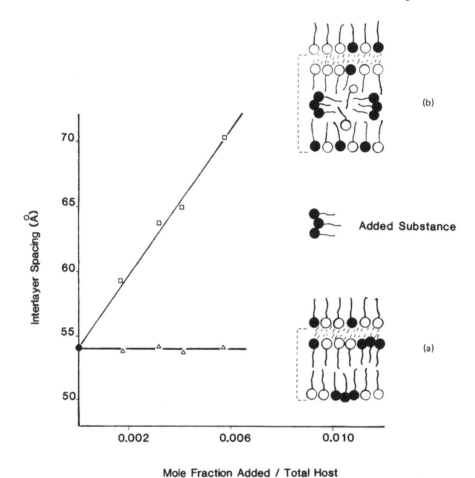

Figure 3.4 Interlayer spacing as a function of mole fraction added: (a) the added substance is located between lipid chains; (b) the added substance is located between the methyl group layers.

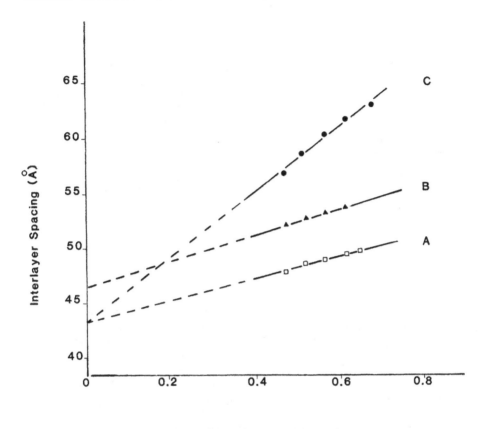

Figure 3.5 Extrapolation to zero water concentration for the curves from Figure 3.

It cannot be overemphasized that this interpretation leads to erroneous results, because the curves in Figure 3.3 reflect the influence of both the water and the added substance. The correct information is obtained only after an extrapolation to zero water concentration has been made (Fig. 3.5). The extrapolated curve C has identical extrapolated interlayer spacing with curve A, whereas curve B shows a higher one. Hence, the correct interpretation is that the added cholesterol is, by no means, located between the methyl group layers in the structure. On the contrary, all cholesterol resides entirely between the lipid chains (Fig. 3.6(b)). In addition, the added

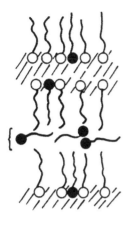   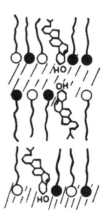

A   B

Figure 3.6 (A) The added phosphatidylethanolamine being localized between the methyl group layers. (B) The added cholesterol brings the added phosphatidylethanolamine into the space between the chains.

cholesterol brings the added phosphatidylethanolamine that was earlier located between the methyl group layers (see Fig. 3.6A) into the space between the chains (see Fig. 3.6B). The high values of the interlayer spacings after addition of cholesterol (see curve C, Fig. 3.3) are a consequence of reduced water penetration into the lipid parts when compared with conditions for curves A and B.

This type of information is definitely useful to clarify the organization−functional relationship for the stratum corneum lipids. The primary source of the information is the x-ray diffraction results, but their interpretation is based on an assumption of no change in order of the amphiphilic chains nor in their tilt. Nuclear magnetic resonance offers a means to confirm the interpretation of the results from the low-angle x-ray diffraction and, in addition, it provides direct information about the order of the individual molecules and groups involved.

B. Order Parameters of Lipid Groups

A deuterated proton site on the lipid molecule gives rise to a quadrupolar splitting of the NMR signal through the interaction with electric field gradients (19—20). The quadrupole splitting, $\Delta\nu$, for an unoriented lamellar dispersion is directly proportional to the order parameter S

$$\Delta\nu = \frac{3}{4} \chi \mid S \mid$$

where χ is the quadrupole coupling constant and S is the order parameter associated with a C-D bond.

For molecules deuterated in different positions, a spectrum will result that comprises a series of splittings centered about a central frequency. The difference in the degree of order in the molecule is reflected in the size of the order parameters and may be expressed as an order profile for which S_i is considered as a function of position in the molecule.

The location between the hydrocarbon chains of the lipids gives an order parameter of about 0.1 to 0.2 for groups attached to a long-chain compound. Compounds located between the methyl group layers show order parameters approximately one-tenth of this value (21).

The NMR spectrum of perdeuterated palmitic acid is a good example. The central peak (Fig. 3.7) is from palmitic acid localized between the central methyl group layers. These palmitic acid molecules show little order, their motion is essentially isotropic. The series of split signals arise from palmitic acid being located in the organized amphiphilic layer. The order parameter for the latter is about 0.15.

This layered structure and its order have the most profound influence on the transport properties of lipids and water.

C. Transport in Layered Structures

The diffusion in liquid crystals is highly anisotropic (22—25). The essential information for the problem of transport across the stratum corneum is that the diffusion *parallel* to the layers is fast; the same magnitude as in a liquid. Conversely, perpendicularly to the layers the diffusion is one or two magnitudes lower.

A layered structure, such as that in the lipid part of the stratum corneum, is not a perfectly organized array of layers parallel to the skin surface but, instead, a series of dislocations always occur as shown schematically in Figure 3.8. Hence, the pertinent diffusion coefficient is the gross one for a partially organized lamellar structure.

The permeability of water through reconstituted stratum corneum was determined earlier (5). The method lends itself to investiga-

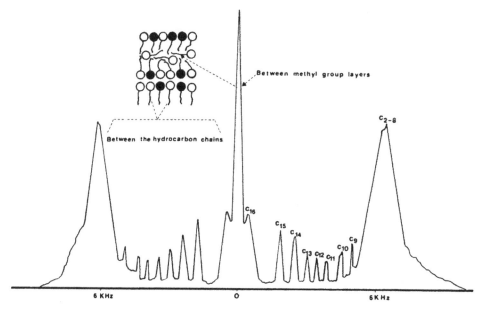

Figure 3.7 The NMR spectrum of perdeuterated palmitic acid added to the system oleic acid—sodium oleate.

tions of any lipid model combination, and we have recently applied it to lipid compounds by use of the following process: Fragments of stratum corneum, obtained either by scraping or from trypsinized skin obtained at autopsy, were lipid-depleted using chloroform/methanol (2:1 vol) and, subsequently, reduced to individual cells with a ground-glass homogenizer. These cells (about 20 mg) were suspended in ether and pipetted onto a phosphate-buffered solution film (pH 7.2) confined within a Teflon O-ring with 1.3 cm diameter. Lipid (1—5 mg in 10 ml ether) was then layered onto the cells, and the films were stored at 0°C for at least 12 hr, after which the reconstituted stratum corneum tablets were used in the permeability studies. The lipid model gave reconstituted stratum corneum tablets with sufficient mechanical properties to be used in the diffusion cell according to Blank (26).

These results, surprisingly, showed the lipid model to give fluxes through the tablet at the same level as the ones found for the extracted lipids (5). Table 2 shows the values from studies in which extracted stratum corneum cells from different persons were

Table 3.2 The Amount of Water Transferred per Unit Time, Expressed as $mg/(cm^2/hr)$ per Unit Membrane Thickness

Reconstructed SC disks	Flux $[mg/(cm^2/hr)]$
Subject A SC with 20% lipid model	0.86 ± 0.05
Subject B SC with 20% lipid model	0.71 ± 0.09
Subject C SC with 5.8% partially saponified FFA	0.99 ± 0.03
Subject C SC with 16% partially saponified FFA	0.87 ± 0.07
Smith's value for calf SC with native lipid	1.0

In each of the above cases, the flux was determined for water penetrating through at least two tablets. The values shown are the mean ± standard deviation.

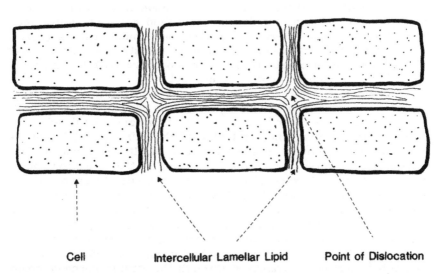

Cell   **Intercellular Lamellar Lipid**   **Point of Dislocation**

Figure 3.8 The current concept of the stratum corneum as a two-compartment system, in which cells can be analogized to bricks, and intercellular lamella to mortar.

combined with the lipid model to give flux values similar to those
found by Smith et al. (5) who used the extracted lipids.

As a matter of fact, the same flux values were also obtained
when the lipids were limited to the partially saponified fatty acids.
Table 3.2 shows us no significant difference between these values and
those from tablets with more complex combinations of lipids present.

These results appear to indicate that structural organization of
the lipids is the important element to prevent fast water transport
through the structure. The difference between the values from a
fatty acid/water-layered structure and the one containing all the lipids of the stratum corneum was insignificant.

The major difference was found when comparing water transport
through layered structures of different fatty acid—soap combinations
without the proteins. The system of water—fatty acids with saturated hydrocarbon chains stood out only in the comparison (27). In
fact, a structure with only saturated hydrocarbon chains gave no
barrier whatsoever to water transport; the values of evaporation
rates were identical with those from an unprotected water surface!
This result agrees with early results on evaporation retardation by
monomolecular layers (28) that showed cracks in a crystalline monolayer to reduce the retardation barrier to water evaporation by 95%.

A comparison of these results with the symptoms of the essential
fatty acid deficiency syndrome is an illustrative exercise. Lack of
unsaturated fats in the diet gives rise to dry cracked skin with increased transdermal transport. In our opinion these results, combined with the diffusion results from the lipid model structure and
the well-known phase diagrams of water—amphiphile systems (29),
show that the lipid hydrocarbon chains in the stratum corneum need
mobility (i.e., total crystallization must be prevented) to function
effectively as a barrier to water transport.

Crystalline packing of the lipids means a brittle structure in
which deformations lead to open cracks: the consistence of a liquid
crystal, on the other hand, is that of soft butter and will tolerate
the necessary bending and stretching of the skin without losing the
basic layered structure. Conversely, our results from evaporation
of water from layered structures (27) revealed that a mixture of
unsaturated and saturated chains results in a slightly enhanced barrier to transport, compared with the values from a structure of unsaturated chains only. This result is interesting in relation to the
recent results by White et al. (16) demonstrating the presence of
both crystalline and liquid chains in the stratum corneum lipids.

With the layered structure as a background, it follows that a
substantial increase of transport through the stratum corneum would
be obtained if the long-range order of the lipid organization were
broken down to form an isotropic liquid. The diffusion now would

not be restricted to paths along the layers, but would be fast in
all three directions. The overall diffusion coefficient is thereby increased more than one order of magnitude when the system becomes
liquid.

However, the complete transition to an isotropic liquid is not
necessary to ensure enhanced transport. Even an increase of the
fluidity (e.g., a small disordering) will give enhanced transport, as
shown by our results (27). This means that substances that increase disorder in a liquid crystalline system are potential transdermal transport enhancers. Such substances are well known in the
colloid chemistry; they are called *hydrotropes*.

Table 3.2 shows structures of selected hydrotropes and of transdermal transport enhancers. The similarity of structure is striking,
and a few experimental methods to determine the action and mechanism of hydrotropes may be of interest for the area of transdermal
transport.

As an example, a hydrotrope in the form of the monosoap of
the 5-carboxy-4-hexyl-2-cyclohexane-1-yl octanoic acid (NaDA).
Its function is well illustrated in Figure 3.9. The curves show the
order parameter of the hydrocarbon chains in the liquid crystal after addition of the hydrotrope or of a surfactant (30). Addition of
the hydrotrope caused a disordering, whereas the added surfactant
caused no change whatsoever.

The reason for the enhanced disorder is found in the conformation of the acid, when present in the liquid crystal (31). The acid
(Fig. 3.10) is present with all of its polar groups at the interface.
The narrow loop and the short chain protruding into the hydrocarbon space of the amphiphile leave a considerable space vacant for
the original hydrocarbon chains to become disordered.

The comparison between molecules that cause enhanced transdermal transport and the hydrotropes is of interest, but it is certainly incomplete. In a previous study (32), it was shown that NADA
(see Fig. 3.9) appears to function as a selective hydrotrope, destabilizing only bilayer systems that are formed from amphiphiles with
relatively small-sized head groups. Hence, NaDA did not enhance
percutaneous permeation (used at 10 wt% concentration) because it
appeared to be ineffective at destabilizing epidermal bilayers formed
from lipids containing medium-sized head groups. The study further implied that a nonirritating transdermal delivery enhancer system is feasible, if a selective hydrotrope was found that would destabilize the bilayers formed by the medium-sized head groups of
epidermal lipids, while not disrupting viable cell membranes that are
formed from large-sized phospholipid head groups. Work in addition
to this preliminary study should prove useful in the search for
unique transdermal enhancers. Note that the essential transdermal

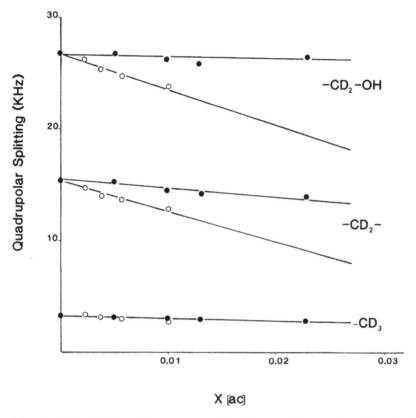

Figure 3.9 Adding the compound in Figure 10 to a lamellar liquid crystal caused disordering (○). Addition of surfactants gave no change (●).

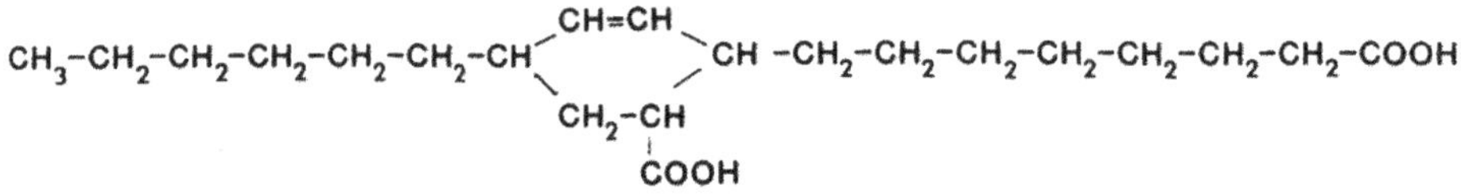

$$CH_3-CH_2-CH_2-CH_2-CH_2-CH_2-CH\underset{\underset{\underset{COOH}{|}}{CH_2-CH}}{\overset{CH=CH}{\diagdown}}CH-CH_2-CH_2-CH_2-CH_2-CH_2-CH_2-CH_2-COOH$$

Figure 3.10 The structure of 5-carboxy-4-hexyl-2-cyclohexane-1-yl octanoic acid.

Table 3.3 Chemical Structures of Selected Hydrotropes and of Selected Transdermal Transport Enhancers

---

hydrotropes

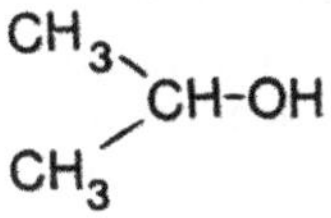

isopropyl alcohol 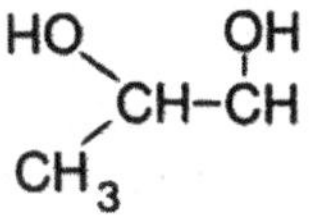 propylene glycol 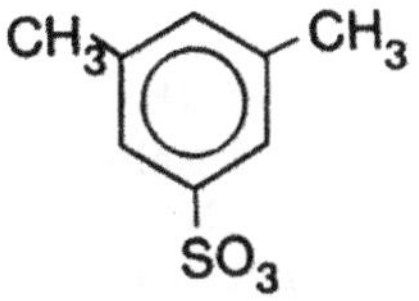 sodium xylene sulfonate

transdermal penetration enhancers

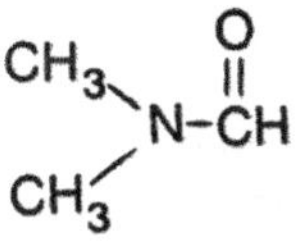

$CH_3(CH_2)_7CH=CH(CH_2)_7CO_2H$

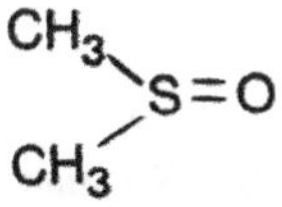

dimethyl formamide dimethyl sulfoxide oleic acid

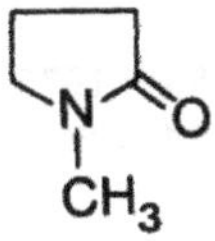

N-methyl-2-pyrrolidone 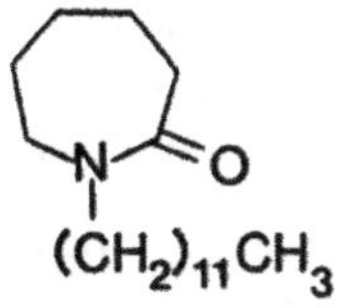   azone

---

enhancer is to facilitate transport of the drug and, thus, the combined influence of the drug molecule and the transport enhancer must be taken into account.

III. SUMMARY

The structure of the lipid component of the stratum corneum was described as a lamellar liquid crystal. It was shown that a change to a predominantly crystalline structure would cause enhanced transdermal transport because of cracks in the structure, and a comparison with the enhanced transport during the essential fatty acid deficiency syndrome was made.

Transdermal transport enhancers act in the opposite direction, and a comparison of their molecular structure was made with that of hydrotropes that cause destabilization of a liquid crystal through enhanced disorder of its hydrocarbon chains.

ACKNOWLEDGMENT

This research was supported in part by Dow Chemical, Midland, Michigan, and by The Upjohn Company, Kalamazoo, Michigan.

REFERENCES

1. A. M. Kliegman, in *The Epidermis*, Chap. 20, (W. Montagna, ed.), Academic Press, New York-London, 1964.
2. P. M. Elias, and D. S. Friend, *J. Cell. Biol.*, *65*:185, 1975.
3. G. E. Burch, and P. Winsor, *Arch. Intern. Med.*, *74*:437, 1944.
4. P. M. Elias, N. S. McNutt, and D. S. Friend, *Anat. Rec.*, *189*:577, 1977.
5. W. P. Smith, M. S. Christenson, S. Nachet, and E. H. Gans, *J. Invest. Dermatol.*, *78*:7, 1982.
6. P. M. Elias, J. Goerke, and D. S. Friend, *J. Invest. Dermatol.*, *69*:535, 1977.
7. G. M. Gray, and R. J. White, *J. Invest. Dermatol.*, *70*:XX, 1977.
8. G. M. Gray, and H. J. Yardley, *J. Lipid Res.*, *16*:435, 1975.
9. E. G. Bligh, and W. J. Dyer, *Can. J. Biochem. Phys.*, *37*: 911, 1959.
10. W. Abraham, P. W. Wertz, and D. T. Downing, *J. Lipid Res.*, *26*:761, 1985.
11. A. W. Ranasinghe, P. W. Wertz, D. T. Downing, and J. C. Mackenzie, *J. Invest. Dermatol.* *86*:187, 1986.

12. P. M. Elias, *J. Invest. Dermatol.*, *80*:44, 1983.
13. P. M. Elias, *Int. J. Dermatol.*, *20*:1, 1981.
14. P. W. Wertz, W. Abraham, L. Landmann, and D. T. Downing, *J. Invest. Dermatol.*, *87*:582, 1986.
15. S. E. Friberg, and D. W. Osborne, *J. Dispersion Sci. Technol.*, *6*:485, 1985.
16. S. H. White, D. Mirejovsky, and G. I. King, *Biochemistry*, *27*: 3725, 1988.
17. J. Seelig, *Q. Rev. Biophys. 10*:353, 1977.
18. S. E. Friberg, H. Suhaimi, and L. B. Goldsmith, in *Stratum Corneum Lipids in a Model Structure.* (in press).
19. A. J. I. Ward, S. E. Friberg, D. W. Larsen, and S. B. Rananavare, *Langmuir 1*:24, 1985.
20. J. H. Davies, *Biochim. Biophys. Acta, 737*:117, 1983.
21. Unpublished results.
22. G. Lindblom, *Acta Chem. Scand.*, *B35*:61, 1981.
23. M. B. Schneider, W. K. Chan, and W. W. Webb, *J. Biophys. Soc.*, *43*:157, 1983.
24. W. C. L. Vaz, R. M. Clegg, and D. Hallman, *Biochemistry*, *24*:781, 1985.
25. G. Chidichimo, D. de Fazio, G. A. Ranieri, and M. Terenzi, *Mol. Cryst. Liq. Cryst.*, *133*:1, 1986.
26. I. H. Blank, J. Moloney, G. E. Alfred, I. Simon, and C. Apt, *J. Invest. Dermatol.*, *84*:188, 1984.
27. S. E. Friberg, and I. Kayali, *J. Pharm. Sci.*, (in press).
28. V. K. LaMer, ed., *Retardation of Evaporation by Macrolayers*: *Transport Processes.* Academic Press, New York, 1962.
29. P. Ekwall, in *Advances in Liquid Crystals.* (G. H. Brown, ed.). Academic Press, New York, 1975.
30. S. E. Friberg, S. B. Rananavare, and D. W. Osborne, *J. Colloid Interface Sci.*, *109*:487, 1986.
31. T. Flaim, and S. E. Friberg, *J. Colloid Interface Sci.*, *97*:26, 1984.
32. D. W. Osborne, *Colloids Surfaces, 30*:13, 1988.

4
Penetration Enhancer Incorporation in Bilayers

ANTHONY J. I. WARD* and REGINA TALLON *University College Dublin, Dublin, Ireland*

I. INTRODUCTION

The use of agents to enhance the penetration of drugs through the skin has been widely studied, although phenomena associated with the modification of the barrier function and the molecular nature of the incorporating molecules remain unclear. An understanding of these interactions, however, is fundamental to the design of topical delivery vehicles and their modes of interaction with the skin.

This chapter explores the use of a simple surfactant system to model some of the features of the lipophilic component of the stratum corneum (SC). For this, the choice of model is determined by the head group nature of those lipids that are predominately neutral (1,2) to about 75% to 80% in the SC compared with about 10% phospholipids. This type of lipid interacts with the aqueous environment through hydrogen-bonding interactions. A similar situation arises with simple surfactants of the alkyl polyoxyethylene glycol ether (CnEOm) type that form associated structures when mixed with water (3). Particular attention will be focused on the lyotropic mesophase formed when the amphiphilic molecules associate into bimolecular arrangements separated by aqueous layers (Fig. 1), known as the *lamellar phase*. Note that this phase comprises randomly dispersed lamellar domains and should not be confused with vesicles and liposomes, although these structures also maintain bimolecular arrangements of the amphiphiles. Indeed, many physicochemical properties are analogous between the two types of systems, but some differences exist in observed behavior (see Sect. II.A) because of differences in isotropy. Furthermore, the lamellar phase does

---

Current affiliation: Clarkson University, Potsdam, New York

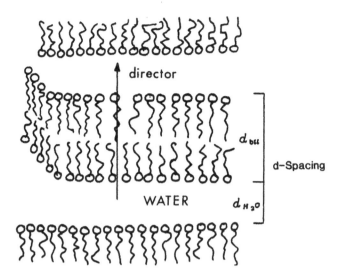

Figure 4.1   The arrangement of surfactant molecules in the lamellar liquid crystal phase.

not suffer from some of the inherent instabilities encountered in vesicles.

The choice of enhancing agent systems is based upon those commonly used and studied; namely, oleyl alcohol, propylene glycol, and dodecyl (rhexahydro-2*H*-azepin-2-dione laurocapram; Azone), which has been regarded as having some potentially novel features (4). The behavior of oleyl alcohol and laurocapram, both singly and in combination with propylene glycol, has been investigated when incorporated in the lamellar phase of n-dodecylpentaoxyethylene glycol ether ($C_{12}EO_5$).

To study this type of system, techniques sensitive to the dimensions of the associated structures and the time scales of the system's dynamics are required. The most powerful techniques available for such studies of anisotropic phases are those of small-angle x-ray diffraction and nuclear magnetic resonance (NMR). Therefore, a brief background and description of these methods will be given as an aid to understanding the presented data.

II.   EXPERIMENTAL BACKGROUND

A.   Nuclear Magnetic Resonance

Nuclear magnetic resonance (NMR) is arguably the most powerful method for investigating the dynamics and structure of associated surfactant and lipid systems. The basis of the power of these techniques is that only the nuclei, which are at resonance, will be directly observed. This allows observation of selected parts and facets of the system, usually by making use of the constituent nuclei of the components (i.e., no perturbation of the system by incorporated foreign "probe" molecules). Substitution of protons by deuterium atoms either in the aqueous fraction or in the lipid component has relatively minor and predictable effects on the system's behavior and has been used to great advantage (5—18) in the investigation of bilayer phases formed by lipids and surfactants in the presence of water.

Unlike the more commonly encountered hydrogen (^{1}H) and carbon (^{13}C) nuclei, the deuterium (^{2}H) nucleus has a spin quantum number, $I = 1$, with an associated quadrupolar moment. This means that there are three nondegenerate nuclear spin energy levels (Fig. 2). In an isotropic environment in which the effects of electric field gradients at the nucleus are averaged to zero, a single resonance is observed. The presence of electrical field gradients in an axially symmetric field, however, leads to changes in the spin energy levels (see Fig. 2), which leaves the $m = 0 \rightarrow +1$ and $m = -1 \rightarrow 0$ transitions with different energies. This has the effect of producing two absorption peaks at resonance separated by a frequency difference $\Delta \nu$, the size of which depends on the size of the interaction with the electrical field gradient as

$$\Delta \nu = 3/4 \; \chi (3\cos^2 \theta - 1) \tag{1}$$

where χ is the *quadrupolar coupling constant* that has a value of 167 kHz for deuterium attached to carbon (19) and θ is the angle between the principal axis, being the vector of the C-D bond and the magnetic field. The case of the nucleus interacting with a field gradient of axial symmetry has been shown to be applicable to those normally studied in lamellar surfactant and lipid systems.

In molecules containing several deuterium atoms, a doublet splitting will be observed for each atom, reflecting the degree of dynamic freedom that atom has in the molecule. Assignment of splittings to particular deuterium atoms in a molecule must be done by observation of selectively deuterated molecules. Deuterium NMR may, therefore, be used to probe the dynamic structure of the chains, as a function of position, composition of phase, and temperature. An *order profile* of the lipid chains, defined by the order parameter, S_{CD}, can be derived from the quadrupolar splittings (Eq. 2) of the ^{2}H nuclei in the different segments.

$$\Delta \nu = 3/4 \, \chi \; |S_{CD}| \tag{2a}$$

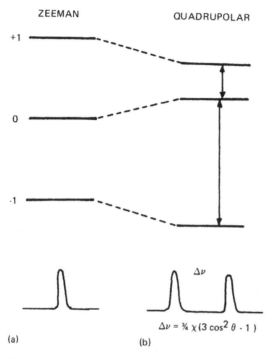

ZEEMAN   QUADRUPOLAR

$\Delta\nu = \frac{3}{4}\chi(3\cos^2\theta - 1)$

(a)   (b)

Figure 4.2   Nuclear spin energy levels for a spin = 1 nucleus (e.g., ^{2}H) showing the allowed transitions (a) for an isotropic and (b) an axially symmetric electrical field gradient.

$$S_{CD} = 1/2\,\overline{(3\cos^2\theta - 1)} \qquad\qquad [2b]$$

Here the bar in Eq. 2b represents the averaging over all possible orientations of the randomly dispersed lamellar domains. Many studies have now been carried out to derive order profiles in lipid and surfactant systems in lyotropic lamellar phases (5—18). The form of the profile (Fig. 3) is remarkably consistent for a wide variety of systems showing highest order (S_{CD} values) in the segments of the amphiphilic chains closest to the head group. A rapid decrease in order, as expressed by S_{CD}, is seen in the methylene segments near the terminal methyl group, which is located at the center of the bilayer. The shape of this type of profile is a result of the packing constraints in the associated bilayer structure, reducing the conformational freedom of the individual chains (17,18).

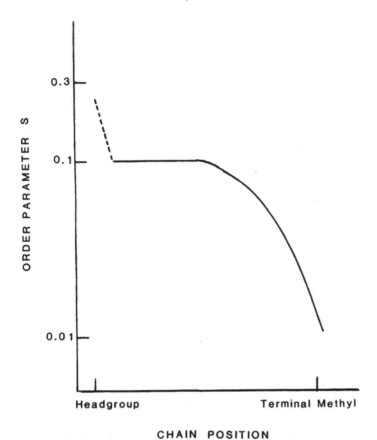

CHAIN POSITION

Figure 4.3 The typical form of order profile of hydrophobe chains observed for surfactants and lipids in lamellar phases.

Perturbations of this structure by incorporation of solubilizates or changes in composition or environment can, in principle, be studied through the resulting effects on the order profile.

Many lyotropic lamellar systems of surfactants and phospholipids may be oriented between flat glass plates (i.e., the effect of the surface field of the glass is to orient bilayers so that the amphiphile chains are perpendicular to the surface). The orientation of the principal axis of the bilayer, called the *director* (see Fig. 1), can be determined relative to the magnetic field. Here, the averaging of the directors in the sample is removed, and the order parameter can be derived from the angular dependence of the quadrupolar splittings (14–16).

B. Small-Angle X-Ray Diffraction

One of the most important methods of identifying and characterizing lyotropic mesophases is x-ray diffraction at low angles (20,21). The well-known Bragg relation, namely,

$$n \lambda = 2d \sin \theta \qquad\qquad [3]$$

where n is the order of reflection of x-rays with wavelength λ from a regular structure with repeat distance d. For *nickel-filtered copper* radiation of wavelength 0.1542 nm, Eq. 3 gives angles of reflection of between 0.4 and 2.5° for the range of repeat lengths commonly encountered in bilayer systems (2–15 nm); hence, the term *small* or *low*-angle x-ray diffraction. Lamellar systems are characterized by a series of reflections in the ratio 1:2:3:4..., corresponding to the one-dimensional repetition with first-order (n = 1) spacing (d) equivalent to the thickness of the water and bimolecular layer (see Fig. 1).

If the partial specific volumes (v) are known, it is possible to calculate the thicknesses of the water layer, d_{H_2O}, and lipid bilayer, d_1, as

$$d_{H_2O} = \upsilon_{H_2O} d \qquad\qquad [4a]$$

$$d_1 = \upsilon_1 d \qquad\qquad [4b]$$

The measured d-spacings are usually plotted as a function of C/(1 − C) where C is the concentration of water (wt%) and the bilayer thickness is obtained by extrapolation to C = 0. The slope of the plot is related to the ratio of the partial specific volumes. Geometric considerations show that the area—polar group (A) in interface of the lamellar phase can be derived from the following relationship:

$$A = \frac{2V_1}{d_1 N} \qquad\qquad [5]$$

where V_1 is the molar volume of the lipid and N is the Avogadro number.

A result of the relatively simple geometric properties of the lyotropic phases is to allow the distribution of components within the system to be inferred from the d-spacings. The experimental d-spacing (d_{exp}) in a lamellar system can be separated into contributions from the bilayer (d_{bil}) and the solvent layer (d_s) as

$$d_{bil} = \phi_{bil} d_{exp} \qquad [6a]$$

$$d_s = d_{exp} - d_{bil} \qquad [6b]$$

where ϕ_{bil} is the volume fraction of the lipid in the system.

A plot of log d_{exp} against $-$log ϕ_{bil} should be linear with unit slope if there is no penetration of the solvent into the bilayer region. If penetration occurs, however, the slope of the plot will increase. The percentage penetration (p) can then be estimated by calculating the difference between the experimental d-spacing and that expected for nonpenetration (d_{cal}) given as

$$d_{cal} = d_0(1 + \phi_p) \qquad [7]$$

$$p(\%) = \left(\frac{d_{cal} - d_{exp}}{d_{cal} - d_0}\right) \times 100 \qquad [8]$$

where ϕ_p is the volume solvent/lipid ratio.

Phase identification rests primarily on the ratios of the spacings of the diffraction lines; thus, unlike the equal spacings of the lamellar lines, the hexagonal phase, for example, exhibits line ratios of $1:(3)^{-1/2}:(4)^{-1/2}:(7)^{-1/2} \ldots$, and the cubic phase often has the ratio $1:(3/4)^{1/2}:(3/8)^{1/2}:(3/11)^{1/2} \ldots$.

III. EXPERIMENTAL

Phase diagrams were constructed by observing the equilibrated state of the samples of different compositions prepared by weight. The presence of anisotropic phases was determined by observation under polarized light microscopy. Thorough mixing of the samples was achieved by vortexing and centrifugation until sample homogeneity was established.

The NMR spectra were obtained at a resonance frequency of 13.71 MHz for deuterium and 22.49 MHz for ^{13}C using a multinuclear pulsed NMR spectrometer (JEOL FX90Q) operating in the Fourier transform mode.

n-Dodecylpentaoxyethylene glycol ether was from Nikko Ltd. (Japan) and used without further purification. n-Alkanes were from Aldrich Ltd. and were passed through alumina before use; deuterium (99.5%) was obtained from Fluorochem Ltd.

IV. RESULTS AND DISCUSSION

A. Surfactant—Water Systems

The phase behavior of binary mixtures of C_nEO_m—water have been extensively studied and presented as temperature-composition diagrams (3). A combination of deuterium NMR and vapor pressure measurements indicated that a simple two-state description of the water in the systems was adequate (22—24). The observed quadrupolar splitting of the water, $\Delta\nu_{obs}$, could then be written simply as

$$\Delta\nu_{obs} = f_b \, \Delta\nu_b + f_{iso} \, \Delta\nu_{iso} \qquad\qquad [9]$$

where f_{iso} represents the fractions of water in the bound or isotropic "free" state and $\Delta\nu_i$ are the corresponding quadrupolar splittings when in state i. Because one of the fractions is regarded as

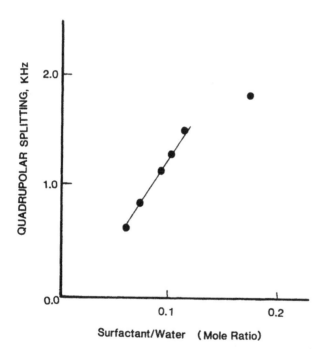

Surfactant/Water (Mole Ratio)

Figure 4.4 The dependence of quadrupolar splitting of water in the lamellar phase of $C_{12}EO_5$—water (3:2 w/w) at 298 K upon the amount of water.

being essentially similar to that of free liquid water, $\Delta\nu_{iso}$ is expected to be zero and Eq. 9 will reduce to

$$\Delta\nu_{obs} = f_b \, \Delta\nu_b \qquad\qquad\qquad\qquad\qquad [10a]$$

$$\Delta\nu_b = 3/4 \, \chi_c S_b \qquad\qquad\qquad\qquad\qquad [10b]$$

Here, χ is the quadrupolar coupling constant and S_b is the order parameter. Equation 10a takes into account the complex averaging through the random dispersion of the lamellar domains. Values of S_b derived from the observed splittings are usually in the range of 10^{-2} to 10^{-3}. A plot of $\Delta\nu_{obs}$ against the mole ratio of surfactant/water (Fig. 4) shows the linear portion predicted from Eq. 10 for the $C_{12}EO_5$—water system studied.

It may be concluded from these observations that the quadrupolar splitting of the water in the lamellar phase depends on

1. The degree of headgroup solvation as manifested by f_b
2. The amount of dynamic order in the bound sites as expressed in the order parameter, S_b

The utility of these facts has recently been made use of in a study of the interaction of commonly used hydrotropic agents with surfactant bilayers (25).

B. Surfactant—Water—Enhancer

The partial phase diagrams of the host surfactant—water system containing laurocapram (Azone), oleyl alcohol, and propylene glycol, respectively (Fig. 5), show that the amount of propylene glycol that can be accommodated is much smaller than the other two oils. It can also be seen that the incorporation of either oleyl alcohol or Azone increases the ability of the lamellar phase to accommodate water, with the maximum water content rising from about 50% in the host system to about 85% in the presence of the enhancer. This is not observed for solubilized propylene glycol for which, in fact, the water capacity of the bilayer phase is reduced on incorporation. Such an enhancement in the water retaining capacity of a bilayer system is of importance in the consideration of penetration enhancement, because it has been shown that the water content of the skin affects delivery rates in several cases (26–28).

Some information about the state of surfactant hydration in the presence of these agents may be derived from the variation of the quadrupolar splittings of the surfactant—$[^2H]H_2O$ system. Studies made of samples with a fixed surfactant/water mole ratio in the

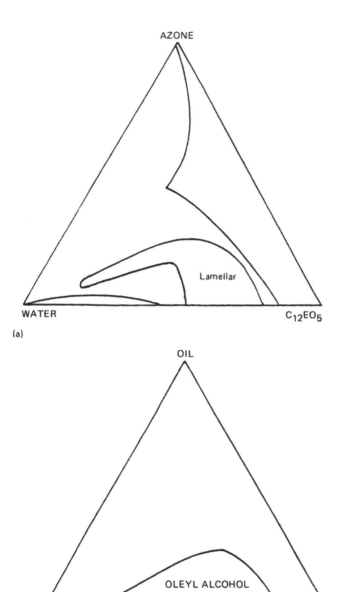

Figure 4.5 (a) Phase diagram at 298 K for the system $C_{12}EO_5$–Azone–water. (b) Partial phase diagrams showing the extent of the lamellar phases of the system $C_{12}EO_5$–water with oleyl alcohol and propylene glycol at 298 K.

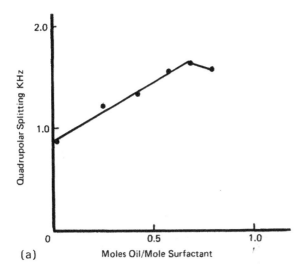

(a)

Figure 4.6 Quadrupolar splitting of $[^2H]H_2O$ in lamellar $C_{12}EO_5-$water (3:2 w/w) at 298 K as a function of solubilizate content: (a) Azone; (b) oleyl alcohol; (c) propylene glycol.

presence of increasing added solubilizate are illustrated in Figure 6. Any changes in $\Delta\nu_{obs}$ must be a result of changes in f_b or $\Delta\nu_i$ because the total water content is constant. The decrease in ΔV_{obs} in the presence of propylene glycol is similar to that found for the $C_{12}EO_4-$water system (25), and can be discussed in terms of changes in the composition of the hydration layer. A replacement of about 2 mol of bound water for each 1 mol of propylene glycol incorporated is implied.

The behavior of the systems containing either laurocapram or oleyl alcohol, however, is quite different with $\Delta\nu_{obs}$ showing large increases with increasing incorporation. This may be interpreted as an increase in the fraction of water molecules contributing to the bound fraction or as an increase in the order parameter of the bound water molecules, or a combination of both phenomena. The observed increase appears to be too large to be a result of an effective increase of f_b alone, if it is assumed that a maximum of 11 mol of water per mole of surfactant may contribute to this fraction. It is more likely that the increase arises, in part, from an increase in the order of the bound water fraction. Consideration of the structures of both oleyl alcohol and Azone reveals them also to be of an amphiphilic disposition, and they would be expected to incorporate between the surfactant molecules forming the bilayers. Such mode of solubilization of long-chained alcohols in lipid and surfactant bi-

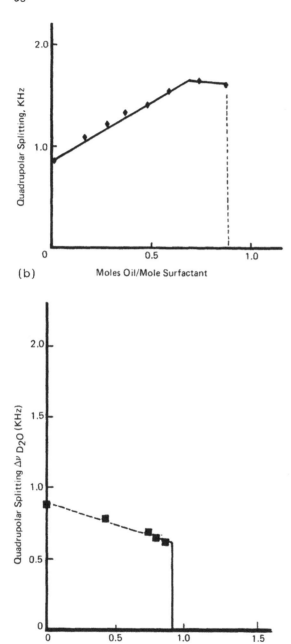

(b)

(c)

Figure 4.6   (Continued).

layers is well known (21) and, generally, leads to increased lamel-
lar-phase stability.

The orientation of the Azone molecules when solubilized in the
bilayer can be inferred from the behavior of the ^{13}C chemical shifts
of the ring carbonyl carbon atoms (Fig. 7). A comparison of the
^{13}C shifts in the lamellar phase with Azone dissolved in polar solvents such as alcohols or polyethylene glycol and in nonpolar media
such as alkanes indicates that the ring is located in the polar environment at the bilayer—water interface.

Inclusion of weakly amphiphilic molecules in the bilayer, therefore, provides further hydrogen-bonding type interactions at the

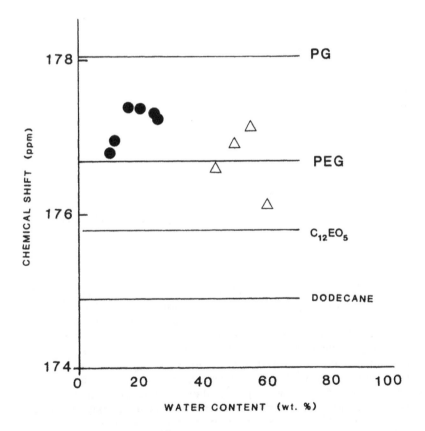

Figure 4.7 Chemical ^{13}C shifts of the carbonyl carbon of Azone in
different media.

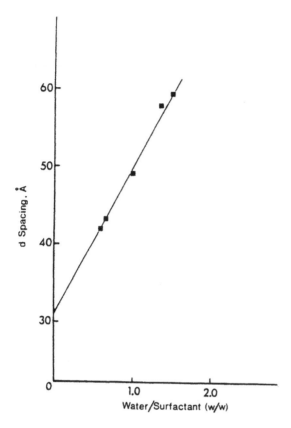

Figure 4.8 Interlamellar layer d-spacings of the system $C_{12}EO_5-$Azone—water.

interface with the aqueous compartment. This fact coupled with the increased structure arising from a greater number of trans config-urations in the hydrophobic region—a necessary result of packing of alkyl chains in this fashion—will lead to more order in the region of the polyoxyethylene chains to which the water molecules are bound. The consequence of this is to increase S_b and $\Delta \nu_b$.

Small-angle x-ray diffraction measurements yielding values of the characteristic d-spacing (Fig. 8) also indicate structural changes in the system as the oil content is varied at fixed mole ra-tios of Azone/water. The expected linear increase in d-spacing is found with increasing water content (Fig. 9), although the slope of the plot decreases with an increasing Azone/surfactant ratio. If the surfactant/water ratio is kept constant, an initial increase in the d-

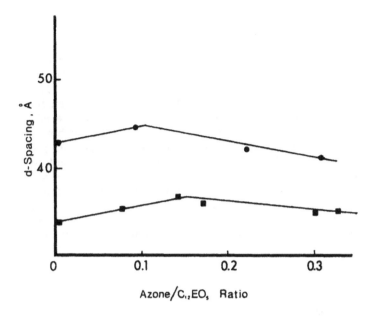

Figure 4.9 Interlamellar layer d-spacings of the system $C_{12}EO_5$–water (surfactant/water ■, 70:30; ●, 55:45 w/w) as a function of Azone content.

spacing compared with the Azone-free system is found, which is followed by a decrease at larger Azone contents. These data, when taken in conjunction with the ^{13}C shift results (see Fig. 7) indicate that Azone does not completely penetrate between the surfactant chains at low Azone concentrations. The degree of penetration increases from about 79% to 100% at 10%(w/w) of added Azone. At higher Azone concentrations, either there is increased water penetration into the bilayers, or extraction of surfactant into the bilayer interior: both of these would give an apparent decrease in the observed d-spacing. The initial increase in the quadrupolar splitting of the water would be expected if the degree of Azone penetration was increasing, as it is the water molecules associated with the EO-groups adjoining the alkyl chains that have the greatest contribution to $\Delta\nu_{obs}$. Thus, increasing Azone penetration leads to a decrease in the area occupied per amphiphile at the EO−hydrophobe interface and increasing order.

C. Incorporation of Mixed Enhancers

The differing influences that propylene glycol and either Azone or
oleyl alcohol have on the stability of the bilayers should be reflected
in the behavior of the system in the presence of binary mixtures
with propylene glycol. Such mixtures have been studied (4,29,30)
as topical delivery vehicles, and synergistic effects have been found
in the delivery of certain drugs. A preliminary study of the effects of incorporating binary mixtures of the type oleyl alcohol—
propylene glycol and Azone—propylene glycol into the lamellar phase
is summarized in Figures 10 and 11. There is an enhancement in
the maximum uptake of the mixture by the bilayers that is greater
than for either of the individual components. The enhancement is
most pronounced for oleyl alcohol—propylene glycol mixtures (about
55% w/w) which compares with about 26% for oleyl alcohol or about
5% for propylene glycol alone. The increased water capacity exists

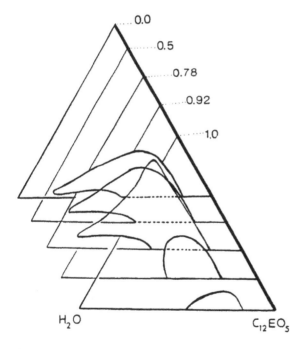

Figure 4.10 Partial phase diagrams of the system $C_{12}EO_5$—water
with solubilized oleyl alcohol—propylene glycol of different PG mole
fractions (numbers at the apex of the diagrams).

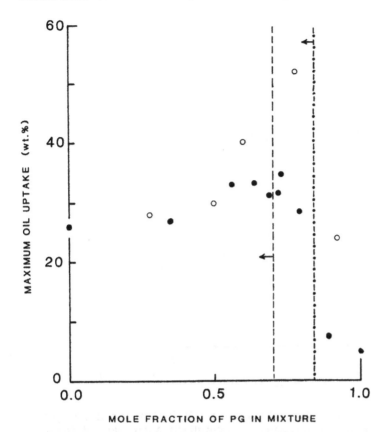

Figure 4.11 Maximum uptake of Azone—PG (closed circles) and oleyl alcohol—PG (open circles) mixtures in lamellar $C_{12}EO_5$—water (3:2 w/w) at 298 K, showing the limit of compositions at which increased water capacity is found; •—••—••—•• oleyl alcohol—PG; — — — — Azone—PG.

only for compositions with propylene glycol contents below those in the range of maximum synergism (i.e., propylene glycol mole fractions less than about 0.75). Therefore, a minimum amount of incorporated oleyl alcohol is required to obtain maximum synergism and water capacity of the bilayers.

Preliminary ^2H NMR measurements of the water in the systems containing solubilized mixtures of Azone—propylene glycol or oleyl alcohol—propylene glycol (Fig. 12) again show changes in slope of

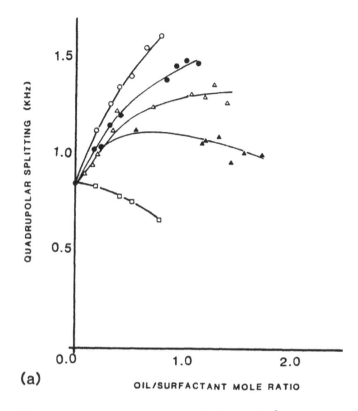

Figure 4.12   Quadrupolar splittings of $[^2H]H_2O$ in lamellar $C_{12}EO_5/$
water at 298 K in the presence of solubilized Azone—PG mixtures
(a) or oleyl alcohol/PG mixtures (b).   Oleyl alcohol mole fraction:
○, 1.0; ●, 0.85; △, 0.65; ▲, 0.52; □, 0.00.

the plots of $\Delta\nu_{obs}$ against solubilizate/surfactant mole ratio at constant water content.   Future small-angle x-ray investigations will
be necessary to determine if the compositions at which these changes
occur will correspond with d-spacings as found for Azone (31).
The initial increase in $\Delta\nu_{obs}$ is common to all of the binary compositions in the region showing synergism.   There is obviously a balance established between the stabilizing (and ordering) effect of the
oleyl alcohol and the destabilizing effect brought about by replacement of bound water molecules by propylene glycol.   In these
terms, there is almost complete balance at the composition of maximum promotion of bilayer phase stability, as denoted by the almost
complete independence of the water quadrupolar splitting from the
amount of the mixture solubilized (see Fig. 12).

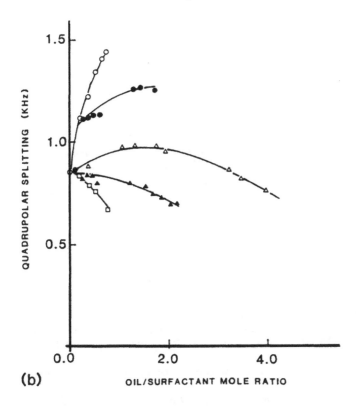

(b)

OIL/SURFACTANT MOLE RATIO

D. Implications for Vehicle Design and Interaction with the Stratum Corneum

A model system of this simplicity cannot, of course, be expected to give a comprehensive description of a system as complex as the stratum corneum. However, these results do give some indications relevant to the design of a topical vehicle and the potential interaction and modification of the skin to which it is applied. It can be seen that components often used in enhancement studies can potentially lead to increased water capacity in the system for which head group hydration is an important factor. Changes are also brought about in the molecular packing and dynamics of the bilayers leading to increased enhancer uptake in the hydrophobic region of the system. This description is in broad agreement with that recently inferred from differential thermal analysis and permeation studies (32).

Second, the model also shows synergistic aspects in both the water and solubilizate retention of the system, both features being characteristic of the real system. Extension of the model to include fatty acid—soap components is being made to incorporate other aspects of skin pH and composition.

ACKNOWLEDGMENT

This work has been part funded by EOLAS, The Upjohn Company, and thanks are due to Professor Stig E. Friberg for making available time on his small-angle x-ray apparatus.

REFERENCES

1. M. A. Lampe, M. L. Williams, and P. M. Elias, *J. Lipid Res.*, *24*:131, 1983.
2. H. J. Yardley, in *Biochemistry and Physiology of the Skin*, Chap. 16 (L. A. Goldsmith, ed.). Oxford University Press, Oxford, Eng., pp. 363—377, 1983.
3. D. J. Mitchell, G. J. T. Tiddy, L. Waring, T. Bostock, and M. P. McDonald, *J. Chem. Soc. Faraday Trans. I*, *79*:975, 1983.
4. R. B. Stoughton, and W. O. McClure, *Drug Dev. Ind. Pharm.*, *9*:725, 1983.
5. J. Charvolin, and B. Mely, in *Magnetic Resonance in Colloid and Interface Science* (H. A. Resing, and C. G. Wade, eds.). ACS Symp. Ser. 34, American Chemical Society, Washington, D.C., 1976.
6. J. Seelig, and A. Seelig, *Q. Rev. Biophys.*, *13*:19,1980.
7. E. Oldfield, in *Techniques in the Life Sciences: Lipid and Membrane Biochemistry* (J. C. Metcalfe, and R. Hesketh, eds.), *B427*:1, 1982.
8. J. H. Davis, *Biochim. Biophys. Acta*, *737*:117, 1983.
9. M. Rance, K. R. Jeffrey, A. P. Tulloch, K. W. Butler, and I. C. P. Smith, *Biochim. Biophys. Acta*, *600*:245, 1980.
10. J. Seelig, *Q. Rev. Biophys.*, *10*:353, 1977.
11. J. Charvolin, in *Lyotropic Liquid Crystals* (S. E. Friberg, ed.), *Adv. Chem. Ser. 152*:101, 1976.
12. T. Ahlnas, O. Soderman, C. Hjelm, and B. Lindman, *J. Phys. Chem.*, *87*:822, 1983.
13. H. Mantsch, H. Saito, and I. C. P. Smith, *Prog. Nucl. Magn. Reson. Spectrosc.*, *11*:211, 1977.
14. J. Seelig, and W. Niederberger, *J. Am. Chem. Soc.*, *96*:2069, 1974.

15. J. Charvolin, P. Manneville, and B. Deloche, *Chem. Phys. Lett.*, *23*:345, 1973.
16. N. Boden, Y. K. Levine, and A. J. I. Ward, *Biochim. Biophys. Acta*, *419*:395, 1976.
17. D. W. R. Gruen, *Prog. Coll. Polym. Sci.*, *70*:6, 1985.
18. S. Marcelja, *Biochim. Biophys. Acta*, *367*:165, 1974.
19. L. J. Burnett, and B. Muller, *J. Chem. Phys.*, *55*:5829, 1977.
20. V. Luzatti, in *Biological Membranes* (D. Chapman, ed.). Academic Press, London, p. 71, 1968.
21. P. Ekwall, *Adv. Liq. Cryst.*, *1*:1, 1975.
22. K. Rendall, and G. J. T. Tiddy, *J. Chem. Soc. Faraday Trans. I*, *80*:3339, 1984.
23. I. G. Lyle, and G. J. T. Tiddy, *Chem. Phys. Lett.*, *124*:432, 1986.
24. M. Carvelle, D. G. Hall, I. G. Lyle, and G. J. T. Tiddy, *Faraday Discuss. Chem. Soc. 81*:1, 1986.
25. A. J. I. Ward, C. Marie, L-A. Sylvia, and M. A. Phillipi, *J. Dispersion Sci. Technol.*, *9*:149—170, 1988.
26. I. H. Blank, *J. Invest. Dermatol.*, *18*:433, 1952.
27. R. J. Scheuplein, and I. H. Blank, *Physiol. Rev.*, *51*:702, 1971.
28. R. J. Scheuplein, *J. Invest. Dermatol.*, *67*:672, 1976.
29. A. Hoelgaard, and B. Mollgaard, *J. Controlled Release*, *2*:111, 1985.
30. P. K. Wotton, B. Mollgaard, J. Hadgraft, and A. Hoelgaard, *Int. J. Pharm.*, *24*:19, 1985.
31. A. J. I. Ward, and R. Tallon, *Drug Dev. Ind. Pharm.*, *14*:1155, 1988.
32. B. W. Barry, in *Skin Pharmokinetics*. (B. Shroot, and H. Schaefer, eds.), Karger, Basel, pp. 121—137, 1987.

5
The Influence of Skin Surface Lipids on Topical Formulations

DAVID W. OSBORNE and DOUGLAS A. HATZENBUHLER *The Upjohn Company, Kalamazoo, Michigan*

I. INTRODUCTION

As both the largest and most visible organ of the human body, the skin is of unequaled importance in portraying an individual's state of being. Therefore, the biology and chemistry of the skin and its appendages is important for both the cosmetic and pharmaceutical industries. The heterogeneous nature of the skin provides one of the difficulties in studying this tissue. Just as the skin itself comprises morphologically different layers, the glands, follicles, and microvasculature of the skin provide zones with different characteristic properties. One of these properties is that a thin, noncontinuous film of lipid is deposited on the skin surface from the sebaceous glands. Although often neglected in the design of a topical formulation, the presence of skin surface lipids can significantly influence the delivery of topical drugs. This influence may be either beneficial to, or detrimental in, obtaining the desired drug delivery characteristics and, thus, should be considered, especially for topicals applied to sebum-rich areas of the body (e.g., forehead, cheeks, and scalp). It is important to realize that the interaction between a drug and the skin surface lipid is not only dependent upon the physicochemical properties of the drug, but also upon the physicochemical nature of the vehicle. Topical agents can be formulated that will target sebum-rich zones of the skin (e.g., the hair follicle) or, alternatively, a topical agent can be formulated to minimize delivery through the hair follicle. Background information on skin surface lipids and a brief description of the approach necessary to formulate sebum-selective vehicles will be the focus of this chapter. The discussion is designed to remain as simplistic as possible in an attempt to introduce concepts that can be studied in more detail by consulting the original research efforts that are referenced.

II. SKIN AND SURFACE LIPID
EXCRETION

The skin is traditionally divided into three major regions: the stratum corneum, the viable epidermis, and the dermis. The outermost of these layers, the stratum corneum, provides a barrier against the permeation of most substances. This layer is composed of nonviable, keratin-filled cells (squames) that are roughly pentagonal plates 0.5 μm thick and 30 to 40 μm across (1). Filling the intercellular space between these cells are bilayer-structured lipids (2). The structure of these lipids has proved to be important to the moisture-retaining ability of the stratum corneum (3,4). The viable epidermis lies below the stratum corneum and consists of stratified keratinizing epithelial cells whose final function is to produce the stratum corneum. This layer does not contain blood vessels, relying on nourishment by cell fluid from the deeper dermis layer. The deepest layer of skin is the dermis, which consists of dense, irregularly arranged connective tissue, and it is nourished directly by blood vessels (5). Combined, these layers form the skin and are connected to the subcutaneous tissue by bundles of collagen fibers.

Embedded in the skin are eccrine sweat glands, apocrine glands, hair follicles, and sebaceous glands. Eccrine sweat glands are simple tubular glands distributed over almost all of the human body. Each gland has a secretory part located below the dermis in the subcutaneous tissue and an excretory duct that ultimately opens directly on the skin surface. These glands produce perspiration and are particularly numerous on the palm of the hand. Apocrine glands produce characteristic body odors and are primarily located in the axilla. The secretion of this gland contains cholesterol, steroids, and proteinaceous substances (6) and contributes substantially to the surface lipid on the axilla, thus, being primarily important to deodorant and antiperspirant formulations. When present, the apocrine gland empties into the hair follicle above the sebaceous gland.

Two major classifications of hair exist on humans: terminal hairs and vellus hairs. Terminal hairs are the courser hairs of the scalp and male trunk hair, and the root of terminal hairs may extend more than 3 mm below the skin surface into the subcutaneous fatty tissue. Vellus hair is the fine, often unnoticed, body hair that populates regions such as the forehead, and extends less than 1 mm into the dermis. The hair follicle comprises the hair, combined with its surrounding root sheath. The sebaceous gland empties directly into the upper third of the hair follicle, and the combination of sebaceous gland and follicle is often referred to as the *pilosebaceous unit*. The sebaceous gland is the major source of skin surface lipid in the adult. A diagram representing both the skin layers and skin appendages is shown in Figure 1.

Because the sebaceous gland is the primary source of the skin surface lipids covering large portions of the anatomy, understanding the properties of this skin appendage becomes important. The sebaceous gland is well defined in the embryo, remains relatively small

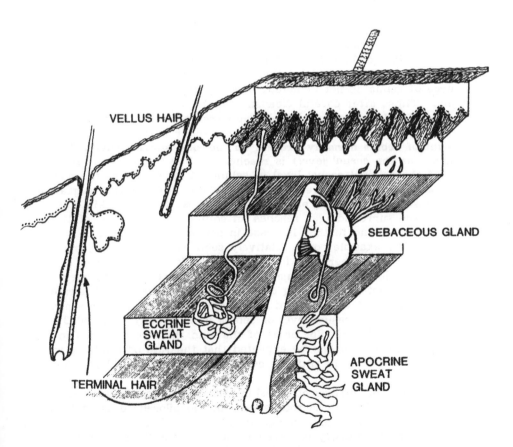

Figure 5.1 Cross-sectional representation of full-thickness thin skin.

during childhood, and becomes drastically enlarged during puberty
(7). In each gland, a common excretory duct is supplied by smaller
ducts that originate in the acini of the gland. As the flattened or
cuboidal peripheral sebaceous cells move toward the center of the
gland, lipid synthesis within the cell increases. The cell swells with
fat until a 100- to 150-fold increase in the cell volume occurs. The
entire cell then ruptures, expelling its contents into the excretory
stream of the gland as sebum. Finally, the sebum passes through
the pilosebaceous follicle and is deposited into the uppermost region
of the hair shaft. Synthesis and discharge of the lipid contained
in the sebaceous gland requires 1 to 3 weeks (7). The discharge
of this lipid has been demonstrated by Kligman to be constant, without regard to either season or the amount of lipid already on the
skin surface (8). However, the quantities of sebum that are normally encountered on unprotected healthy skin depend more upon
the rates of flow (or refatting process) of sebum than upon the absolute rate of production of sebum in the sebaceous gland. This
results because there is a large reserve of fully synthesized sebum
contained in the *follicular reservoir* (9), that is, the upper portions
of the hair follicle and the orifice to the sebaceous gland. This
source of sebum is not depleted, even by repeated solvent extrac-
tions. Thus, after careful cleansing of the skin, sebum contained
in the follicular reservoir initially appears to flow out onto the skin
surface at a constant rate over a several-hour period until the
amount of sebum normally present on the skin surface (defined by
Kligman as the *casual level*) is reached. After the skin has re-
gained the original casual level of sebum, it appears that the normal loss process (i.e., reabsorption into the skin, further migration
to less sebum-rich areas of the skin, and loss from the skin surface to the surroundings) reaches an equilibrium with the refatting
processes, and no further increase in sebum concentration is observed. For example, on a relatively sebum-rich skin area, such
as the forehead, where the follicular reservoir is estimated to contain a few milligrams of sebum per cubic centimeter, it has been
shown (10) that the refatting of the skin surface occurs at a rate
of from about 0.1 to 2.1 µg of sebum per square centimeter a minute, and that the casual level of sebum is essentially restored 3 or
4 hr after defatting. For areas of the skin that are less rich in
sebaceous glands, it can be expected that the rate of refatting will
be slower and that the ultimate casual level of sebum will also be
lower.

   Once on the surface of the skin, the sebum has already been
chemically modified by microorganisms in the pilosebaceous unit and
becomes mixed with lipids of epidermal origin. Hence, the term
sebum is more precisely reserved to describe the lipid contained in
the sebaceous gland, and the term *skin surface lipid* is used to describe the lipid mixture on the skin surface.

III. COMPOSITION OF SKIN SURFACE LIPIDS

Skin surface lipid primarily contains diglycerides, cholesterol, fatty acids, triglycerides, wax esters, cholesterol esters, and squalene (7). The amounts of these components and their variations with anatomical location have also been investigated. Table 1 lists the results obtained by Greene et al. (11). When the specific structures of the lipids in these general categories are examined, the skin surface lipid becomes significantly more complex. Not only do the chains of the lipids exhibit different lengths and degrees of saturation, but they also contain odd-numbered-carbon chains, uncommon in biological systems, and significant branching (12). The amount that the lipid fractions of human sebum exhibit these characteristics is given in Table 2 (12).

The complexity of the chemically diverse mixture of lipid that eventually reaches the skin surface is a result of equally diverse influences on the sources of this lipid. The two sources already mentioned are always associated with skin surface lipids and warrant further discussion. The sebaceous gland is considered the primary, if not exclusive, source of squalene. Classic work by Nicolaides and Rothman (13) demonstrated that the sebaceous glands contain an incomplete enzyme system, compared with the epidermis, rendering the sebaceous gland incapable of finishing the synthesis of cholesterol from squalene. For this same reason, the source of cholesterol is distinctly epidermal. Use of this fact explained the bulk differences in the anatomical variation in human skin surface lipids. As seen in Table 1, the forehead, which contains 400 to 800 sebaceous glands per square centimeter (7), produces skin surface lipid rich in squalene and low in cholesterol. This is because the skin surface lipid of the forehead is primarily of sebaceous origin. Conversely, the surface lipid of leg skin is high in cholesterol (9.4%) and low in squalene (6.2%) because of the greatly reduced sebaceous contribution (11).

The source of the fatty acids found on the skin surface is both epidermal and indirectly sebaceous. Elias et al. (14) has characterized human epidermal lipids (Table 3) and found the free sterols/ free fatty acid ratio for the face to be 0.9. Assuming that all of the cholesterol characterized in Table 1 for the forehead is of epidermal origin, then only about 6% of the free fatty acid on the forehead skin surface is of epidermal origin. The remainder is the hydrolysis product of triglycerides that originate in the fatty acid-free sebum (15). Kligman showed that the bacterium *Corynebacterium acnes*, which inhabits the follicular canal, is responsible for this hydrolysis (16). Variations in the bacterial flora among different individuals is used to explain why comparisons between individuals exhibit dramatic differences in the amount of fatty acid in their

Table 5.1 Anatomical Variation in the Amount and Composition of Human Skin Surface Lipids in Adult Male Subjects[a] 3 hr after Washing with Soap and Water

Composition (wt%)	Forehead	Cheek	Chest	Back	Side	Arm	Leg
Triglycerides	30.3	30.9	32.7	34.5	39.2	30.7	24.2
Diglycerides	2.3	2.9	2.3	2.7	1.9	1.5	1.8
Free fatty acids	24.1	22.4	28.1	24.9	23.4	36.4	37.8
Wax esters	27.0	24.3	21.7	21.9	17.7	15.8	12.9
Squalene	12.3	13.5	10.7	10.7	8.5	6.9	6.2
Cholesterol	1.2	2.0	1.6	2.1	3.9	4.1	9.4
Cholesterol esters	2.9	3.9	2.9	3.0	5.3	4.4	7.5
Total lipid[b] (g/cm^2)	160	104	59	38	29	30	19

[a]The data are averages for three subjects, each examined on two occasions.
[b]Graphic extrapolation indicated that the epidermal lipid contributed an average of 4 g/cm^2.
Source: From Ref. 11.

Table 5.2 Fatty Acid Composition of Each Lipid Fraction Taken

Chain type (%)	Triglyceride	Cholesterol ester and hydrocarbon	Free fatty acid
Branched chains	15.0	13.6	12.4
Straight chains	84.9	86.4	87.5
Odd-carbon-number chains	22.5	26.2	21.8
Even-carbon-number chains	77.4	73.8	78.1
Saturated chains	70.9	56.2	69.7
Unsaturated chains	29.9	43.8	30.2

Source: From Ref. 12.

Table 5.3 Composition of Human Epidermal Lipid

Component	Wt%
Polar lipids	4.9
Cholesterol sulfate	1.5
Neutral lipids	74.8
free sterols	14.0
free fatty acids	19.3
triglycerides	25.2
sterol or wax esters	5.4
squalene	4.8
n-alkanes	6.1
Sphingolipids	18.8

Source: From Ref. 14.

skin surface lipid, although for a given individual, the fatty acid
component of the skin surface lipid remains relatively constant with
time. Thus, each individual shows an inverse relationship between
the amount of free fatty acids and the triglycerides characteristic of
their skin surface lipids.

A hydrolysis reaction is the source of one of the important
structural components of the epidermal lipids. The enzyme sphingomyelinase hydrolyzes sphingomyelin to ceramide and phosphorylcholine (17). This enzyme degrades all cellular phospholipids before
the cells move from the granular layer of the epidermis into the
stratum corneum. The ceramides then become one of the main components of the layered intercellular lipids found between the terminally differentiated cells of stratum corneum. The other hydrolysis
product, phosphorylcholine, does not fill the stratum corneum spaces
but, rather, the phosphorus is thought to be reabsorbed and re-
utilized by the epidermis. This is an explanation for why the skin
surface lipids are void of phospholipids (18).

In addition to the substances listed in Table 1, small quantities
of other compounds have been found in sebum. Detectable amounts
of androgens are present in skin surface lipids (19); however, the
amounts of these steroids are considered too small to claim that the
sebaceous glands play a notable role in the total body excretion of
androgens. Paraffinic hydrocarbons, such as pristane, also have
been shown to be of sebaceous origin (2). Their presence is speculated to be linked to diet.

The influence of diet on skin lipids is another factor that adds
to the chemical complexity of skin surface lipids. Fasting causes a
50% reduction in the synthesis of fatty acids, wax esters, and triglycerides, although not influencing squalene production (19).
Likewise, the absence of dietary essential fatty acids (linoleic acid)
results in a stratum corneum with a changed appearance and a significantly reduced barrier function (2). Nicolaides (21) showed that
feeding 1-[14C] octadecane to the rat led to the presence of the
labeled substance in the skin. Although a complete understanding
of skin metabolism is not yet available, it is important to remember
that some of the components found in the skin surface lipids are
present simply because the skin was the most efficient route of
"waste" removal.

When examining the components found in skin surface lipids,
the possibility that the trace substances are from exogenous sources
must be considered. The obvious sources are cosmetics, topical
pharmaceuticals, toiletries, and environmental pollutants. Although
most studies that use skin surface lipids collected from human do-
nors try to eliminate these contaminants, the reservoir function of
the stratum corneum makes it particularly difficult to remove totally
some substances from the skin surface. Because chemicals can be

readily retained in the stratum corneum, even after 16 days (22), the total elimination of exogenous lipophilic materials becomes difficult.

Methods used to obtain skin surface lipids for studies usually involve application of a solvent to the anatomical site of interest. This can be done by using cotton swabs (23), solvent cups (24), and head soaks (25). Nonsolvent methods exist that take advantage of the absorbent properties of bentonite clay (26), and the use of absorbent paper (7,27) as well as ground-glass surfaces (28) has been reported. The amounts and composition of the lipids are then determined by chromatography. Other spectroscopic (29) and thermal (3) techniques have also been used.

IV.   FALLACIES CONCERNING SKIN SURFACE LIPIDS

In 1963, Kligman, in the article *The Uses of Sebum?* (31), challenged a number of previously held concepts concerning the human integument. Despite the vintage of this essay, some of these fallacies that Kligman refuted still surface in the literature and, thus, are worthy of restating. Two of these fallacies are that sebum is antibacterial and antifungal and that sebum contains a vitamin D precursor. The vitamin D precursor was found to be of epidermal origin, whereas the antibacterial and antifungal properties were considered to be primarily artifacts of in vitro testing methods. Kligman concluded this article with the statement: "Human sebum seems to be useless. Since hair has become vestigial over most of the surface of the human body, sebaceous glands are probably obsolescent appendages." Although this final statement seems extreme, it provides a balance to the previously overemphasized benefits of skin surface lipids.

Another area of confusion in the literature is the question of whether or not there was a feedback mechanism (the concept that the glands shut down sebum production when a certain level of surface fat is achieved) controlling the sebaceous glands. Kligman and Shelly (32) demonstrated that, under carefully controlled conditions, it was clearly possible to demonstrate a large buildup of skin surface lipids, well above the level that was speculated to cause shutdown of the glands' production. This work, together with a better understanding of the holocrine nature of the sebaceous glands (1), has fully discredited the feedback hypothesis. Note, however, that even though the rate of formation of sebum is constant, from the practical point of view of, "how much sebum will a topical formulation encounter on normal healthy skin?", the sebum flow and refatting processes on the surface of the skin are a more direct concern. Moreover, it is also well established that the amount of sur-

face lipid present on healthy, unprotected skin will plateau a few
hours after the skin has been thoroughly washed (33).

The rate of sebum production does not show seasonal changes
(9), although it is again possible to demonstrate seasonal variations
in perceived oiliness of the skin, primarily because of the amount
of perspiration and other factors that affect the sebum refatting and
redistribution processes.

V. MODEL SKIN SURFACE LIPIDS

One approach to determining the effect that skin surface lipids have
on topical formulations is to model either the properties or composition of skin surface lipids by using readily available materials. This
avoids the laborious task of pooling lipids from volunteers and provides the investigator with convenient quantities of lipid material.
Skin surface lipid models are of interest to not only cosmetic and
pharmaceutical industries but, also, to the detergent industry, for
which they are used to compare the abilities of laundry formulations
to clean soiled fabrics. Because of this interest from various factions of the industrial research community, a number of model skin
surface lipids are cited in the literature. However, most of these
models were designed for specific studies, thus, limiting their prac-
ticality in evaluating cosmetic and pharmaceutical topical agents.
The compositions of some of these model lipids will be given, and
the advantages and disadvantages of each model will be discussed,
in this section.

The Friberg—Osborne model skin surface lipid (34) given in Table 4 is based on the chemical composition elucidation studies that
have been carried out by various investigators. In this model, each
of the major chemical factions of human sebum are represented by
a purified commercially available chemical. For instance, the triglycerides are represented by triolein, whereas the wax esters are
represented by oleic acid palmitic ester. This model has two major
advantages: it chemically mimics natural sebum, and it is a single-
phase liquid that is easy to work with at ambient temperatures.
This second advantage is also the model's strongest disadvantage,
because the physical properties of sebum (melting point about 35°C)
are not obtained. This model was initially used to describe the
phase behavior of a simplified cosmetic system when mixed with the
model lipid. It is in this capacity that a single-phase liquid model
becomes essential for practical studies.

Another model, which is based on the chemical components
found in natural skin surface lipids, is given in Table 5. Again,
the major advantage to this model is that it chemically mimics the
skin lipids. At room temperature, this mixture is a waxy solid that

Table 5.4 Composition of the Friberg–
Osborne Model Skin Surface Lipid

Component	%
Oleic acid	16.5
Myristic acid	1.9
Triolein	41.8
Oleic acid palmitic ester	20.3
Cholesteryl oleate	3.0
Pristane	2.8
Squalene	12.2
Lecithin	1.5

Source: From Ref. 34.

Table 5.5 Model Skin Surface Lipid

Component	%
Palmitic acid	10
Myristic acid	4
Oleic acid	6
Tripalmitin	20
Triolein	20
Palmitic acid palmitic ester	10
Oleic acid palmitic ester	10
Cholesterol	3
Cholesteryl palmitate	1
Cholesteryl oleate	1
Squalene	15

Table 5.6 Composition of Spangler's Sebum Model

Component	%
Olive oil	20.0
Coconut oil	15.0
Palmitic acid	10.0
Stearic acid	5.0
Oleic acid	15.0
Paraffin wax	10.0
Squalene	5.0
Spermaceti	15.0
Cholesterol	5.0
Total	100.0

Table 5.7 Composition of the Gordon Sebum Model

Component	%
Hydrocarbon oil (medium-viscosity lubricating oil)	25
Tristearin	10
Arachis oil	20
Stearic acid	15
Oleic acid	15
Cholesterol	7
Octadecanol	8

Table 5.8 Composition of the Morris
Skin Surface Lipid Model

Component	%
Squalene	8
Tristearin	23
Triolein	23
Stearic acid	15
Oleic acid	15
Cholesterol	8
Octadecanol	8

contains considerable crystalline material, even after repeated melt-ing and cooling cycles. It is the presence of crystalline material, even at 60°C, that limits the usefulness of this model because it is difficult to manipulate samples of this nonhomogeneous mixture at ambient temperatures.

Spangler's sebum model (Table 6) dates back to 1964 and com-prises a mixture of natural fats and oils (35). After melting, mix-ing, and then cooling these components, a homogeneous waxy mate-rial results that mimics well certain physical properties of skin sur-face lipids. This model has the chemical advantage of comprising multiple components that have a wide range of branching and of degrees of saturation. Unfortunately, the chemical properties of vegetable oils are unlikely to be similar to those of human sebum, and the rationale for the materials used appears to be based on ob-taining a wax at ambient temperatures. This model has the added problem of crystal formation after approximately 1 month on the shelf.

The Gordon model (Table 7) resembles that of Spangler, but it was designed for the double-label radiotracer studies that are used to compare the detergency of a series of detergent systems (36). The advantages of this doubly labeled system also apply to topical formulations; namely, the selectivity of a formulation can be inves-tigated, while the number of experiments is minimized. The com-mercial availability of the radiolabeled material dictated the composi-tion of this model. The very similar Morris model (Table 8) was originally used to determine retention of an artificial oily soil on cotton and polyester—cotton durable-press fabrics after laundering (37). This model has the slight advantage that all of the compo-nents are readily available.

VI. INTERACTIONS BETWEEN SKIN SUR-
FACE LIPIDS AND TOPICAL
FORMULATIONS

Research that is concerned with the interaction of topical agents
and skin surface lipids can be divided into two major areas. The
first is the research that is concerned with targeting sebum-rich
zones of the skin (e.g., hair follicles and sebaceous glands). This
is of obvious importance to acne formulations. The second is research concerned with structural changes that occur in the topical
vehicle after mixing with skin surface lipid. This area is most important when formulating topical agents that will be used on lipid-
rich regions of the body, such as the forehead or scalp.

The initial observation that it is possible to "target" delivery
of drug substances to the hair follicle appears to have been made
by investigators at Unilever (38). Their work noted that if a vehicle with low polarity (i.e., matching the polarity of sebum) and
low viscosity was used, greater amounts of drug could be detected
near the base of the hair follicle. Further work in our laboratories
(Hatzenbuhler, unpublished results) has demonstrated that this appears true, even for drug molecules that are inherently too polar
to be readily soluble in sebum. Thus, for localized delivery into
the region of the hair follicle, sebum appears to present an important barrier, somewhat analogous to that of the stratum corneum
barrier (although structurally totally different) across the remainder of the skin. Importantly, the much greater surface area of
the stratum corneum makes this route of delivery generally dominant
for the systemic delivery of topically applied drugs (39). However,
this does not negate the foregoing observations for local delivery
to sebum-rich regions such as the hair follicle.

The technique most widely used for the measurement of local
drug distributions in tissue is autoradiography; a method that uses
radiolabeled drug to measure the concentration of drug throughout
thin sections (typically only a few microns thick) of the tissue.
The sectioning process has been described as highly technique-de-
pendent for any type of tissue, and it is particularly difficult for
the study of skin tissue. Thus, one should carefully reproduce
results before making final judgment on their validity. Classic
methodology in this technique is provided in Ref. 40.

An alternative method to autoradiography, which can also be
used to specifically measure the amount of drug delivered into sebum-rich regions, has been proposed by Elias (41). This technique
uses different solvents to differentially extract sebum, rather than
the lipids of skin surface origin. By comparing the relative amounts
of drug present in extracts of skin surface lipids with extracts of

sebum, an estimation of the relative efficiency that two vehicles show for delivering a drug into the sebum-rich areas may be made.

The basis for understanding a vehicle's effect on the delivery into sebum-rich areas, such as the hair follicle, appears to be fully explained by conventional solubility properties. Hildebrand solubility coefficients (42) appear adequate to predict this performance. The Hildebrand coefficients for model sebum compositions demonstrate that sebum is an overall nonpolar, oily material (with a Hildebrand coefficient of approximately $7.5-8$ cal$^{1/2}$/cm$^{3/2}$). Many topical vehicle components, such as water, propylene glycol, and ethanol, with Hildebrand coefficients of 23.4, 14.0, and 12.55, respectively, are too polar to be readily soluble in sebum (typically ± 2 units on the Hildebrand coefficient indicates miscible materials). Therefore, for many relatively polar drugs, if the vehicle is not specifically designed to solubilize the sebum reservoir in the hair follicle, there will be little chance to effectively deliver the drug into the deeper portions of the hair follicle. This may be utilized to either avoid delivering a local excess of drug or, perhaps more significantly, to attempt to localize drug delivery into the pilosebaceous unit.

The determination of interactions between skin surface lipids and topical agents involves utilizing various physicochemical techniques. These techniques include contact angles, solubility parameters, and phase behavior determination. Contact angles are used to provide information on the ability of a formulation to wet the skin. These measurements can be done in vivo and in vitro. In vitro studies can use stratum corneum sheets that have been separated from cadaver skin with a trypsin solution. Natural or model skin surface lipids can then be added in known amounts to the stratum corneum sheets. This technique has shown that contact angle measurements can differentiate between relatively polar vehicles that show superior wetting on stratum corneum sheets and less polar vehicles that more effectively wet a sebum film deposited on the stratum corneum (Hatzenbuhler, unpublished results). This work demonstrates that the difference in the polarity of the protein surface of the stratum corneum and the polarity of sebum is sufficiently large to show observably different vehicle performance. Detailed clinical or cosmetic evaluation of this performance difference apparently has not been reported. However, use of solubility parameters in cosmetics formulations was thoroughly described in a recent article by Vaughan (43), in which he lists the solubility parameters for over 150 cosmetic materials.

VII. CONCLUDING REMARKS

As described, the skin and its appendages combine to form a complex, heterogeneous network of epithelial tissue and glands that produces both the proteins and lipids that form the skin's barrier. One aspect of this network, the skin surface lipids, can influence the topical products that are applied to the skin. This influence may be either beneficial or detrimental, depending upon the formulator's awareness of the physicochemical properties of skin surface lipids and the desired drug delivery characteristics. By matching the polarity of the formulation to the polarity of the skin surface lipids, the hair follicle can be targeted for drug delivery, even if the drug is inherently too polar to be readily soluble in sebum or skin surface lipid. The reader is further encouraged to consult the review articles by Strauss et al. (7) and Wheatly (44) for more detailed consideration of skin surface lipids.

REFERENCES

1. G. F. Odland, in *Biochemistry and Physiology of the Skin*, Vol. 1 (L. A. Goldsmith, ed.). Oxford University Press, New York, pp. 3–63, 1983.
2. P. M. Elias, *Int. J. Dermatol.*, *20*:1, 1981.
3. G. Imokawa, and M. Hattori, *J. Invest. Dermatol.*, *84*:282, 1985.
4. D. W. Osborne, and S. E. Friberg, *J. Dispersion Sci. Technol.*, *8*:173, 1987.
5. A. H. Ham, *Histology*, 7th ed. J. B. Lippincott, Philadelphia, pp. 593–623, 1974.
6. J. N. Labows, K. J. McGinley, and A. M. Kligman, in *Principles of Cosmetics for the Dermatologist* (P. Frost, and S. N. Horwitz, eds.). C. V. Mosby, St. Louis, pp. 89–97, 1982.
7. J. S. Strauss, D. T. Downing, and F. J. Ebling, in *Biochemistry and Physiology of the Skin*, Vol. 1 (L. A. Goldsmith, ed.). Oxford University Press, New York, pp. 569–595, 1983.
8. A. M. Kligman, and W. B. Shelley, *J. Invest. Dermatol.*, *30*: 99, 1958.
9. A. M. Kligman, J. J. Leyden, and K. J. McGinley, in *Principles of Cosmetics for the Dermatologist*, Chap. 3. C. V. Mosby, St. Louis, 1982.
10. D. Saint-Leger, and E. Cohen, *Br. J. Dermatol.*, *113*:551, 1985.
11. R. S. Greene, D. T. Downing, P. E. Pchi, and J. S. Strauss, *J. Invest. Dermatol.*, *54*:240, 1970.
12. H. Kosugi, and N. Ueta, *Jpn. J. Exp. Med.*, *47*:335, 1977.
13. N. Nicolaides, and S. Rothman, *J. Invest. Dermatol.*, *24*:125, 1955.
14. M. A. Lampe, A. L. Burlingame, J. Whitney, M. L. Williams, B. E. Brown, E. Roitman, and P. M. Elias, *J. Lipid Res.*, *24*: 120, 1983.

15. N. Nicolaides and G. C. Wells, *J. Invest. Dermatol.*, *29*:423, 1957.
16. R. R. Marples, D. T. Downing, and A. M. Kligman, *J. Invest. Dermatol.* *56*:127, 1971.
17. P. A. Bowser, and G. M. Gray, *J. Invest. Dermatol.* *70*:331, 1978.
18. S. H. Emara and H. S. El-Mokaddem, *Z. Hautkr.* *54*:641, 1979.
19. M. E. Karunakaran, P. E. Pochi, J. S. Strauss, E. A. Valerio, H. H. Wotiz, and S. J. Clark, *J. Invest. Dermatol.*, *60*:121, 1973.
20. H. J. O'Neill, L. L. Gershbein, and R. G. Scholz, *Biochem. Biophys. Res. Commun.* *35*:946, 1969.
21. N. Nicolaides, *Lipids*, *1*:87, 1966.
22. R. B. Stoughton, *Arch. Dermatol.*, *91*:657, 1965.
23. K. K. Jones, M. C. Spencer, and S. A. Sanchez, *J. Invest. Dermatol.*, *17*:213, 1951.
24. N. Nicolaides, R. E. Kellum, and P. V. Woolley III, *Arch. Biochem. Biophys.*, *105*:634, 1964.
25. D. T. Downing, A. M. Stranier, J. S. Strauss, *J. Invest. Dermatol.*, *79*:226, 1982.
26. S. J. Lewis, A. R. Shalita, and W. Lee, *J. Invest. Dermatol.*, *71*:370, 1978.
27. J. S. Strauss and P. E. Pochi, *J. Invest. Dermatol.*, *36*:293, 1961.
28. D. Saint-Leger, C. Berrebi, C. Duboz, and P. Agache, Arch. Dermatol. Res., *265*:79, 1979.
29. P. Bore, N. Goetz, and J. C. Caron, *Int. J. Cosmet. Sci.*, *2*:177, 1980.
30. W. J. Cunliffe, J. N. Kearney, and N. B. Simpson, *J. Invest. Dermatol.*, *75*:394, 1980.
31. A. M. Kligman, in *Advances in Biology of Skin*, Vol. 4, *The Sebaceous Glands* (W. Montagna, R. A. Ellis, and A. F. Silver, eds.). Permagon Press, Oxford, pp. 110—124, 1963.
32. A. M. Kligman and W. B. Shelley, *J. Invest. Dermatol.*, *30*: 99, 1958.
33. D. Saint-Leger and J. L. Leveque, *Br. J. Dermatol.*, *106*: 669, 1982.
34. S. E. Friberg and D. W. Osborne, *J. Am. Oil Chem. Soc.*, *63*:123, 1986.
35. W. G. Spangler, *J. Am. Oil Chem. Soc.*, *41*:300, 1964.
36. B. E. Gordon, J. Roddewig, and W. T. Shebs, *J. Am. Oil Chem. Soc.*, *44*:289, 1967.
37. M. A. Huisman and M. A. Morris, *Text. Res. J.*, *41*:657, 1971.
38. T. Rutherford and J. G. Black, *Br. J. Dermatol.*, *81*(suppl. 4):75, 1969.

39. R. J. Scheuplein, *J. Invest. Dermatol.*, *48*:79, 1967.
40. W. E. Stumpf, and L. J. Roth, *Autoradiography of Diffusible Substances*, Academic Press, New York, 1969.
41. P. M. Elias, Personal communication.
42. J. H. Hildebrand, and R. L. Scott, *The Solubility of Nonelectrolytes*. Reinhold, New York, 1949.
43. C. D. Vaughan, *J. Soc. Cosmet. Chem.*, *36*:319, 1985.
44. V. R. Wheatly, *Physiol. Pathophysiol. Skin*, *9*:2829, 1986.

6

Immunology of the Skin

DAVID H. LYNCH *Immunex Corporation, Seattle, Washington*

LEE K. ROBERTS *University of Utah School of Medicine, Salt Lake City, Utah*

I. INTRODUCTION

The skin, by weight, is the largest organ of the body. Human skin serves to provide several important functions. These include maintaining physical protection (barrier function) against external agents and dessication; receiving sensory stimuli from the environment; regulating body temperature and water balance; excreting a variety of substances; participating in metabolic pathways (e.g., initiation of vitamin D synthesis and subcutaneous fat metabolism); and serving as a compartmentalized component of the immune system to provide protection against certain pathogens, toxins, and neoplasia. To protect the host from foreign materials and organisms, an extremely complex relationship has evolved between the skin and the immune system. Many of the cell types in skin synthesize a variety of bioactive compounds, many of which have profound effects, not only on local inflammatory responses and skin-associated immune responses but, also, on systemic immune responses.

The development of strategies to transdermally deliver pharmacologically active compounds is an exciting new area of research with far-reaching implications. However, transdermal delivery of pharmacologically active compounds also presents a number of challenging problems because the skin is normally fairly impervious to even relatively small molecules. Thus, from the pharmacologic point of view, one of the major requirements for effective transdermal delivery is to accentuate the rate of movement of compounds across the many different layers of cells and connective tissue that collectively form skin. However, transdermal delivery of a number of different compounds has resulted in the generation of significant hypersensitivity or inflammatory responses. For example, although transdermal delivery of arachalene (a compound that is useful in the treatment of Alzheimers disease) results in the sustained delivery of pharmacologically beneficial concentrations of the drug, it results in the generation of symptomatic contact hypersensitivity (CH) responses (Krueger, Pershing, and Roberts, personal communication). Furthermore, subsequent oral administration of this compound elicits a "memory" CH response. Thus, the induction of an immune response to a transdermally delivered compound may result in a condition in which it is no longer feasible to achieve delivery by conventional routes.

Historically, during the early development phases of transport systems, little attention was paid to the potential immunological consequences of transdermal drug delivery. A further complication of the issue is the reticence on the part of a number of companies and research groups to report the deleterious effects of transdermal transport systems in the scientific literature, thus, making systematic evaluation of many of the problems associated with transdermal delivery systems extremely difficult, if not impossible. This chapter will present an overview of the immunobiology of the skin and its interactions with systemic immune responses as an aid in consideration of the potential immunological ramifications of transdermal drug delivery systems. It is hoped that this will lead to development of rational approaches to obviate potentially deleterious side effects that may arise as a result of these delivery systems.

II. STRUCTURE AND HISTOLOGICAL OR-
GANIZATION OF THE SKIN

Like other organs, the skin is well organized into histologically defined tissues to facilitate the performance of its various functions. The structural and functional relationships of the skin for its immunological role are discussed in the following:

The skin is divided into two main layers, the surface epithelium, termed *epidermis*, and the underlying connective tissue layer, termed the *dermis*. Beneath the dermis is the *hypodermis*, which is a layer of loose connective tissue consisting mainly of subcutaneous adipose tissue. The hypodermis is loosely connected to underlying deep fascia, aponeurosis, or periosteum. A schematic representation of the skin is presented in Figure 1.

The epidermis is a keratinizing, stratified squamous epithelium composed of several distinct cell populations. The keratinocytes represent more than 90% of the cells within the epidermis. These are the keratinizing squamous cells of the self-renewing epithelium

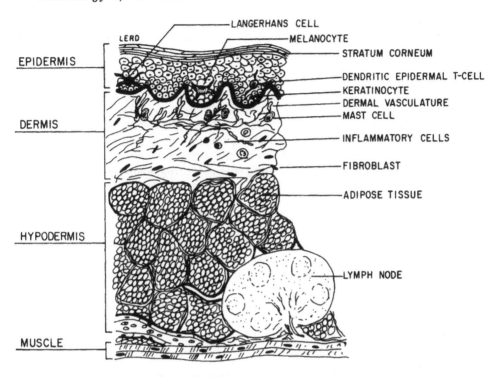

Figure 6.1 Schematic representation of a histological cross-section of the skin.

which terminally differentiate into the nonviable, flattened corneocytes of the stratum corneum. The stratum corneum is the outermost layer of the epidermis. The interlocking platelike structure of the highly keratinized corneocytes provides the major barrier function for the skin.

Melanocytes, Merkel cells, Langerhans cells, dendritic epidermal T cells, and epidermotropic lymphocytes represent the remainder of the minor epidermal cell populations. *Melanocytes* are pigmented dendritic epidermal cells that function to provide protection against ultraviolet radiation (UVR)-induced skin damage through the melaninization, or tanning, process. *Merkel cells* do not, as yet, have a defined function, although it has been suggested that they serve as neurosensory receptors. *Langerhans cells, dendritic epidermal T cells*, and *epidermotropic lymphocytes* are distinct immunocompetent cell populations within the epidermis. Their phenotypic characteristics and functional properties are outlined in greater detail in the following section.

The dermis is a dense connective tissue layer that supports the epidermis. The bulk of the dermis consists of bundles of collagen and elastic fibers that form a reticular network to support the vasculature, nerves, lymphatics, and adnexal structures (hair, sweat glands, and such). *Fibroblasts*, which secrete the reticular fibers, represent the major cell population within the dermis. *Mast cells*, which are in close proximity to the blood vessels, represent the major inflammatory immunocompetent cell population of the dermis. These cells synthesize and secrete a number of active inflammatory mediators. Other immunocompetent and inflammatory cells within the dermis include: resident antigen-presenting cells and transient inflammatory—lymphoid cells (e.g., polymorphonucleocytes, monocytes, and lymphocytes).

Distinct immunocompetent cell populations do not reside within the subcutaneous tissue layer of the skin. However, this layer of the skin does support blood vessels, lymphatics, and lymph nodes that are important structural components of the immune system. Transient inflammatory and lymphoid cells are also observed in the hypodermis.

III. INFLAMMATION AND IMMUNE RE-
SPONSES IN THE SKIN

To understand the immunobiology of the skin, it is important to distinguish between inflammation and immunity, because the consequences and approaches to circumventing the two are different. *Inflammation* is a basic nonspecific response localized in various tissue sites in all multicellular animals. A variety of stimuli are capable of triggering an inflammatory response, which is characterized by a series of interacting and interdependent physiologic reactions. The initial phase of an inflammatory reaction is marked by edema resulting from an increase in both the blood supply to the local site and the permeability of the capillary walls. Phagocytic cells (primarily neutrophils and monocytes) leave the circulation and accumulate at these sites. This is accompanied by the release of a large number of soluble mediators produced by lymphoid cells, macrophages, and neutrophils, as well as by the clotting, kinin, and complement systems. The inflammatory response is a complicated series of reactions that sets the stage for the repair of damaged tissues by stimulating collagen biosynthesis and the proliferation of both connective tissue cells and small blood vessels.

The hallmarks of *immunologically mediated responses*, on the other hand, are specificity and memory. The lymphoid cells, which compose the immune system, are separated into two broad categories: *B cells*, which can differentiate into plasma cells that produce antigen-specific antibodies; and *T cells*, which are further subdivided into a number of subsets based upon their function. These include cytotoxic T lymphocytes (CTL), delayed-type hypersensitivity cells (T_{DTH}), T-helper (T_h) cells that aid in the generation of either cell-mediated immune responses or antibody-mediated immune responses, and suppressor T (T_s) cells that act as regulatory cells for both cell-mediated and antibody-mediated immune responses. Functionally associated with the lymphocytes are antigen-presenting cells (APC), such as macrophages and dendritic cells, which act to "process" and "present" antigens to the various subpopulations of B and T cells. The APC are required for the induction, generation, and regulation of immune responses. Antigens are "presented" to antigen-specific T lymphocytes in the context of major histocompatibility complex-encoded cell-surface molecules that are expressed by APC (1). The induction and elicitation phases of the immune response that are initiated by the APC causes both the activation and the clonal expansion of the antigen-specific T cells.

Mature T cells are continually trafficking between the circulation and various different secondary lymphoid organs such as the spleen, peripheral lymph nodes, mesenteric lymph nodes, and Peyer's patches (2). The continual surveillance of different organ systems in the body by these recirculating lymphocytes thereby provides an enhanced protection against not only pathogens and toxins, but also against neoplastically transformed somatic cells. This continual immunosurveillance is of particular importance for those organ systems, such as the skin and the gastrointestinal tract, that are in constant contact with the numerous harmful agents encountered in the external environment.

The immunosurveillance capabilities of the immune system are further enhanced by the compartmentalization of subsets of T lymphocytes into defined circulatory circuits. For example, recent studies have demonstrated the preferential homing of a subset of T cells to the Peyer's patches, mesenteric lymph nodes, and lamina propria of the intestine, a feat mediated by a distinct lymphocyte-binding molecule expressed only within the microvasculature of these tissues (3). These data indicate the existence of a population of lymphoid cells that are restricted to gut-associated lymphoid tissue. Although much work remains to be done, data from a variety of sources also indicate the existence of an immunological circuit, restricted to the skin, that has been termed *skin-associated lymphoid tissue* (SALT). Compartmentalized immunological circuits thereby increase the probability of an interaction between an antigen-specific T cell and an antigen-bearing APC by directing effector cells to the anatomical sites(s) of antigen deposition (4,5).

The immunological effects of antigenic exposure are manifold, with the simultaneous induction of both positive effector responses (which may be cellular or antibody-mediated) and negative regulatory responses. In addition to the necessary cellular interactions, a variety of lymphokines (soluble mediators) are required to elicit an immune response. Thus, the ultimate response detected is the result of a complex series of interactions between these two arms of the immune response. Furthermore, the amount and route of antigen exposure also play a major role in determining which of these two arms of the immune response is preferentially induced.

IV. SKIN-ASSOCIATED LYMPHOID TISSUES

In an anatomical sense, the skin is uniquely situated and faces perhaps the most diverse array of antigenic stimuli, as well as physical and chemical insults, of any organ system in the body. For example, the skin is bombarded daily with ultraviolet (UV) radiation, a physical environmental agent that is known to be capable of causing neoplastic transformation of cells. Recognition of the unusual set of immunological requirements placed upon the skin led Streilein to hypothesize that a special relationship exists between the skin and the immune system (6). Thus, it has been proposed, and a substantial amount of evidence supports, the existence of SALT that provide skin with a unique and vitally important surveillance mechanism.

The SALT is hypothesized as being an integrated immunosurveillance system comprised of (1) antigen-presenting cells that reside within the epidermis; (2) a population of recirculating T lymphocytes that are distinctly epidermotropic; and (3) a unique anatomical environment that can affect the induction, generation, and regulation of immune responses both locally and systemically. Support for the concept of SALT includes (1) the identification of a population of cells (Langerhans cells) that can act as APC (7); (2) identification of functional subsets of T cells that exhibit a definite epidermotropism (8); (3) the demonstration that antigen-specific T cells can recognize antigen within the skin (9); (4) characterization of a unique population of dendritic epidermal T cells within the skin (10—12); (5) the demonstration that epidermal T cells are immunocompetent in situ (13,14); (6) the elucidation of unique immunoregulatory circuits that operate within the skin (15—17); (7) the demonstration that keratinocytes can be induced to express major histocompatibility complex-encoded class II molecules (18); (8) the identification and characterization of a number of different keratinocyte-derived cytokines that can have profound effects on the regulation of immune responses, T-cell differentiation, and hem-

atopoiesis (19–28); and (9) the observation that a number of physical (e.g., ultraviolet radiation, heat, surgical trauma) and chemical
(e.g., steroids, arachidonic acid, transdermal delivery systems)
agents can have profound effects on the generation of immune responses to antigens first encountered through the skin (29–32).
Thus, by affecting any of the various components of SALT the normal immunological capacity of the skin can be changed, with the potential for altering not only local, but also systemic, immune responses.

The peripheral secondary lymphoid tissues that drain the skin
provide the necessary immunological environment for the induction
of an immune response. Antigens first encountered through the
skin are transported to the draining lymph nodes by epidermal
Langerhans cells and macrophages through the afferent lymphatics
(33). In contrast, more than 90% of the lymphoid cells that enter
the lymph nodes do so from the blood stream. The entry of T cells
from the circulation into peripheral lymph nodes is mediated and
regulated by specific interactions between the T cells and anatomi-
cally distinct sites of the cuboidal endothelial cell-lined postcapillary
venules, the so-called high endothelial venules (HEV; 34). After
binding to HEV, lymphocytes are able to extravasate from the circulation and enter the parenchyma of the lymph node (35).

Events that affect skin can have profound effects on the SALT
circuit. For example, Spangrude et al. found that the number of
radiolabeled murine T cells that "homed" to peripheral lymph nodes
rose from 11.1% in control mice to 21.8% in mice exposed to 5000 J/
m^2 per day of UVB radiation over a period of 6 days (36). Furthermore, this alteration in the lymphocyte trafficking pattern was
observed to persist for up to 2 months. The increase in the numbers of T cells homing to peripheral lymph nodes in UV-irradiated
mice appears to be due to many factors. For example, increases in
HEV structure and function have been observed in the peripheral
lymph nodes of mice exposed to relatively modest doses of UVB radiation (37). This may well be due to an increase in the number of
activated APC in the peripheral lymph nodes as well as an increase
in the production of cytokines, such as interleukin-1 (IL-1), by
UV-irradiated keratinocytes (38). The situation is further accentu-
ated by the effects of increased production of prostaglandins during
the inflammatory response induced by UV exposure. The increased
release of prostaglandin produces an efferent lymphatic blockade
that, in turn, leads to increased retention times for recirculating
lymphocytes in the lymph nodes (39). It is important to note that
the changes in lymphocyte recirculation patterns produced by short-
term exposure to UV radiation are similar to those caused by cutaneous administration of antigen (40). Whether similar changes are

induced by various transdermal delivery systems (either by themselves or in conjunction with the compounds being delivered) is now
unknown and deserves serious consideration.

V. IMMUNOCOMPETENT CELLS WITHIN
   THE SKIN

We will now focus on those skin cell populations that either have
been shown to play, or are suspected of playing, a role in skin-associated immune responses. The experimental model of allergic contact dermatitis, referred to as contact hypersensitivity (CH), will
be discussed in detail because this type of skin reaction represents
a major obstacle faced by the pharmaceutical industry in developing
successful transdermal drug delivery systems. The roles played by
individual cell types in the afferent, or induction, phase (which involves T-cell antigen-priming by APC) and the efferent, or elicitation, phase (which involves the activation and migration of effector
T cells to sites of antigen deposition) of an immune response will be
outlined. The kinetics and mediators ascribed to CH responses will
be compared with those involved in antigen nonspecific inflammatory
reactions.

A. Langerhans Cells

Langerhans cells (LC) account for 3% to 5% of the cells in the epidermis (41). These cells constitutively express high levels of ATP-
ase activity, receptors for the Fc portion of immunoglobulin and
the complement protein C3b, as well as major histocompatibility gene
complex-encoded class II molecules (termed Ia in mice and HLA-D in
man; 41). Human LC also express the CD4 (T4) and CD1 (T6) T-
cell differentiation antigens (42,43). Birbeck granules, which are
detected by transmission electron microscopy, serve as the definitive
markers of LC (44). With use of ATPase or immunohistochemical
staining techniques, LC are morphologically identified by light microscopy as evenly spaced dendritic cells. An example of murine
LC stained for cell surface class II molecules by an indirect immuno-
peroxidase method is presented in Figure 2.
     Functionally, LC are very effective antigen-presenting cells (41).
In vitro and in vivo studies have shown that LC are capable of acting as APC for the induction and elicitation of humoral and cell-mediated immune responses. Their location in the suprabasal cell layer of the epidermis indicates that LC are the primary sentinels for
the processing and presentation of antigens that are introduced
through the skin to the immune system of the host.
     Contact hypersensitivity reactions are mediated by antigen-spe-
cific T cells in response to epicutaneously encountered contact sen-

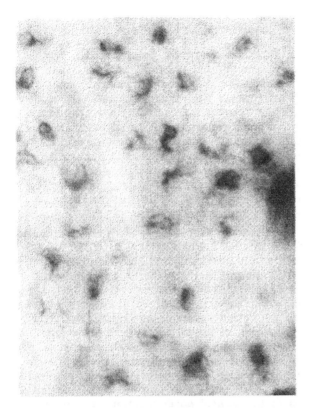

Figure 6.2 Murine Langerhans cells. Indirect immunoperioxidase staining was used to visualize Ia-expressing Langerhans cells in an epidermal sheet removed from a BALB/c mouse. Monoclonal antibodies against *I-A* and *I-E* class II molecules were used as primary antibodies for identifying Langerhans cells.

sitizing agents (45). A variety of contact sensitizers have been identified. They include metals, plant oils, organic dyes and chemicals, preservatives, cosmetics, medicaments, and petrochemical products (46). Although this appears to represent a highly diverse group of compounds, contact sensitizers display a similar chemical property in that they function as haptens. *Haptens* are classified as small (< 2000 molecular weight), highly reactive compounds that form antigens by establishing chemical bonds with soluble proteins or cell surface molecules.

By employing the contact sensitizing agent dinitrofluorobenzene (DNFB), it has been shown that LC carry DNFB from the epidermis, through afferent lymphatic vessels to the draining peripheral lymph

nodes (9). While in transit in the afferent lymphatics, the LC are
histologically identified as veiled cells. Once in the draining lymph
nodes, the LC initiate the afferent phases of a CH immune response
by presenting the cell surface-bound antigen to responsive clones
of T cells (41). The antigen-stimulated T cells undergo clonal expansion and seed the peripheral lymph nodes with memory cells that
are capable of mediating the intense CH responses observed in allergic contact dermatitis after subsequent challenge with the immu-
nizing contact sensitizer. Thus, under normal conditions, any agent
that is introduced through the skin (assuming it has the properties
of a foreign antigen) can induce or elicit an immune response through
the actions of the LC to stimulate an effector T-cell response.

    Although immune responses (whether cell-mediated immunity or
antibody production) are generally attributed to the activities of
the effector cells, it must be appreciated that the immune system is
highly regulated. The induction of T_S-cell responses by most antigens serves to regulate the intensity and duration of an immune response. It has been shown that distinct populations of antigen-specific T_S cells are capable of inhibiting the afferent and efferent
phases of a CH response (47,48). Thus, the balance between the
activities of the T_{CH}-cell and T_S-cell populations dictates whether
or not an allergic contact dermatitis reaction will occur. It has
been suggested that a population of epidermal APC that are distinct
from LC may selectively elicit antigen-specific T_S-cell responses
(15). Likewise, a soluble antigen absorbed through the skin that
gains access into the blood may also selectively stimulate a T_S—cell-
dominated immune response (50). This may account for the elicitation of T_S-cell responses to the topical administration of high doses
of contact sensitizing agents (51). The immunological conditions
and mechanisms responsible for the selective elicitation of T_S—cell-
mediated responses are outlined in detail in Section VI of this chapter.

B. Dermal Mast Cells

The dermal mast cell population plays a central role in the efferent
phases of CH responses, inflammatory reactions to skin irritants,
and immediate-type hypersensitivity responses (52—54). Mast cells
reside in close approximation to the dermal capillaries. These cells
synthesize and store in cytoplasmic granules a number of vasoactive
compounds (e.g., serotonin and histamine; 55). When activated,
the mast cells release the contents of these storage granules into
the interstitium where the soluble mediators are able to activate endothelial cells to cause increased vascular permeability. The actions of the mast cell products thus account for the edema and ery-

thema that accompany inflammatory and immune skin reactions. The increased vascular permeability also allows the movement of inflammatory and immune cells into the dermis.

Loveren and co-workers have reported that certain antigen-primed T cells produce factors that activate mast cells during the early phases of the elicitation of a CH response (56). As the response progresses, the increase in vascular permeability at sites of antigen deposition allows the influx of antigen-specific T_{CH} cells. The T_{CH} cells secrete a variety of chemotactic, cell-activating, and migration inhibitory factors that sequester inflammatory cells at the antigen-reactive skin site. The reaction proceeds to effect clearance of the antigen from the skin. The response subsides with the appearance of the T_S−cell-mediated regulatory response.

Mast cells mediate antigen-nonspecific inflammatory reactions and immediate-type hypersensitivity responses in a similar manner. The appearance of inflammatory cells within the skin is directly associated with the increased vascular permeability produced after mast cell degranulation. In these two reactions, mast cells are activated by either the direct or indirect effects of the skin irritant (inflammatory), or direct activation by allergen binding to IgE on the surface of the mast cells (immediate-type hypersensitivity) (54). Although the appearance of the skin reaction produced by either inflammatory or immune cells is similar, the mechanisms that are involved are distinct and, thus, different approaches should be taken to circumvent the elicitation of a specific skin reaction produced by a given compound introduced through the skin (i.e., does the compound act as an irritant, contact sensitizer, or allergen?).

C. Epidermotropic Blood-Borne Cells

Lymphocytes, monocytes, and polymorphonuclear cells are the mediators of the various skin reactions. The influx and concentration of these cells in a skin site are responsible for the urticaria observed at inflammatory and immune reactive sites. We have already discussed how these cells gain entry into the skin. However, it should be appreciated that certain populations of T cells display epidermotropic properties (57). When activated, these cells tend to selectively recirculate to skin sites. Mycosis fungoides cells represent a neoplasia of T cells that have retained this property (58). Thus, it appears that in addition to possessing antigen-specific activity through the expression of the T-cell antigen receptor (Ti/CD3 complex), epidermotropic T cells also express receptors for skin endothelium-expressed cell adhesion molecules which allows them to selectively bind to, and extravasate across, the dermal vasculature. These types of adhesion molecules have been identified

on lymphocytes that recirculate between mesenteric lymph nodes and
the lamina propria of the small intestine (59,60).

D. Dendritic Epidermal T Cells

Several years ago a population of murine dendritic epidermal cells
(DEC) was identified that expressed the T-cell differentiation antigen Thy 1 (10—12). These cells appear to possess natural killer
cell-like activity (61). Furthermore, it has been suggested that
these cells are responsible for the induction of T_S-cell responses to
skin-associated antigens (62). However, their exact functional significance is unknown. Recently, it has been reported that cloned
dendritic epidermal T-cell lines express the gamma/delta form of the
T-cell antigen receptor complex (63). These cloned cells do not express either CD4 or CD8 cell-surface proteins, which are differential markers of mature functional T-cell subsets (61). It is possible
that these cells represent epidermotropic population of a minor sub-
population of T cells found in the peripheral blood that have similar
phenotypic properties (64). Alternatively, it is also possible that
dendritic epidermal T cells may be immature cells that have under-
gone limited extrathymic T-cell maturation. This is consistent with
the data that indicate that the epidermis possesses similar morphological and limited hormonal characteristics of thymic epithelium (65).
Whether DEC play a role in the generation or regulation of effector
functions of skin-associated immune responses is now unknown.
What is clear is that further analysis is required to determine the
immunobiological role of this particular cell population.

E. Keratinocytes

Under normal conditions, the LC are the only cells within the epidermis that express class II cell surface determinants (41). This
is consistent with their known APC function, since class II molecules are important for antigen presentation and the cellular interactions required for the induction of an immune response. Curious-
ly, it has been shown that keratinocytes in the skin of patients
with certain skin diseases are induced to express HLA-DR antigens
(18,66—69). Similar results have been obtained experimentally
where murine keratinocytes can be induced to express Ia (Fig. 3).
Interestingly, those conditions in which the keratinocytes are induced to express class II antigens are generally associated with the
infiltration of lymphoid cells into the skin. For example, γ-interferon (γ-IFN) can induce keratinocytes to express class II antigens,
which suggests that the release of this lymphokine by activated T
cells may be responsible for inducing expression of these cell-surface molecules (72). However, experiments in mice suggest that

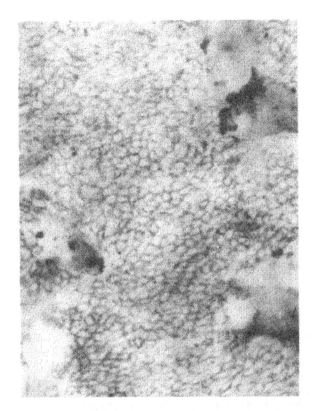

Figure 6.3 Induced expression of Ia molecules by murine keratino-cytes. *I-A* and *I-E* reactive monoclonal antibodies were used in an indirect immunoperoxidase-staining procedure to visualize Ia-expressing keratinocytes in the epidermis of BALB/c mice treated with high doses of recombinant γ-interferon (10^6 units/day IP for 6 days).

stimuli aside from γ-interferon may also be capable of inducing keratinocyte expression of cell-surface class II molecules.

That keratinocytes are induced to express class II antigens suggests that these cells may possess some immunological potential. Numerous studies have been undertaken to determine whether or not keratinocytes expressing class II molecules can function as APC (75–77). The results of these studies indicate that these cells possess very limited, if any, APC activity. For example, γ-interferon-treated human HLA-DR$^+$ keratinocytes function as stimulators of allogeneic T cells only when interleukin-2 (IL-2) is exogenously supplied (75). Furthermore, recent studies have indicated that class

II$^+$ keratinocytes are incapable of processing soluble antigen (76–77). Thus, it is not surprising to find that Ia$^+$ keratinocytes are capable of presenting only "preprocessed" antigenic fragments to certain cloned antigen-specific T-cell lines. We have obtained similar results using the T-cell mitogen MAS which is prepared from supernatants of cultures of *Mycoplasma arthritidis* (78–79). The MAS mitogen does not require processing (79) and is restricted in its presentation to T cells by the I-E encoded class II molecule (78). In our study the I-E-expressing keratinocytes were capable of presenting MAS to murine T cells (Ghadessi et al., manuscript in preparation). Thus, the data indicate that class II$^+$ keratinocytes display extremely limited APC activity.

The general lack of APC activity displayed by HLA-D/Ia-expressing keratinocytes apparently is not due to their inability to produce the appropriate lymphokines. Keratinocytes synthesize and secrete both the soluble and the membrane-bound forms of interleukin-1 (IL-1) (22,38). Expression of class II antigens and production of IL-1 appear to be the primary requisites of a functional APC (80). Furthermore, keratinocytes have been shown to produce a wide array of lymphokines, including IL-3, IL-6, granulocyte–monocyte colony-stimulating factor (GM-CSF), tumor necrosis factor-α (TNF-α), and prostaglandins (24,25,81–84). Several of these keratinocyte-derived lymphokines are produced in increased amounts after appropriate stimulation (e.g., exposure to UV radiation; 22, 38). The fact that these lymphokines have potent immunomodulatory activities indicates that keratinocytes can actively influence an immune response. Further investigation is required, however, to differentiate between localized and systemic effects of keratinocyte-derived cytokines.

Although class II antigen-expressing keratinocytes do not appear to function to any significant degree as effective APC, recent studies suggest that these cells may play alternative immunological roles. Roberts et al. have reported that Ia expression by murine keratinocytes is strongly correlated with the intensity and duration of CH responses as well as the migration of LC into the epidermis (85,86). This suggests that these cells may function to direct (either attracting or sequestering) lymphoid cell movement in the skin. This concept is supported experimentally by the results of studies in which it was observed that cloned epidermotropic T cells display increased migrational activity toward Ia$^+$ keratinocytes (87). It was also recently reported that Ia$^+$ keratinocytes may, in fact, function to selectively elicit antigen-specific T$_S$-cell responses (88). Thus, these cells may have an important immunoregulatory function. Further evaluation of class II antigen-expressing keratinocytes is clearly required to determine their exact immunological role.

VI.   EFFECTS OF AGENTS THAT DEPLETE SKIN OF LANGERHANS CELLS ON IMMUNE RESPONSES TO ANTIGENS ENCOUNTERED THROUGH THE SKIN

The best-characterized system in which alterations of functions of some of the cells in the SALT circuit can profoundly alter systemic immune response uses UVB (280—320 nm) radiation as the modulatory agent.   The results of studies by both Toews et al. (89) and Lynch et al. (31) have shown that exposure of skin with even low doses of UVB radiation causes a rapid and dramatic reduction in the density of LC in the exposed skin sites.   For example, Toews et al. found the density of $ATPase^+$ LC in murine epidermis obtained from shaved abdominal skin sites that were exposed on four consecutive days to 100 J/m^2 of UVB radiation had decreased from 800 cells/mm^2 (in nonirradiated skin) to less than 50 cells/mm^2 (89).   Contact hypersensitivity responses to DNFB applied to the UV-irradiated skin sites were markedly depressed relative to responses elicited in either nonirradiated mice or in UVB-irradiated mice that were sensitized to DNFB through a nonexposed skin site.   Depletion of $ATPase^+$ LC by UVB irradiation persisted for as long as UV treatments were continued, but returned to normal densities within 2 weeks after the cessation of treatments (31).   Furthermore, a direct correlation was observed between the density of "returning" $ATPase^+$ LC and the capacity of irradiated skin to promote sensitization to DNFB.   The results of these studies have been extended to humans by Cooper et al. and Slot et al., who found that UV exposure of skin caused a marked reduction in the density of LC that correlated with a depression in the ability to elicit CH response to epicutaneously applied contact-sensitizing agents (90,91).

In addition to being unable to induce CHS responses through UV-irradiated skin sites, application of contact sensitizers to UVB-exposed murine skin results in a condition in which the mice are refractory to resensitization through nonirradiated skin sites (31,89). The refractory state is antigen-specific, and the results of later studies by Elmets et al. demonstrate that the antigen-specific unresponsiveness induced by topical application of contact sensitizers to UV-exposed skin is mediated, at least in part, by T_S cells, which function to downregulate the afferent, but not the efferent, phases of the CH response (92).

Insight into the mechanisms by which the T_S cells are induced is provided by the results of the much earlier studies by Macher and Chase (9,93).   Their investigations revealed that if the site of sensitization was surgically excised 12 hr after application of a reactive hapten that only 4% of the sensitized animals developed detectable CH responses, as compared with 80% of the animals in the control group in which the sensitization sites were not excised. Excision of the sensitization sites 24 or 48 hr after sensitization resulted in the induction of CH responses in 14% and 61% of the animals, respectively. Animals that were unresponsive to the contact sensitizer after excision of the sensitization site were also refractory to subsequent immunization. Of major importance was the observation that most of the contact sensitizer left the site of application through the venous outflow within minutes of sensitization. As little as 5% was found remaining 24 hr after application. Moreover, intravenous administration of the contact sensitizer resulted in the generation of specific CHS unresponsiveness. These data suggest that epicutaneous application of reactive chemicals to normal skin results in the simultaneous induction of both positive and negative immune responses, and that antigen must persist at the sensitization site for an appropriate period to activate an effector response.

This scenario has received considerable support from the results of a number of different investigators who have demonstrated that positive and negative immune responses can be induced either simultaneously, or in isolation, depending upon the immunization protocol used. For example, pretreatment of animals with cyclophosphamide (to eliminate suppressor cell precursors) before sensitization by conventional skin-painting techniques results in enhanced CH responses (94). Subcutaneous immunization with haptens or intravenously injected haptenated, unfractionated epidermal cells has also induced both positive and negative immune responses (95−97). This should not be interpreted to mean that both types of immune responses are *always* induced simultaneously. For example, skin-painting with threshold levels of a contact sensitizer leads to induction of CH responses in the apparent absence of T_S-cell responses (98). Conversely, skin-painting with supraoptimal doses of hapten preferentially induce T_S-cell generation (51). It remains to be determined which of these situations will be mimicked with each individual antigen that is used in a transdermal delivery system.

VII. CONCLUSION

Transdermal delivery systems potentially offer both therapeutic and economic advantages over many of the more conventional drug delivery systems currently in use. However, it is also becoming increasing clear that there are still major problems associated with transdermal delivery systems that must be overcome for these delivery modalities to gain widespread applicability and use.

In addition to determining what types of immune responses may be induced to transdermally delivered drugs, it is also important to evaluate the consequences of immune responses to antigens that enter the system through a "coat-tail" effect. For example, induction of a suppressor response to antigens expressed by pathogens (e.g., bacteria) that normally reside on the skin may render the individual unable to mount effective immune responses to that agent and may result in an increased susceptibility to disease caused by that pathogen. Nor is it clear what the effects of transdermal carrier systems will have on the expression of other immunologically and physiologically active molecules (such as IL-1, IL-3, IL-6, γ-IFN, histamine, GM-CSF, TNF, prostaglandins), on the expression of cell surface class II molecules on either epidermal LC or keratinocytes, on the ability of LC to process and present antigens to T cells, or on the function of the T cells (DEC) which normally reside in the skin. Thus, the effect of transdermal delivery systems on the functions and effects (both locally and systemically) on all of these aspects of the immune response capabilities of the skin also deserve scrutiny and careful evaluation.

The major obstacle to be hurdled is not in the development of effective transport systems but, instead, is in understanding the interaction of these systems on both local and systemic immune responses. Should either local or systemic immune responses be induced by a transdermally delivered compound, there are certainly a number of approaches that can be taken to resolve the problem. One approach would be to modify either the vehicles used in the delivery system or the amounts of the compound being delivered, to obviate the induction of undesirable immune response. A second approach would be to chemically modify the drug being delivered in such a way that the compound (or its metabolites) are no longer immunogenic, yet still retain pharmacological activity. Finally, it might be possible to temporarily modify localized SALT circuits in such a way that compounds delivered through appropriate skin sites no longer result in the induction of deleterious immune responses.

Clearly, none of the aforementioned potential approaches are trivial in the amount of effort that will be expended in their implementation. It also seems to be an inescapable conclusion that if rational strategies are to be developed to evaluate the exact nature of immunologically mediated inflammatory responses induced by transdermally delivered compounds, then we must first possess a significantly enhanced understanding of immune response in the skin. This goal can only be achieved by devoting increased efforts toward understanding the immunobiology of the skin, its interaction with systemic immune responses, and by improving the dissemination of the knowledge obtained by the different research groups.

ACKNOWLEDGMENT

The authors would like to thank Janet Swapp, Carol Peck, and Patricia Tibola for their excellent secretarial assistance in the preparation of this manuscript.

This work was supported by EPA Contract 68-01-7288 and National Institutes of Health Grant AM33543.

REFERENCES

1. D. Hansburg, E. Heber-Katz, T. Fairwell, and E. Apella, *J. Exp. Med.*, *158*:25, 1983.
2. J. L. Gowans, and E. J. Knight, *Proc. R. Soc. Biol.*, *59*:257, 1964.
3. P. R. Streeter, E. L. Berg, B. T. N. Rouse, R. F. Bargatze, and E. C. Butcher, *Nature 331*:41, 1988.
4. G. E. Woloschak, and T. B. Tomasi, *CRC Crit. Rev. Immunol.*, *4*:1, 1983.
5. J. W. Streilein, *J. Invest. Dermatol.*, *71*:167, 1978.
6. J. W. Streilein, *J. Invest. Dermatol.*, *71*:167, 1978.
7. G. Stingl, K. Tamaki, and S. I. Katz, *Immunol. Rev.*, *53*:149, 1980.
8. R. L. Edelson, *J. Am. Acad. Dermatol.*, *2*:89, 1980.
9. E. Macher, and M. W. Chase, *J. Exp. Med.*, *129*:103, 1969.
10. D. A. Chambers, *Br. J. Dermatol.*, *113*:24, 1985.
11. E. Tschachler, G. Schuler, J. Hutterer, H. Leibl, K. Wolff, and G. Stingl, *J. Invest. Dermatol.*, *81*:282, 1983.
12. P. R. Bergstresser, R. E. Tigelaar, J. H. Dees, and J. W. Streilein, *J. Invest. Dermatol.*, *81*:286, 1983.
13. J. B. Solomon, *Transplanatation*, *1*:327, 1963.
14. C. F. Barker, and R. E. Billingham, *Transplantation*, *14*:525. 1972.
15. R. D. Granstein, *J. Invest. Dermatol.*, *84*:206, 1985.
16. J. W. Streilein, *J. Invest. Dermatol.*, *80*:12s, 1983.
17. W. Patak, D. Rozycka, P. W. Askenase, and R. K. Gershon, *J. Exp. Med.*, *151*:362, 1980.
18. J. Smolle, *Acta Dermmato-Venereol.*, *65*:9, 1985.
19. G. Lisby, C. Avnstorp, and G. L. Wantzin, *Int. J. Dermatol.*, *26*:8, 1987.
20. T. A. Luger, and J. J. Oppenheim, *Adv. Inflamm. Res.*, *5*:1, 1983.
21. D. N. Sauder, T. V. Arsenault, D. Stetsko, and C. B. Harley, *J. Invest. Dermatol.*, *88*:515, 1987.
22. T. S. Kupper, A. O. Chua, P. Flood, J. McGuire, and U. Gubler, *J. Clin. Invest.*, *80*:430, 1987.

23. A. C. Chu, J. A. K. Patterson, G. Goldstein, C. L. Berger, S. Takezaki, and R. L. Edelson, *J. Invest. Dermatol.*, *81*: 194, 1983.

24. M. Danner, and T. A. Luger, *J. Invest. Dermatol.*, *88*: 353, 1987.

25. T. S. Kupper, M. Horowitz, F. Lee, D. Coleman, and P. Flood, *J. Invest. Dermatol.*, *88*: 501, 1987.

26. J. F. Nicolas, D. Kaiserlian, M. Dardenne, M. Faure, and J. Thivolet, *J. Invest. Dermatol.*, *88*: 161, 1987.

27. T. Schwarz, A. Urbanska, F. Gschnait, and T. A. Luger, *J. Immunol.*, *138*: 1457, 1987.

28. L. C. Gahring, A. Buckley, and R. A. Daynes, *J. Clin. Invest.*, *76*: 1585, 1985.

29. B. Berman, D. S. France, G. P. Martinelli, and A. Hass, *J. Invest. Dermatol.*, *80*: 168, 1983.

30. W. L. Morison, and M. L. Kripke, *Cell. Immunol.*, *85*: 270, 1984.

31. D. H. Lynch, M. F. Gurish, and R. A. Daynes, *J. Immunol.*, *126*: 1892, 1981.

32. L. A. Rheins, and J. J. Nordlund, *J. Immunol.*, *136*: 867, 1986.

33. I. Silberberg-Sinakin, G. Thorbecke, R. L. Baer, S. A. Rosenthal, and V. Berezowsky, *Cell. Immunol.*, *25*: 137, 1976.

34. N. D. Anderson, A. O. Anderson, and R. G. Wyllie, *Immunology*, *31*: 455, 1976.

35. G. J. Spangrude, B. A. Braaten, and R. A. Daynes, *J. Immunol.*, *132*: 354, 1984.

36. G. J. Spangrude, E. J. Bernhard, R. S. Ajioka, and R. A. Daynes, *J. Immunol.*, *130*: 2974, 1983.

37. W. E. Samlowski, H. T. Chung, D. K. Burnham, and R. A. Daynes, *Regional Immunol.*, *1*: 41, 1988.

38. L. Gahring, M. Baltz, M. B. Pepys, and R. Daynes, *Proc. Natl. Acad. Sci. USA*, *81*: 1198, 1984.

39. H. T. Chung, W. E. Samlowski, D. K. Kelsey, and R. A. Daynes, *Cell. Immunol.*, *101*: 571, 1986.

40. W. G. Kimpton, A. Walsh, D. C. Poskitt, and H. K. Muller, *Int. Arch. Allergy Appl. Immunol.*, *74*: 40, 1984.

41. G. Stingl, K. Tamaki, and S. I. Katz, *Immunol. Rev.*, *53*: 149, 1980.

42. V. Groh, M. Tani, A. Harrer, K. Wolff, and G. Stingl, *J. Invest. Dermatol.*, *86*: 115, 1986.

43. F. Fithian, P. Kung, G. Goldstein, M. Rubenfeld, C. Fenoglio, and R. Edelson, *Proc. Natl. Acad. Sci. USA*, *78*: 2541, 1981.

44. M. S. Birbeck, A. S. Breathnach, and J. D. Everall, *J. Invest. Dermatol.*, *37*: 51, 1961.

45. R. L. Baer, and J. L. Turk, in *Biochemistry and Physiology of the Skin*. (L. A. Goldsmith, ed.). Oxford University Press, New York, pp. 921–937, 1983.

46. J. D. Wilkinson, and R. J. G. Rycroft, in *Textbook of Dermatology* (A. Rook, D. S. Wilkinson, F. J. G. Ebling, R. H. Champion, and J. L. Burton, eds.). Blackwell Scientific, Oxford, pp. 435–532, 1986.

47. G. L. Ashershon, and M. Zembala, *Proc. R. Soc. Lond. B.*, *187*:329, 1974.

48. P. Phanuphak, W. Moorehead, and H. N. Claman, *J. Immunol.*, *113*:1230, 1974.

49. R. D. Granstein, *J. Invest. Dermatol.*, *84*:206, 1985.

50. Y. Nakano, and K. Nakano, *Immunology*, *42*:111, 1981.

51. M-S. Sy, S. D. Miller, and H. N. Claman, *J. Immunol.*, *119*:240, 1977.

52. R. K. Gershon, P. W. Askenase, and M. W. Gershon, *J. Exp. Med.*, *142*:732, 1975.

53. A. Schwartz, P. W. Askenase, and R. K. Gershon, *J. Immunol.*, *118*:159, 1977.

54. G. O. Solley, G. J. Gleich, R. E. Jordon, and A. L. Schroeter, *J. Clin. Invest.*, *58*:408, 1976.

55. S. I. Wasserman, in *Biochemistry and Physiology of the Skin*. (L. A. Goldsmith, ed.), Oxford University Press, New York, pp. 878–898, 1983.

56. H. Loveren, R. E. Ratzlaff, K. Kato, R. Meade, T. A. Ferguson, G. Iverson, C. A. Janeway, and P. W. Askenase, *Eur. J. Immunol.*, *16*:1203, 1986.

57. M. L. Rose, D. M. V. Parrot, and R. G. Bruce, *Cell. Immunol.*, *27*:36, 1976.

58. R. L. Edelson, *J. Am. Acad. Dermatol.*, *2*:89, 1980.

59. S. Jalkansen, R. F. Bagatze, J. de los Toyos, and E. C. Butcher, *J. Cell. Biol.*, *105*:983, 1987.

60. P. R. Streeter, E. L. Berg, B. T. N. Rouse, R. F. Bargatze, and E. C. Butcher, *Nature*, *331*:41, 1988.

61. N. Romani, G. Stingl, E. Tschachler, M. D. Witmer, R. M. Steinman, E. M. Shevach, and G. Schuler, *J. Exp. Med.*, *161*:1368, 1985.

62. M. Bigby, T. Kwan, and M. S. Sy, *J. Invest. Dermatol.*, *89*:495, 1987.

63. F. Koning, G. Stingl, W. M. Kokoyama, HJ. Yomada, W. L. Maloy, E. Tschachler, E. M. Shevach, and J. E. Coligan, *Science*, *236*:834, 1987.

64. L. L. Lanier, N. A. Federspiel, J. J. Ruitenberg, J. H. Phillips, J. P. Allison, D. Littman, and A. Weiss, *J. Exp. Med.*, *165*:1076, 1987.

65. B. F. Haynes, *Clin. Res.*, *34*:422, 1986.

66. P. G. Natali, C. D. Martino, V. Quaranta, M. R. Nicotra, F. Frezza, M. A. Pellegrino, and S. Ferrone, *Transplantation*, *31*:75, 1981.
67. U. M. Tjernlund, *Arch. Dermatol. Res.*, *261*:81, 1978.
68. A. Scheynius, and U. Tjernlund, *Scand. J. Immunol.*, *19*:141, 1984.
69. S. Aiba, and H. Tagami, *Br. J. Dermatol.*, *3*:285, 1984.
70. I. A. Lampert, A. J. Suitters, and P. M. Chisholm, *Nature*, *293*:149, 1981.
71. A. J. Suitters, and I. A. Lampert, *Br. J. Exp. Pathol.*, *63*: 207, 1982.
72. T. Y. Basham, B. J. Nickoloff, T. C. Merigan, and V. B. Morhenn, *J. Invest. Dermatol.*, *83*:88, 1984.
73. B. D. Jun, G. G. Krueger, and L. K. Roberts, *J. Invest. Dermatol.*, (in press).
74. L. K. Roberts, L. C. Gahring, S. E. Wiedmeier, G. G. Krueger,and R. A. Daynes, *J. Invest. Dermatol.*, *86*:503, 1986.
75. V. B. Morhenn, and B. J. Nickoloff, *J. Invest. Dermatol.*, *89*: 464, 1987.
76. S. M. Breathnach, S. Shimada, Z. Kovac, and S. I. Katz, *J. Invest. Dermatol.*, *86*:226, 1986.
77. A. A. Gasparini, and S. I. Katz, *J. Immunol.*, 140:2956, 1988.
78. D. H. Lynch, M. F. Gurish, B. C. Cole, and R. A. Daynes, *J. Immunol.*, *131*:1702, 1983.
79. M. C. Bekoff, B. C. Cole, and H. M. Grey, *J. Immunol.*, *139*:3189, 1987.
80. E. A. Kurt-Jones, and W. Virgin IV, *J. Immunol.*, *135*:3652, 1985.
81. V. A. De Leo, H. Horlick, D. Hanson, M. Eisinger, and L. C. Harber, *J. Invest. Dermatol.*, *83*:323, 1984.
82. T. Ruzicka, J. F. Walter, and M. P. Printz, *J. Invest. Dermatol.*, *81*:300, 1983.
83. J. A. Chodakewitz, T. S. Kupper, and D. L. Coleman, *J. Immunol.*, *140*:832, 1988.
84. D. N. Sauder, D. Wong, R. McKenzie, D. Stetsko, D. Harnish, V. Tron, B. Nickoloff, T. Arsenault, and C. B. Harley, *J. Invest. Dermatol.*, *90*:605, 1988.
85. L. K. Roberts, G. J. Spangrude, R. A. Daynes, and G. G. Krueger, *J. Immunol.*, *135*:2929, 1985.
86. L. K. Roberts, G. G. Krueger, and R. A. Daynes, *J. Immunol.*, *134*:3781, 1985.
87. T. Shiohara, and M. Nagashima, *J. Invest. Dermatol.*, *88*:518, 1987.
88. A. Gaspari, M. Jenkins, and S. L. Katz, *J. Invest. Dermatol.*, *90*:562, 1988.

89. G. B. Toews, P. R. Bergstresser, and J. W. Streilein, *J. Immunol.*, *124*:445, 1980.

90. K. D. Cooper, P. Fox, G. Neises, and S. I. Katz, *J. Immunol.*, *134*:129, 1985.

91. W. B. Slot, J. R. Taylor, and J. W. Streilein, *J. Invest. Dermatol.*, *88*:519, 1987.

92. C. A. Elmets, P. R. Bergstresser, R. E. Tigelaar, P. J. Wood, and J. W. Streilein, *J. Exp. Med.*, *158*:781, 1983.

93. E. Macher, and M. W. Chase, *J. Exp. Med.*, *129*:81, 1969.

94. L. Polak, and J. L. Turk, *Nature*, *249*:654, 1974.

95. M-S. Sy, S. D. Miller, J. W. Moorhead, and H. N. Claman, *J. Exp. Med.*, *149*:1197, 1979.

96. M. E. Sunday, B. Benacerraf, and M. F. Dorf, *J. Exp. Med.*, *153*:811, 1981.

97. K. Tamaki, H. Fujiwara, and S. I. Katz, *J. Invest. Dermatol.*, *76*:275, 1978.

98. S. Sullivan, J. W. Streilein, P. R. Bergstresser, and R. E. Tigelaar, *J. Invest. Dermatol.*, *82*:446, 1984.

II

DRUG FORMULATION SELECTION
AND TESTING FOR TOPICAL USE

7

Computational Methods for Prodrug or Drug Analogue Selection Optimized for Percutaneous Delivery

DAVID W. OSBORNE   *The Upjohn Company, Kalamazoo, Michigan*

I. INTRODUCTION

The introduction of transdermal patch systems capable of delivering therapeutic systemic levels of clonidine, nitroglycerin, estradiol, and scopolamine (1,2) has caused the realization that the skin is an alternative route for drug administration. For the topical formulator, an important result of the work conducted on patch systems is that models are available that can predict the rate at which a drug can cross the stratum corneum barrier. Although topical formulations will seldom try to maintain systemic drug levels, localized delivery of a drug will require that the stratum corneum barrier be crossed. Thus, it is important to know the flux value of the drug across the stratum corneum. By having this information early during development of the formulation, timely discussions of drug concentration within the formulation can be completed. The need for addition of absorption enhancers to the formulation can also be determined.

Recently four empirical models were evaluated to determine their ability to predict percutaneous absorption of drugs (3). Two of the predictive models evaluated were developed by Berner and Cooper (4), the third was taken from a publication by Michaels et al. (5), and the fourth was proposed by Albery and Hadgraft (6,7). In this previous study, ten drugs, having a wide range of physical properties, were evaluated and their resulting predicted transdermal flux values compared with experimental values. These permeation equations are useful for predicting the flux of a drug through the skin barrier when values for the water solubility, partition coefficient, and molecular weight of the drug are known. However, all of the models examined occasionally predicted transdermal fluxes that were unacceptably high or low when compared with experimentally determined data. It was hoped that insight into these discrepancies could be gained by obtaining additional experimental transdermal flux values from the literature for comparison with predicted values. Thus, an appropriate combination of the predicted values from each of the models might provide a more reliable overall predicted flux.

Another application for these computational methods becomes apparent by realizing that a reasonable transdermal flux value for a drug can be predicted with knowledge of the molecular weight and two readily calculable physicochemical drug properties (8,9). Because group contribution methods allow the calculation of both partition coefficient and water solubility, it becomes possible to predict a transdermal flux value for any liquid drug molecule whose structure can be drawn, regardless of whether or not the drug has actually been synthesized. For crystalline drugs, knowledge of the melting point is required in addition to the structure of the molecule. This could be of particular value when designing analogues to existing drugs or prodrugs in hopes of obtaining increased systemic breakthrough after topical administration. Indeed, by calculating flux values over a full range of water solubilities and partition coefficients, particularly useful insight into how to alter drug structure can be obtained.

II. DESCRIPTION OF THE MODELS

Because the models have been described in detail in both the original publications and in the previous technical update, only a brief description will be presented here. Model 1, the two—parallel-pathway model for skin permeation proposed by Berner and Cooper, (4) separates permeation through the skin's barrier function into two paths, a *polar* (aqueous) path and a *nonpolar* (lipophilic) path. The diffusion constants for the polar, D_p, and lipophilic, D_L, pathways are calculated using the equations

$$D_p = (3.8 \times 10^{-5})e^{-0.016(M)} \tag{1}$$

$$D_L = (1.7 \times 10^{-5})e^{-0.016(M)} \tag{2}$$

where M is the molecular weight of the drug, and the resulting diffusion constants are given in cm^2/hr. Because the fluxes of the polar and lipophilic pathways are considered additive, the total flux

(J) of the drug through the stratum corneum can be stated in terms of the diffusion coefficients from Eqs. 1 and 2, as follows:

$$J = (A_p D_p + A_L PD_L) C_w /L \qquad [3]$$

where A_p and A_L are the area fractions of the polar and lipophilic pathways, respectively; P is the partition coefficient of the nonpolar phase; C_w is the drug's water solubility (in mg/ml); and L is the thickness of the stratum corneum (in cm). Berner and Cooper suggest that $A_p = 0.1$ and $A_L = 0.9$. The octanol/water partition coefficient is used for P, and a value of 0.0015 cm is suggested for the thickness of the stratum corneum. Using these values for the parameters, J will have the units $mg/cm^2 \cdot hr$.

Berner and Cooper have also proposed a three—parallel-pathway model (4), which combines not only the polar and lipophilic pathways, but also an oil—water multilaminate pathway (2). Use of the same values for I, D_L, D_p, and A_p as for the two—parallel-path model, and decreasing the value for AL from 0.9 to 0.5, the equations to calculate the upper (J^U_{max}) and lower (J^L_{max}) bounds of the maximum flux through the stratum corneum are given in Eqs. 4 and 5. The values for J^U_{max} and J^L_{max} have been averaged in this evaluation to provide a single value for the three—parallel-path model.

$$J^U_{max} = \frac{2(C_w)e^{-0.016(M)}}{15} \left\{ \frac{[85(P) + 190][38 + 153(P)]}{(228 + 238(P)} \right\} \qquad [4]$$

$$J^L_{max} = \frac{(C_w)e^{-0.016(M)}}{15} \left[85(P) + 38 + \frac{5 \times 10^{-3}(P)}{380 + 170(P)} \right] \qquad [5]$$

In the heterogeneous structural model (model 3) described by Michaels and associates (5), the barrier function is treated as a dispersion of hydrophilic protein in a continuous lipid matrix through which the drug migrates by dissolution and Fickian diffusion. Given this reasonable assumption about the structure of the stratum corneum, the simplified flux equation is where P is the lipid/protein partition coefficient, which in the original paper, is set equal to the mineral oil/water partition coefficient. The octanol/water partition coefficient has also been successfully used.

$$J = 0.27(P)(C_w) \left[\frac{1160 + 3.4 \times 10^{-3}(P)}{160 + 2(P)} \right] \qquad [6]$$

The last model (4) was adapted from a theoretical description of percutaneous absorption published by Albery and Hadraft in 1979 (6,7).
In steady-state applications, the total fluxes can be predicted by
using the following equation:

$$J = \frac{36C_w}{2.82 + (29.6/P)}$$
[7]

III.   COMPARISON OF PREDICTED AND EX-
PERIMENTAL FLUX VALUES

Table 1 lists the molecular weight, partition coefficient, water solubility, experimental flux, and predicted fluxes for 50 compounds
(5,10−20).   When compiling a list of experimental values, the most
striking result is the huge value range that is found on the few
occasions in which values are determined by more than one investigator.   Partition coefficients and water solubilities can vary by 2
orders of magnitude, whereas experimentally determined transdermal
flux values can vary by 3 to 5 orders of magnitude.   With transdermal flux measurements, such variability is usually the result of
using substantially different experimental techniques (see Chapt.
12).   In many respects, the "ballpark" predictive nature of these
empirical models is desirable over "quick-and-dirty" in vitro transdermal experiments, especially if the purity of radiolabel is questionable.   The average predicted flux value is provided for comparison purposes.

IV.   USE OF THE PREDICTIVE MODELS

A few examples should demonstrate the usefulness of these predictive models.   First, if the drug or drugs to be evaluated have been
physicochemically characterized, then partition coefficient and water
solubility data will be available.   For 13 substituted melamines and
related compounds (Table 2), these properties have been determined (21).   This information, combined with the molecular weights,
directly provides predicted transdermal flux values.   If either maximum or minimum systemic breakthrough was desired, then these
predicted values would aid in the decision of which compound to
test in vitro.   However, it is important to remember that for analogues that are not expected to be metabolized to the same parent
compound (i.e. analogues that are not prodrugs) the activity of
each species must be considered.   If an analogue has a tenfold higher flux rate, but is 100-fold less potent, then it obviously will not

be the drug of choice. Once again, the balance between localized effect and systemic breakthrough is important.

The physicochemical characterization of a drug may not be completed early enough during drug development efforts to have both the partition coefficient and water solubility data available when evaluation by computational methods is desired. If partition coefficients are not available, then they can be calculated using various group contribution methods (8). Because these computational techniques are becoming more readily available commercially, they will not be further discussed here. High-performance liquid chromatography (HPLC) methods are also cited in the literature that correlate retention time to partition coefficients (22,23). These methods have the advantage that small amounts of the drug material are required to determine the partition coefficient. Thus, partition coefficients can be experimentally determined by HPLC when only a small quantity of the drug is available, or partiton coefficients can be calculated based upon drug structure alone.

Knowledge of the partition coefficient and melting point of crystalline compounds can be utilized to estimate the water solubility of the compound by use of the method of Valvani and Yalkowsky (9). Specific equations for various classes of organic compounds are given in Table 3. This water-solubility value combined with the calculated or experimental partition coefficient can then be utilized to predict flux values as shown previously. Table 4 gives predicted transdermal flux values for a series of benzodiazepines, calculated using partition coefficients and melting points (24). This is an example of how predicted flux values for a series of analogues can be combined with pharmacokinetic information (25) to select the drug candidate most likely to provide the most efficacious topical treatment. The use of these models for a series of timolol prodrugs (26) is given in Table 5. Because these compounds are expected to be metabolized to the same active species, the flux values are the primary concern. However, the rate of hydrolysis may also be important for localized delivery, since the prodrug could be "swept" by the targeted site of delivery before conversion to the desired species, if the rate of hydrolysis was sufficiently slow.

V. CONTOUR PLOTS

Although the models described in this chapter may prove useful in evaluating existing drugs for use as topicals, a more fundamental utility exists for these computational methods; namely, a set of qualitative guidelines can be established that will aid the synthetic chemist in designing drug analogues and prodrugs. In the past, vague guidelines existed indicating that increases in lipophilicity re-

Table 7.1 Comparison of Predicted and Experimental Transdermal Flux Values (mg/cm^2 · hr)

Compound (Ref.)	MW	PC	Cw (mg/ml)	Exp Flux	Model 1	Model 2	Model 3	Model 4	Avg
Acetaminophen (10)	151	2.95	0.012[a]	4.6	0.033	0.032	0.064	0.032	0.038
Alprazolam[b]	309	19	0.063	0.5	0.088	0.063	1.9	0.51	0.52
Atropine (5)	289	0.006	2.4	0.02	0.061	0.081	0.028	0.017	0.05
Benzoic Acid (10)	122	89	3.4(11)	720	439	286	280	38	270
Benzyl Alcohol (10)	108	10.5	40(11)	1,060	779	589	727	251	590
Caffeine (10)	194	0.955	22(11)	1.5	12	14	40	23	20
Chlorpheniramine (5)	275	0.46	1.6	3.5	0.14	0.18	1.4	0.85	0.6
Clindamycin Phosphate[b]	505	166	11	11	5.8	3.8	1160	130	260
Diethylcarbamazine (5)	199	0.064	800	100	106	142	100	61	110
Digitoxin (5)	765	0.014	0.01	0.00013	1.3×10^{-7}	1.7×10^{-7}	0.00027	0.00017	0.00009
Dextromethorphan (10)	271	13,489	0.017[a]	10	30	19	2.8	0.21	14
Diazepam (10)	285	631	0.05(14)	4.7	3.4	2.2	7.0	0.6	3.0
Ephedrine (5)	165	1.0	50	300	45	53	97	55	61
Estradiol (5)	272	12	0.003	0.016	0.0048	0.0036	0.061	0.020	0.019
Fentanyl (5)	337	200	0.2	2.0	1.9	1.2	22	2.3	5.7
Flurbiprofen	244	0.064 (15)	0.04(15)	20(16)	165	100	66	5.4	89

5-Flurouracil (10)	130	0.120	13.9(18)	3.2	6.5	8.7	3.2	2.0	5.9
Furosemide (10)	331	29	0.54a	0.21	0.81	0.55	22	5.0	5.9
Griseofulvin (10)	353	151	0.014 (12)	0.24	0.076	0.049	1.4	0.16	0.35
Hydralazine HCl (10)	161	0.0295	44.2(11)	20	9.5	13	2.6	1.6	7.8
Hydrocortisone (10)	363	41	0.28(11)	0.42	0.35	0.24	14	2.8	3.7
Ibuprofen (10)	206	3235	0.036 (15)	430	366	234	6.0	0.45	170
Idolyl-3-acetic acid (10)	175	26	1.56(12)	11	25	18	60	14	27
Indomethacin (10)	358	1200	0.016(12)	0.25	0.64	0.40	2.4	0.20	0.80
Isosorbide dinitrate (10)	236	16	1.09(11)	4.8	4.1	3.0	28	8.3	9.4
Ketoprofen (10)	254	1000	2.97(12)	12	520	330	431	37	330
Methotrexate (17)	454	0.005	11.8	0.034	0.021	0.028	0.11	0.071	0.053
Methyl salicylate (10)	152	288	0.67(11)	1,350	173	110	82	8	97
Minoxidil (10)	209	17	1.9b	0.81	11.8	8.4	52	15	19
Mitomycin C (MMC) (13)	334	0.41	0.94	0.32	0.030	0.039	0.75	0.45	0.26
Benzyl MMC (13)	434	38.6	0.65	3.2	0.25	0.16	33	6.4	8.0
Benzoyl MMC (13)	448	77.9	0.0045	0.0072	0.0028	0.0018	0.35	0.049	0.081
Benzylcarbonyl MMC (13)	462	34.5	1.0	0.3	0.22	0.14	47	9.6	11

Table 7.1 (Continued)

Compound (Ref.)	MW	PC	Cw (mg/ml)	Exp Flux	Model 1	Model 2	Model 3	Model 4	Avg.
Benzyloxycarbonyl MMC (13)	478	113	0.25	2.5	0.14	0.089	23	2.9	5.2
Propyloxycarbonyl MMC (13)	430	32.7	0.14	0.12	0.05	0.033	6.4	1.3	1.6
Pentyloxycarbonyl MMC (13)	458	279	0.27	1.7	0.50	0.32	32.9	3.2	7.4
Nonyloxycarbonyl MMC (13)	514	3670	0.00013	0.012	0.0013	0.00083	0.020	0.0016	0.0049
Morphine sulfate (10)	285	0.0174	64(11)	0.19	1.8	2.5	2.2	1.3	2.0
Naproxen (10)	230	1510	0.016 (12)	4.8	6.2	3.9	2.4	0.2	3.3
Nicotinic acid (10)	123	0.63	0.003 (11)	2.2	0.0038	0.0046	0.0037	0.0021	0.0038

Nitroglycerin (5)	227	10.0	1.3	13	3.6	2.8	23	8.0	8.0
Piroxicam (10)	331	1.12	0.57(12)	0.70	0.040	0.046	1.2	0.69	0.41
Ouabain (5)	585	0.00026	10	0.008	0.0022	0.0029	0.0051	0.0031	0.0032
Salicylamide (10)	137	7.76	0.02(11)	53	0.18	0.14	0.28	0.11	0.17
Salicylic acid (10)	138	174	0.002 (11)	1,900	0.39	0.25	0.21	0.024	0.22
Scopolamine (5)	303	0.026	75	3.8	1.6	2.2	3.8	2.3	2.4
Testosterone (10)	288	2040	0.039 (15)	34	21	14	5.9	0.48	11
Theophylline (18)	180	0.66	7.5	1.08	3.9	4.8	9.6	5.6	5.7
Timolol (19)	316	81	8	60	42	27	630	89	160
Triamcinolone acetonide (10)	434	339	0.04(20)	0.18	0.13	0.085	5.1	0.48	1.2
Correlation for plot of ln(theoretical) vs. ln(experimental)					0.88	0.88	0.73	0.73	

aCalculated using equations in Table 3.
bPersonal communications - various sources.

Table 7.2 Predicted Transdermal Flux Values for a Series of Substituted Me-amines Calculated from Experimentally Determined Octanol/Water Partition Coefficients and Water Solubilities (mg/ml)

Drug Structure

Drug/R_1	$R_2 = R_3$	MW	PC	CW	Calc. flux[a]
NMe_2	NMe_2	210	350	0.091	8
NHMe	NMe_2	196	67	2.16	65
NH_2	NMe_2	182	15.7	0.374	4
NEt_2	NMe_2	238	2800	0.071	23
N(OH)Me	NMe_2	212	41	0.904	17
$N(CH_2CH_2OH)Me$	NMe_2	240	92	3.22	91
$N(CH_2CH_2OH)_2$	NMe_2	270	39.5	8.25	120
Cl	NMe_2	201	204	0.61	38
$N(CH_2COOH)Me$	NMe_2	254	0.065	200	16
$N(CH_2CH_2OOCCH_2CH_2COOH)Me$	NMe_2	340	0.56	260	96
$SCH_2CH(COOH)NH_2$	NMe_2	286	0.45	2.24	0.7
NEtMe	NEtMe	252	8000	0.05	34
$N(CH_2OH)Me$	$N(CH_2OH)Me$	258	2.5	9.04	15

[a]Average of the four models ($\mu g/cm^2 \cdot hr$).
Source: From Ref. 21.

Table 7.3 Equations for Prediction of Water Solubility

Fundamental equations

liquid solutes

$$\log S_w = -1.072 \log PC + 0.672$$

crystalline solutes: rigid nonelectrolyte drug molecules

$$\log S_w^c \doteq -\log PC - 0.01 MP + 1.05$$

Applied equations

rigid polycyclic aromatic hydrocarbons, e.g., naphthalene

$$\log S_w^c = -0.88 \log PC - 0.01 MP - 0.012$$

mono- and multifunctional halobenzenes, e.g., 1,2-dichlorobeneze

$$\log S_w^c = -0.99 \log PC - 0.01 MP + 0.72$$

steroid hormones, e.g., progesterone

$$\log S_w^c = -0.88 \log PC - 0.01 MP + 0.08$$

hexamethylmelamines, e.g., hexamethylmelamine

$$\log S_w^c = -0.904 \log PC - 0.007 MP + 0.07$$

alkyl *para*-substituted benzoates, e.g., hexyl-*p*-aminobenzoate

$$\log S_w^c = -1.14 \log PC - 0.005 MP + 0.633$$

alkyphenols

$$\log S_w^c = -0.997 \log PC - 0.008 MP + 1.43$$

Source: From Ref. 9.

Table 7.4 Predicted Transdermal Flux Values for a Series of Benzodiazepines Calculated Using Experimentally Determined Partition Coefficients and Melting Points

Drug	MW	PC	Melting point (°C)	Calc. CW[a]	Calc. flux[b]
Diazepam	285	309	130	0.089	3.8
Oxazepam	287	97	205	0.044	1.0
Lorazepam	321	73	167	0.15	2.9
Temazepam	301	62	120	0.49	8.9
Desmethyldiazepam	271	54	216	0.054	0.98
Triazolam	343	48	234	0.050	0.74
Midazolam	326	43	115	0.82	11
Bromazepam	316	30	237	0.066	0.75
Chlordiazepoxide	300	28	236	0.068	0.75
Alprazolam	309	18	228	0.12	0.96
Clobazam	301	9	181	0.65	3.1

[a]Using equation for rigid polycyclic aromatic hydrocarbons in Table 3.
[b]Average of the four models ($\mu g/cm^2 \cdot hr$).
Source: PC from Ref. 24; MP from Ref. 11.

sulted in increased permeability and that major changes in molecular weight were required to produce significant differences in diffusivity (27). Although these generalities were very useful, more detailed insight into the relationship between a drug's physical properties and percutaneous permeation can be obtained by use of contour plots of the predicted flux that are plotted as a function of the partition coefficient versus water solubility at a fixed molecular weight.

Contour plots for molecular weights of 50, 100, 300, and 500 are given in Figures 1 and 2. For molecular weights of 500 to 1000, the contours were indistinguishable on this scale from the plot for molecular weight 500. As seen, for each molecular weight, the predicted flux values smoothly increase as both water solubility and partition coefficient increase. Although drugs, such as the prostaglandins and some antibiotics, have both relatively high water solubilities and partition coefficients, generally speaking, the upper

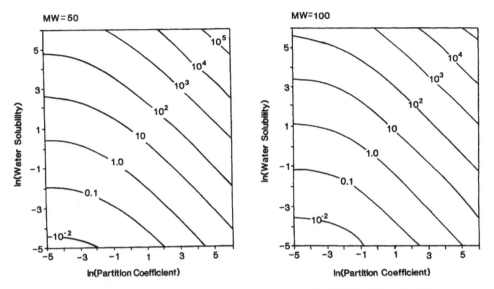

Figure 7.1 Plot for drugs of molecular weight (MW) 50 and 100. Contour lines are values of predicted transdermal flux in units of $\mu g/cm^2 \cdot hr$. The ordinate and abscissa are natural logarithms of water solubility (mg/ml) and octanol/water partition coefficients, respectively.

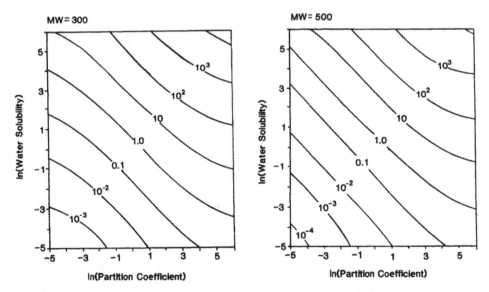

Figure 7.2 Plot for drugs of molecular weight 300 and 500. Contour lines are values of predicted transdermal flux in units of $\mu g/cm^2 \cdot hr$. The ordinate and abscissa are natural logarithms of water solubility (mg/ml) and octanol/water partition coefficients, respectively.

Table 7.5 Physical Data and Predicted Transdermal Flux Values of Timolol

Ester	R	MW	PC	Melting point (°C)	Calc CW^a	Calc. flux
O-Acetyl	$-COCH_3$	394	13	204	3.1	18
O-Propionyl	$-COCH_2CH_3$	408	42	188	1.4	18
O-Butyryl	$-CO(CH_2)_2CH_3$	422	120	159	1.0	22
O-Isobutyryl	$-COCH(CH_3)_2$	422	160	175	0.53	13
O-Valeryl	$-CO(CH_2)_3CH_3$	516	470	158	0.32	9.5
O-Pivaloyl	$-COC(CH_3)_3$	436	480	147	0.34	11
O-2-Ethylbutyryl	$-COCH(C_2H_5)_2$	530	1800	138	0.14	4.8
O-3,3-Dimethylbutyryl	$-COCH_2C(CH_3)_3$	530	1200	174	0.90	30
O-Hexanoyl	$-CO(CH_2)_4CH_3$	530	2200	177	0.046	1.6
O-Octanoyl	$-CO(CH_2)_6CH_3$	558	46000	170	0.0027	0.18
O-Cyclopropanoyl	$-COcC_3H_5$	500	54	173	1.9	28
O-1'-Methylcyclopropanoyl	$-COcC_3H_4-1'-CH_3$	514	160	175	0.64	15

O-2'-Methylcyclopropanoyl	$-COcC_3H_4-2'-CH_3$	514	180	162	0.77	19
O-Cyclobutanoyl	$-COcC_4H_7$	514	230	160	0.63	16
O-Cyclopentanoyl	$-COcC_5H_9$	528	560	173	0.20	6.1
O-Cyclohexanoyl	$-COcC_6H_{11}$	542	2000	181	0.047	1.6
O-Benzoyl	$-COC_6H_{11}$	536	350	181	0.27	7.6
O-2-Methylbenzoyl	$-COC_6H_4-o-CH_3$	550	1000	164	0.14	4.5
O-4-Methylbenzoyl	$-COC_6H_4-p-CH_3$	550	1200	202	0.048	1.6
O-2-Methoxylbenzoyl	$-COC_6H_4-o-OCH_3$	566	320	127	1.1	30
O-4-Methoxybenzoyl	$-COC_6H_4-p-OCH_3$	566	450	203	0.13	3.8
O-2-Acetoxylbenzoyl	$-COC_6H_4-o-OCOCH_3$	594	16	136	18	120
O-2-Benzoyloxymethylbenzoyl	$-COC_6H_4-o-CH_2OCOC_6H_5$	656	23000	152	0.0097	0.38
O-2-Aminobenzoyl	$-COC_6H_4-o-NH_2$	551	320	170	0.38	10
O-2-Methylaminobenzoyl	$-COC_6H_4-o-NHCH_3$	565	1100	167	0.12	3.9
O-3-Thienyl	$-COC_4H_3S$	510	190	150	0.95	23

ᵃUsing equations for rigid nonelectrolyte drug molecules given in Table 3.
Source: From Ref. 26.

right corner of the contour plot is seldom attainable. Most notable
on these plots is the shape of the contours and changes in this
shape as molecular weight is increased. Changing the physical
properties of a drug in a manner that results in movement parallel
to a flux contour line will result in neither an increase nor decrease
in the predicted transdermal delivery of the drug. Conversely,
changing the physical properties of a drug in a manner that results
in movement perpendicular to a flux contour line will provide the
greatest change in predicted delivery. For example, increasing the
partition coefficient from 0.05 to 0.37 (lnPC = -3 to -1), while
maintaining the water solubility at 2.72 mg/ml (lnS = 1) results in
a 5.3-fold increase in predicted flux for a drug of molecular weight
300, but only a 2.4-fold increase in the predicted flux for a drug
of molecular weight 100.

A good example of moving along a contour line is seen in the
experimental transdermal flux results for the mitomycin C (MMC)
analogues listed in Table 1. Note that the partition coefficient of
pentyloxycarbonyl MMC is twofold greater than the partition coefficient for benzyloxycarbonyl MMC, whereas the water solubilities and
molecular weights are approximately the same. The experimental
results show that the in vitro percutaneous absorption of these substances are essentially the same, just as anticipated based upon the
contour plot for molecular weight 500. Alternatively, benzyl MMC
and propyloxycarbonyl MMC can be compared. Both compounds have
essentially the same molecular weights and lipophilicity, but benzyl
MMC has about fourfold greater water solubility than propyloxycarbonyl MMC. Here, the increased water solubility of the analogue
increases drug flux by greater than an order of magnitude. Again,
this increase in flux of the drug would have been anticipated from
noting the perpendicular crossing of the contour line in Figure 2.

VI. ADDITIONAL CONSIDERATIONS

Although these contour plots are of obvious utility, it must be re-
membered that these computational methods provide predicted flux
values that are generally within an order of magnitude of the experimental values determined using in vitro transdermal methods.
It is fully expected that such computational methods may not be appropriate for topical drugs, such as steroids, that uniquely interact
with the bilayer-structured epidermal lipids (see Chap. 3) (28–30),
or for drugs that alter the barrier properties of the skin. This
second consideration can be seen in Table 1 when considering data
for the keratolytic, salicylic acid. Thus, experimental values
10,000-fold greater than the predicted values result. It is also
noteworthy that each of these models assumes that diffusion is stratum corneum-controlled, rather than dermally controlled.

Also it is important that the compounds initially investigated to establish the empirical parameters were nonelectrolytes. Thus, drug molecules that will dissociate should be evaluated with reservation. As seen in Table 1, the computational methods can be predictive for drugs that dissociate. However, predictions for compounds that fall outside of the assumptions of the models must be considered suspect. Likewise, use of these techniques to predict the percutaneous delivery of polypeptides is probably unfounded. Octanol/water partition coefficients, for polypeptides are often larger than would be expected compared with non—hydrogen-bonding solvent/water partition coefficients. The use of mineral oil/water, or alkane/water partition coefficients may be more predictive of the actual percutaneous flux values. Finally, remember that only the two models by Berner and Cooper utilize the molecular weight of the compound. For compounds of molecular weight above 1000, the resulting flux values from these two models might be more predictive of experimental results.

REFERENCES

1. A. F. Kydonieus, and B. Berner, eds., Transdermal Delivery of Drugs, Vols. 1—3. CRC Press, Boca Raton, Fla., 1987.
2. D. C. Monkhouse, and A. S. Huq, *Drug Dev. Ind. Pharm.*, *14*:183, 1988.
3. D. W. Osborne, Pharm. Manufact., *3*(4):41, 1986.
4. B. Berner, and E. R. Cooper, in *Transdermal Delivery of Drugs*, Vol. 2 (A. F. Kydonieus, and B. Berner, eds.). CRC Press, Boca Raton, Fla., pp. 41—62, 1987.
5. A. S. Michaels, S. K. Chandrasekaran, and J. E. Shaw, *AICHE J.*, *21*:985, 1975.
6. W. J. Albery, and J. Hadgraft, *J. Pharm. Pharmacol.*, *31*:129, 1979.
7. W. J. Albery, and J. Hadgraft, *J. Pharm. Pharmacol.*, *31*:140, 1979.
8. J. T. Chou, and P. C. Jurs, in Physical Chemical Properties of Drugs (S. H. Yalkowsky, A. A. Sinkula, and S. C. Valvani, eds.). Marcel Dekker, New York, pp. 163—199, 1980.
9. S. C. Valvani, and S. H. Yalkowsky, in Physical Chemical Properties of Drugs (S. H. Yalkowsky, A. A. Sinkula, and S. C. Valvani, eds.). Marcel Dekker, New York, pp. 201—229, 1980.
10. G. B. Kasting, L. Smith, and E. R. Cooper, in *Pharmacology and the Skin*. Vol. 1, *Skin Pharmacokinetics* (B. Shroot and H. Schaefer, eds.). Karger, Basel, pp. 138—153, 1987.

11. *The Merck Index*, 10th ed. Merck & Co., Rahway, N.J., 1983.
12. S. H. Yalkowsky, S. C. Valvani, W-Y Kuu, and R-M Dannen-
    felser, *Arizona Database of Aqueous Solubility*, 2nd ed., copy-
    right 1987 by S. Yalkowsky.
13. E. Mukai, K. Arase, M. Hashida, and H. Sezaki, *Int. J.
    Pharm.*, 25:95, 1985.
14. A. MacDonald, A. F. Michaelis, and B. Z. Senkowski, in *Ana-
    lytical Profiles of Drug Substances*, Vol. 1. (K. Florey, ed.).
    Academic Press, New York, pp. 79–99, 1972.
15. S. H. Yalkowsky, S. C. Valvani, and T. J. Roseman, *J.
    Pharm. Sci.* 72:866, 1983.
16. S. A. Akhter, and B. W. Barry, *J. Pharm. Pharmacol*, 37:27,
    1985.
17. S. M. Wallace, J. O. Runikis, and W. D. Stewart, *Can. J.
    Pharm. Sci.* 13:66, 1978.
18. K. B. Sloan, S. A. M. Koch, K. G. Siver, and F. D. Flowers,
    J. Invest. Dermatol., 87:244, 1986.
19. R. Cargill, K. Engle, G. Rork, and L. J. Caldwell, *Pharm.
    Res.*, 3:225, 1986.
20. K. Florey, in *Analytical Profiles of Drug Substances*, Vol. 1.
    (K. Florey, ed.). Academic Press, New York, pp. 397–421,
    1972.
21. A. J. Cumber, and W. C. J. Ross, *Chem. Biol. Interactions*,
    17:349, 1977.
22. R. Kaliszan, *Quantitative Structure-Chromatographic Retention
    Relationships*, John Wiley & Sons, New York, pp. 232–278,
    1987.
23. W. J. Lambert, and L. A. Wright, *J. Chromatog.*, 464:400, 1989.
24. D. J. Greenblatt, R. M. Arendt, D. R. Abernethy, H. G.
    Giles, E. M. Sellers, and R. I. Shaker, *Br. J. Anaesth.* 55:
    985, 1983.
25. D. J. Greenblatt, M. Divoll, D. R. Abernethy, H. R. Ochs,
    and R. I. Shader, *Drug Metab. Rev.*, 14:251, 1983.
26. H. Bundgaard, A. Buur, S.-C. Chang, and V. H. L. Lee,
    Int. J. Pharm. (Amst.), 46:77, 1988.
27. J. L. Zatz, CTFA Scientific Monograph Series No. 2., CTFA,
    Washington, D.C., 1983, pp. 29–45.
28. D. W. Osborne, COLL, Abstract 072, 191st ACS National Meet-
    ing, New York City, New York, April 13–18, 1986.
29. C. Ackermann, and G. L. Flynn, *Int. J. Pharm. (Amst.)*, 36:
    61, 1987.
30. C. Ackermann, G. L. Flynn, and W. M. Smith, *Int. J.
    Pharm. (Amst.)*, 36:67, 1987.

8
Kinetic Considerations in the Design of Surfactant-Based Topical Formulations

ANTHONY J. I. WARD* *University College Dublin, Dublin, Ireland*

I. INTRODUCTION

The development of formulations using surfactants as vehicles for transdermal drug delivery is currently of great interest. Much of the attraction of this type of system lies in the acceptable rheological and wide-ranging physicochemical properties that can be obtained. The objects of system design are to achieve a suitable reservoir dosage and controlled release by percutaneous absorption. Contact between such vehicles and the skin inevitably entails modification of the skin's barrier function, either by hydration changes or structural changes in the molecular arrangements in the stratum corneum. This is a result of absorption of components either from a vehicle into the skin or vice versa. The thermodynamic equilibrium properties of the system, as always, will control the ultimate state of the system; however, the time scales of the various kinetic processes can be of more importance for consideration in practical applications because controlled or sustained release over a period is usually a major requirement.

Recently (1−10), attention has been focused on the kinetics of the transfer of components across oil−water interfaces in systems containing surfactant aggregation structures. The amount of available data is small, however, when compared with the corresponding literature relating to equilibrium properties. An understanding of the mechanisms found for mass transfer across interfaces in such systems is becoming clearer as a result of such studies. The purpose of this paper is to review the available information in the context of what it tells us about processes, such as solubilization, in-

---

Current affiliation: Clarkson University, Potsdam, New York

terfacial liquid crystal formation, and diffusion, that control the overall kinetic behavior of systems containing surfactant aggregates.

II. EXPERIMENTAL BACKGROUND

A. Passive Drop-on-Fiber

One of the problems of quantitatively examining systems, such as emulsions, to obtain the kinetic and contained mechanistic informa- tion has been that of simultaneously defining the volume and inter- facial area of the system as a function of time. Studies of formula- tion stability to determine the factors involved in their time evolu- tion are usually somewhat qualitative and system-specific. An at- tempt to overcome some of the inherent difficulties in such measure- ments has been the development of the passive drop-on-fiber (3,5) technique. This defines the system by fixing in space one droplet of emulsion size using an inert cylindrical fiber. Provided the dis- tortions from gravity are negligible, the shape of the axis-symmetric drop is purely determined by capillary forces. An analysis of this type of system (3) shows that the rate of solubilization into solution is related in a simple fashion to the relative dimensions of the drop- let and fiber. Both the volume (V) of the drop and the interfacial area (A) can be obtained and used to calculate the rate from the general relationship:

$$\text{Rate} = \frac{1}{A} \frac{dV}{dt} \qquad [1]$$

The way in which the experiment has been performed has mainly been concerned with systems in which the amount of oil in the drop is extremely small compared with the amount required to saturate the micelles. In this respect, the data obtained are concerned main- ly with the *initial* kinetic process so that contributions from micelles containing oil can be neglected.

B. Rotating Disk

Cases in which the solubilizate under investigation is in the form of a solid that can be fashioned into a shape allows the use of methods in which the material is suspended in the solubilizing medium. Dis- solution of the solubilizate is then followed analytically as a function of time, with the particular technique being dependent upon the nature of the system. Radioisotopes, for example, have been used (1,2) with fatty acid solubilizates. An advantage of this technique is that it potentially allows the effects of rheology on the kinetics to be determined by spinning the solubilizate at different rates.

III. RESULTS AND DISCUSSION

A. Insoluble Oil—Aqueous Micellar Surfactant

Application of the drop-on-fiber technique (3,5,7,8) to the study of initial solubilization kinetics of highly water-insoluble oils into aqueous surfactant solutions has yielded consistent mechanistic descriptions. Pseudo—zero-order kinetics (Fig. 1) were observed (3,5–8) for single-component oils solubilizing into different surfactants above the critical micellization concentration (CMC), whereas the rates were not observable below the CMC (Fig. 2). This behavior was considered (3) in terms of the following possible diffusion mechanisms:

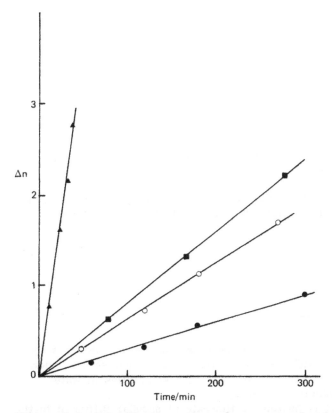

Figure 8.1 Solubilization stage of n-nonane droplets solubilizing into surfactant solutions (1% w/w) as a function of time [Δn is the drop radius/fiber radius compared at zero time and time (t)]; solid triangles, $C_{12}EO_6$; solid square, DDAPS; open circle, dodecyltrimethylammonium bromide, and closed circle, SDS.

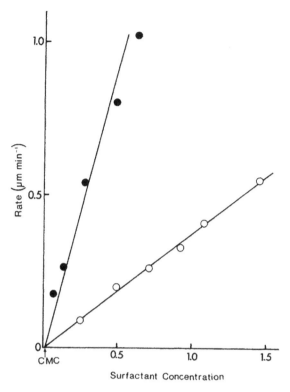

Figure 8.2 Solubilization rate of oils as a function of $C_{12}EO_6$ concentration; open circles, tetradecane; solid circles, n-nonane.

1. Diffusion of oil into water is related to the water solubility of the oil $C_{o,w}$ as

$$\text{Rate} = \frac{D_{o,w} C_{o,w}}{\delta} \qquad [2]$$

where $D_{o,w}$ is the diffusion coefficient of the oil and δ is the diffusion layer thickness (Fig. 3). This mechanism was shown to be significant (3) only for oils with solubilities in water greater than about 10^{-6} wt%.

2. A process that is limited by the rate of micellar diffusion to the oil—water interface with mass transfer of oil to micelle.

The rates of solubilization predicted on this mechanism are also too high compared with the experimentally observed values for wa-

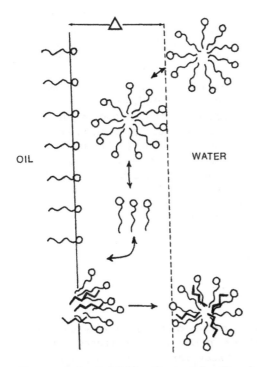

OIL   WATER

Figure 8.3 Solubilization mechanism involving micellar association-dissociation within a diffusion layer of thickness, Δ.

ter-insoluble oils. It was argued that this was because the dissoci-ation of the micelle, being a precursor to adsorption with consequent desorption of the mixed micelle-containing oil, did not occur for every excursion of micelles into the interfacial region (3). A factor describing the probability of the micelle dissociating within the vi-cinity of the oil—water interface, leading to solubilization, has to be included (3) giving the relation

$$\text{Rate} = \frac{\Delta}{2\tau} V_0 (b/a) (C - CMC) \qquad\qquad [3]$$

where Δ is the thickness of the interfacial region (see Fig. 3), which has dimensions of the order of the micelle diameter; τ is the time interval between micellar dissociations that lead to adsorption and concerted solubilization steps; V_0 is the molar volume of the solubilizate, and b/a is micellar capacity for the solubilizate (i.e., moles oil per mole surfactant).

Equation 3 has been used to describe the behavior of alkanes solubilizing into aqueous micellar solutions of various surfactant types (3,5-8). The term (b/a) is found to be proportional to the *equilibrium* solubilization capacity of the micelle for the homologous series of *n*-alkanes (7). Comparisons among oils of differing architecture (5) indicated that the constant of proportionality between the equilibrium and kinetic values of the solubilization capacity was different, possibly reflecting differences in the oil packing in the micelle. A further development upon the packing requirements of the micelle-solubilizate aggregate has been made (8,11) in the consideration of solubilization from binary oil mixtures. It was shown (11) that, within the restriction of no preference in the solubilization of the oil components (i.e., the composition of the solubilizate mixture remaining the same as the bulk contacting oil phase), relationships of the following form should apply.

$$b_1 = x_1 \{(x_1 b_1^0 + x_2 b_2^0) - x_1 x_2 P\} \qquad [4a]$$

$$b_2 = x_2 \{(x_1 b_1^0 + x_2 b_2^0) - x_1 x_2 P\} \qquad [4b]$$

where b_i^0 are the solubilization capacities of the pure oils and the x_i represents the oil mole fractions in the binary mixtures. The value of the factor P can vary between zero and $(1 + b_1^0/b_2^0)$, the former value representing ideal mixing of the solubilizate in the micelle interior. Comparison of data determined from kinetic experiments with values determined at equilibrium show agreement in the value of P required to fit the data in some cases (8), whereas, in others, different values are needed in each case. The solubilization of oil from dodecane-tetradecane mixtures into micelles of the nonionic surfactant *n*-dodecylhexaoxyethylene glycol ether, $C_{12}EO_6$, (12) shows good agreement (Fig. 4a) with a P value of 1.3 (i.e., very close to ideal mixing). In contrast, solubilization from similar mixtures into a micellar zwitterionic surfactant, *n*-dodecyldimethylammonium propane sulfonate, shows (Fig. 4b) that, although the equilibrium data require P = 0, the kinetic data is fitted only if the maximum possible value of P is used. This discrepancy may be a result of micelle shape changes in the first stages of the solubilization process that are not considered in the derivation of Eq. 4. A further manifestation of micellar shape changes that occurs in the initial stages of solubilization is the nonlinear temperature dependencies observed for the rates in the region of the phase-inversion temperature of some nonionic surfactants (5,8).

One consequence of the proposed mechanism is that the rate is essentially independent of stirring in the system, unlike the passive

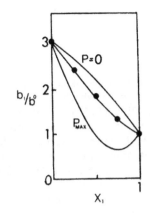

(a)

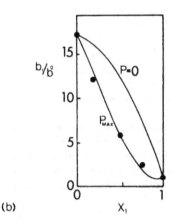

(b)

Figure 8.4 Composition dependence of solubilization for (a) $C_{12}EO_6-$ $C_{12}H_{26}-C_{16}H_{34}$ and (b) $DDAPS-C_{12}H_{26}-C_{16}H_{34}$ at 298 K.

diffusion process described by Eq. 2. Here, the value of δ is decreased and the rate increased by increased stirring (1,2). A similar conclusion has been reached more recently from the description of interfacial kinetics in experiments involving contacting phases with different states of surfactant aggregation (9,10). A linear increase in volume of the aqueous micellar phase with time in contact with a lamellar phase was found. If the rate of mass transfer from the micellar phase had been diffusion-limited, a square root of time-dependence of the layer thickness would have been expected. The restructuring to produce surfactant monomers from the micellar dissociation in passage into the lamellar phase was regarded as being the rate-limiting step.

B. Soluble Oil—Aqueous Micellar Surfactant

Oils that have sufficient water solubility are more likely to produce kinetics that are dependent upon mechanisms in which a diffusion step is rate-determining. Preliminary studies (7,13) of systems containing benzene, as either a single component or a component of a binary mixture, have indicated that this is true. Benzene, for example, is soluble to the extent of about 1500 ppm in water, and any oil that has a water solubility greater than about 10^{-5} wt% may be regarded as soluble in this context. Oil transport across the liquid—liquid interface is essentially that of solubility and is governed by the passive diffusion Eq. 2. Rates determined by the drop-on-fiber experiment are for systems under conditions of minimal stirring and, therefore, do not represent rates that are at the diffusion limit (i.e., the diffusion layer thickness, δ, being on the order of the dimensions of the oil molecule). Some rates typically found for soluble oils dissolving in water are presented in Table 1.

Interestingly, the diffusion layer thicknesses derived in these experiments are similar to those found in membrane processes in vivo. Furthermore, the ratio of the rates is the same as the ratio of the oil solubilities in water, as required by Eq. 2.

Equation 2 shows that the rate is independent of the surfactant concentration, inasmuch as it does not affect the activity of the oil in the aqueous phase. A surfactant concentration scan of the rate of benzene in contact with aqueous solutions of sodium dodecyl sulfate (Fig. 5) shows no increase until the total surfactant concentration is above the CMC. This may be understood in terms of the relative sizes of the oil "sinks" provided by the bulk water and the micelles; thus, only when the volume of micelles available for solubilization is comparable with, or greater than, that from the water solubility will there be an observable effect from the surfactant on the observed rate. The concentration at which this occurs will obviously depend on the size of the water solubility, CMC, and micelle volume. The larger the micellar volume and lower the CMC in a system with oil of a given water solubility, the lower will be the total surfactant concentration at which micellar solubilization will have an important contribution. Factors such as the size of the surfactant hydrophobe, therefore, will be important, as will any factor that influences the CMC of the surfactant. Similarly, because nonionic surfactants tend to have larger solubilization capacities for oils than do ionic surfactants of similar molecular volume, this effect should occur at lower surfactant concentrations in formulations based on nonionics.

Table 8.1 Typical Rates for Soluble Oils Dissolving in Water

Oil	Rate (m · min^{-1})	Diffusion layer thickness (μm)
Benzene	2.4×10^{-6}	50
Toluene	0.7×10^{-6}	55
p-Xylene	0.18×10^{-6}	55

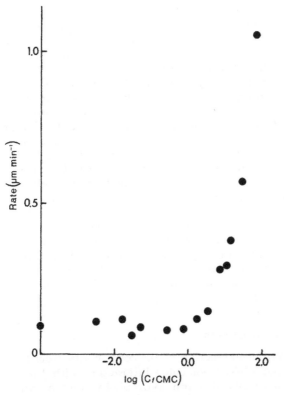

Figure 8.5 Solubilization rate of benzene as a function of cetyltri-methylammonium bromide concentration (normalized to the CMC) at 298 K.

C. Interfacial Liquid Crystal Formation

Early work (14,15) on systems with a fatty acid or soap constituent indicated the importance of interfacial liquid crystal formation in solubilization and associated detergency processes. This was further emphasized in studies of solubilization kinetics in such systems using a rotating disk apparatus (1,2). The mechanisms found in these situations involved the relative temperature of the system compared with that required to produce interfacial liquid crystal formation, diffusion of components in and out of the interfacial layer, and the degree of stirring. Experiments to determine the pathways of phase formation in surfactant—oil—water systems have shown the important role played by the formation of lamellar phases in interfacial mass transfer processes (9,10). Furthermore, the rate at which components diffuse in and out of this interfacial layer, when considered in the context of the phase behavior of the system, determines the overall observed kinetics. An example has been given in which the formation of a layer of interfacial water effectively stopped the transport of the oil component in the system (9). Here, although the bulk composition of the total components indicated that a micellar phase should be formed, this state was, in fact, never reached because the kinetic pathway led to the formation of an effective barrier, to transport in the unstirred system; namely, the water layer formation. These considerations must be taken into account in situations in which multicomponent systems are involved and interfaces formed. An important interface to consider, in this context, is that of the vehicle and skin containing unequal water contents on either side of the interface. The occlusive effect of the vehicle will lead to an increase in the water content of the stratum corneum. Similarly, there will be diffusion of the water component of the vehicle into the interfacial region. The tendency of transport of water from one side to the other will depend upon the relative thermodynamic activities that, in turn, will vary relative to each other, depending on the rates at which their concentrations are being established. This will have an influence upon the transport of any hydrophilic or polar components across the interface. The same kind of considerations apply to the nonpolar components of both the vehicle and the stratum corneum.

D. Differential Solubilization Effects

An example of the differential extraction of oil components from the skin has been demonstrated (16). Molecules interacting with the stratum corneum lipids, primarily through hydrophobic forces and located within the lipid bilayers, were extracted the most rapidly. Those molecules that were more amphiphilic and interacted more in-

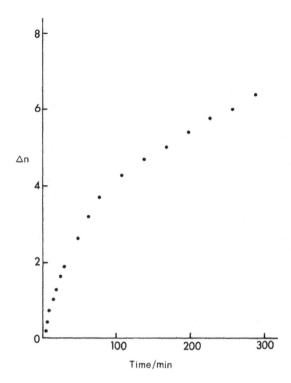

Figure 8.6 Solubilization stage of a benzene—cyclohexane mixture (7:3 w/w) into aqueous SDS (0.032 M) at 298 K.

timately with the bilayer took much longer to be removed. This implies that the nature of the oil component location within the surfactant association structure is also of importance in the solubilization and related processes. Demonstration of such preference in the solubilization process in simple micellar solutions is found in extraction from binary mixtures of benzene and cyclohexane into either water or surfactant solutions (6). The kinetic profile derived from a drop-on-fiber experiment (Fig. 6) is no longer linear, implying that the composition of the oil drop is also time-dependent. A bulk-scale experiment, in which the composition of the oil in the micellar phase was monitored by gas chromatography (13), showed the composition to increase to a ratio of about 9:1 from an inital 1:1 bulk-oil composition within the first hour of contact between the bulk oil and micellar phases. Here, the process is determined by the diffusion rates into water, which are in a ratio of approximately 30:1 in favor of benzene. Similar effects have recently been found in the

initial solubilization kinetics from binary mixtures of *n*-alkanes (13),
for which the mechanism is that discussed in Section I.

Such observations for the formulation of topical agents are important because they raise the question of the relative tendencies of
molecules to leave (or enter) the vehicle. This is determined by the
nature of the molecules and the molecular properties of the surfactants used to provide the vehicle. For solute species for which the
rates are determined primarily by diffusional processes, this, in a
first approximation, reduces to a consideration of their relative solubilities. However, the location of the solubilizate within an aggregated structure must also be considered because it will have a bearing on the rate at which it is released or solubilized. Molecules
that intercalate between the surfactant molecules, with an orientation
toward the aggregate—solvent interface, that form the aggregated
structure, whether simple micellar or lyotropic liquid crystal, require a greater energy to overcome the constraints of their environment. A consequence of this is a slower rate of release than that
of nonpolar components held in the essentially liquid hydrocarbon
interior of such aggregated surfactant structures. The rate of accumulation at, or transport across, the vehicle—stratum corneum interface of the different components in a typical multicomponent system will be determined, to a certain extent, differentially as result
of their locations within the system.

E. Conclusions

The time evolution of the vehicle—stratum corneum interface will undoubtedly be such that steady-state penetration of a drug from the
vehicle will not be instantaneous. It is usual to express these nonequilibrium effects in terms of a *lag time* (Fig. 7). Conventionally,
this has been developed in terms of the passive diffusion equation
written as

$$Q_t = \frac{DP\Delta C_\upsilon}{\delta} (t - \tau) \qquad [5]$$

where Q_t is the total amount absorbed in time t, D the solute diffusion coefficient (ΔC_υ) is the difference in the solute concentration
between the vehicle and tissue, P is the solute partition coefficient
between vehicle and skin, and δ is the thickness of the stratum
corneum—vehicle interfacial region. The lag time has been shown to
be equal to $\delta^2/6D$. Generally the value of δ has been taken as the
thickness of the stratum corneum using the measured lag times to
obtain values for the diffusion coefficients. The considerations presented here, however, indicate that such a simple passive descrip-

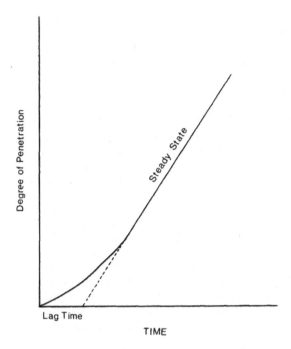

Figure 8.7 Idealized drug—penetration-time profile for diffusion through skin.

tion of the diffusion barrier is limited, because the active interaction of components at the vehicle—stratum corneum interface will effectively alter the value of δ. Exact description of the time evolution of δ depends on a knowledge of the effects discussed earlier. This is important in view of the large range of lag times that have been observed (minutes to several days) in studies of drug penetration from topical vehicles.

IV. SUMMARY

The work outlined shows some of the insights to be obtained from a controlled and well-defined experimental approach to the study of interfacial kinetics. The importance of interfacial surfactant association structures and their rearrangements has been demonstrated. Such considerations are also important in many processes, such as liquid chromatography (17), potential low-energy cost oil separations, as well as multivarious physiological processes.

The importance of the phase behavior in determining kinetic behavior (1,2,9,10), or vice versa, has also been demonstrated. It should be noted that a combination of local adsorption phenomena and development of interfacial concentrations dependent on the relative rates of ingress or diffusion away can lead to an interfacial phase formation that is unexpected from the total concentrations of the components in the system. Thus, for example, the observation of interfacial liquid crystal formation at the oil—water interface in systems in which the total oil is less than 1% of the amount required to saturate the surfactant micelles with which the oil is in contact in initial kinetics experiments.

Both of these considerations (i.e., time scale and phase evolution) will affect the performance of a topical vehicle formulation. Humectant aspects of an applied cream, for example, will raise the water concentration at the vehicle—skin interface, thereby altering the local concentrations of the system. The effect of this will be determined by the phase behavior of the system in response to increased water and the rates at which the components diffuse into and out of the interfacial region. One possible scenario would be that for which the rate of ingress of components from the vehicle was much slower than the rate of water buildup. Here, an essentially impenetrable barrier to the passage of a hydrophobic drug could be established.

In general terms, for the rate of steady-state penetration of a drug to be attained (see Fig. 7), the effects discussed will be manifested in terms of the lag time (τ). The range of lag times encountered is from minutes to several days; thus, they are of clinical significance, and an understanding of the factors underlying them is of major importance in the design of topical systems.

REFERENCES

1. J. A. Shaeiwitz, A. F.-C. Chan, E. L. Cussler, and D. F. Evans, *J. Colloid Interface Sci.*, *84*:47, 1981.
2. C. Huang, D. F. Evans, and E. L. Cussler, *J. Colloid Interface Sci.*, *82*:499, 1981.
3. B. J. Carroll, *J. Colloid Interface Sci.*, *79*:126, 1981.
4. Y. C. Chiu, Y. C. Han, and H. M. Cheng, in *Structure/ Performance Relationships in Surfactants*. (M. J. Rosen, ed.). ACS Symposium Series, p. 23, 1984.
5. B. J. Carroll, B. G. O'Rourke, and A. J. I. Ward, *J. Pharm. Pharmacol.*, *34*:287, 1982.
6. A. J. I. Ward, M. C. Carr, and J. Crudden, *J. Colloid Interface Sci.*, *106*:558, 1985.

7. A. C. Donegan, and A. J. I. Ward, *J. Pharm. Pharmacol.*, *39*: 45, 1987.

8. B. G. C. O'Rourke, A. J. I. Ward, and B. J. Carroll, *J. Pharm. Pharmacol.*, *39*:865, 1987.

9. S. E. Friberg, M. Mortensen, and P. Neogi, *Separation. Sci. Technol.*, *20*:285, 1985.

10. P. Neogi, M. Kim, and S. E. Friberg, *Separation. Sci. Technol.*, *20*:613, 1985.

11. B. J. Carroll, *J. Chem. Soc. Faraday Trans. I*, *82*:3205, 1986.

12. P. Faulkner, M. Sc. Thesis, Natl. University of Ireland, 1985.

13. A. J. I. Ward, unpublished results.

14. A. S. C. Lawrence, *Disc. Faraday Soc.*, *25*:51, 1958.

15. A. S. C. Lawrence, A. Bingham, C. B. Capper, and K. Hume, *J. Phys. Chem.*, *68*:3470, 1964.

16. G. Imokawa, and M. Hattori, *J. Invest. Dermatol.*, *84*:282, 1985.

17. D. W. Armstrong, T. J. Ward, and A. Berthod, *Anal. Chem.*, *58*:579, 1986.

9

The Use of Phase Behavior and Laboratory Robotics for the Optimization of Pharmaceutical Topical Formulations

DAVID W. OSBORNE   *The Upjohn Company, Kalamazoo, Michigan*

I. INTRODUCTION

Formulators seldom have the time required to optimize topical pharmaceutical formulations over complete ranges of both active and inert ingredients. Traditional cream, lotion, and ointment formulations are usually taken from cosmetic formulatories. The drug is
added at the desired level to the heated ointment base or emulsion
oil phase (water phase for hydrophilic drugs) and the resulting
formulation is observed for physical stability and appropriately assayed for potency, preservative challenge, etc. Since we assume
the vehicle was optimized prior to being included in the formulatory,
no changes are made unless the formulation shows indications of
phase separation, or chemical incompatibility with the drug. While
this process should work well for the formulator, anyone who has
followed this procedure knows that drugs tend to either destabilize
the vehicle or have inadequate solubility in the vehicle. Even
worse physical stability problems can also occur, such as slow crystallization from a super saturated solution, or crystallization of a
more stable polymorph of the drug. These latter occurrences are
particularly troublesome because the resulting changes in product
properties may not become apparent until months after manufacture.
Optimization of a pharmaceutical topical formulation can help to minimize the likelihood of encountering these problems.

Typically, a formula is considered acceptable when critical product characteristics (appearance, potency, viscosity, content uniformity, etc.) have been identified and shown to be maintained for
the shelf life of the product. The acceptable formulations are then
further scrutinized by considering raw material costs and processing

restraints until a final formulation is selected. It is important that the critical product characteristics be obtainable using raw materials whose properties span the supplier's specification range. This is of particular importance if one or more of the ingredients is a natural product that may tend to have extensive lot-to-lot variability. If the formulation is not tested using at least three different lots of each ingredient, then costly problems may arise during manufacture of the product. Likewise, the ranges for both active and inert ingredient amounts that are tolerable to the formulation should be determined. If more than one drug concentration is anticipated, then the ranges of inert ingredients should be checked at each drug concentration. It becomes quickly apparent that the optimization of even a five-component formulation is a substantial undertaking. Finally, the critical and noncritical process variables must be defined once an optimal formulation is selected. Although this last step is usually associated with scaleup, laboratory scale experiments can help define processing steps such as holding times, holding temperatures, and the effect of heating and cooling cycles.

Obviously, the formulator is faced with a tremendous challenge to actually optimize a topical formulation. To meet this challenge, an efficient, well-planned series of laboratory investigations must be performed. These experiments can be divided into three steps:

1. A preliminary formulation is selected and component ranges are determined by conducting phase behavior studies.
2. A statistically valid experiment is designed utilizing these component ranges and any other process variables considered important.
3. Responses from the experimental trials are empirically modeled and optimized.

The last two steps in this process have been the focus of much attention for solid-dosage forms (1,2) and, more recently, for disperse systems (3). However, these optimization techniques require that the responses be smooth and continuous throughout the variable ranges. For topicals, smoothness and continuity must be established for the composition ranges before optimization. Thorough characterization of the formulation by determining the phase behavior of the system is a way of establishing smoothness and continuity throughout the composition range. Thus, selection of component ranges by phase behavior determination will be the main focus of this chapter.

II. DETERMINATION OF COMPONENT RANGES

The first step in the optimization of a topical drug delivery system is selecting a preliminary formulation and determining the component ranges to be optimized by conducting phase behavior studies. Because these studies require preparation of a large number of samples, laboratory robotic techniques can be used to automate the quantitative dispensing and mixing of the components. Although the data that result from evaluating each of the samples can be represented by various techniques, plotting the data to compose a binary or ternary phase diagram is the method used in this chapter. Representing the data with binary or ternary diagrams is the most efficient way of presenting all of the equilibria for systems containing only pure components. For complex systems typical of industrial applications, binary and ternary diagrams provide a method of mapping (characterizing) the "phases" whose component composition range must be known. Thus, the general usefulness of binary and ternary phase diagrams warrants the following detailed description of how to assimilate the information provided by these diagrams. This tutorial on phase diagrams will be followed by a general, yet detailed description of the laboratory robotic techniques that are useful in preparing the samples required to complete the phase diagrams. Remember, adequate characterization of any disperse system is required to guarantee smoothness and continuity throughout the composition range to be optimized.

A. Phase Behavior

Disperse systems are unique among pharmaceutical dosage forms because the physicochemical interactions between the components dominate the stability of the system. Small changes in the component composition can completely destabilize or dramatically improve the stability of the dispersion. A well-characterized example would be the stable emulsion region for the water—p-xylene—decaoxyethylene glycol nonyphenol ether system. In this system the stability of the emulsion dramatically increased when the surfactant concentration was changed from 3 to 4 wt% (4). This dramatic change in physical stability was related to the formation of lamellar liquid crystalline bilayers around the oil droplets. The formation of a third phase (e.g., a liquid crystalline phase) is now a common method of increasing emulsion stability (5). Other examples of small compositional changes affecting product properties would include the effect of pH on clay suspensions or the effect of solids content on the melting point of a suppository.

  To understand dispersion stability, it must be realized that changes in stability are the result of changes in the phase behavior

of the system. For surfactant systems, sufficient scientific investigations have been completed to begin to correlate emulsion (4,5),
foam (6), and gel (7) stability with the equilibrium that exists between the thermodynamically stable micellar and liquid crystalline
phases. Although the techniques for representing phase behavior
described in this chapter are not limited to surfactant systems, they
have been primarily applied to this area of study. Nontraditional
formulations may generally be less dependent upon emulsifiers than
current topical creams; however, binary and ternary phase representations are useful in characterizing any of a variety of systems.
Most topical formulations are composed of materials that fall into one
of three groupings: polar components, nonpolar components, and
other components (e.g., amphiphilic, polymeric, therapeutic, bio-
technology products). For any formulation whose components can
be divided into two or three such groupings, binary or ternary
plots of the phase behavior can be extremely useful in characterizing the physicochemical interactions between the components of the
system. These physicochemical interactions between the individual
components ultimately lead to the stability or lack of stability characteristic of pharmaceutical dispersions.

1. Binary Systems

Binary diagrams are typically plotted on Cartesian coordinate grids
in which the composition is plotted along the abscissa and temperature is plotted along the ordinate. Phase boundaries are plotted on
this grid to separate single-phase regions from multiple-phase regions. An idealized example of a two-component liquid system having an upper critical solution point is shown in Figure 1. The left
vertical axis represents the phase behavior of pure A over the temperature range of interest, whereas the right vertical axis is for
pure B. Both A and B are liquids for this temperature range at
the given pressure. As expected, the solubility of B in A increases as the temperature is increased. This change in solubility
is shown by the left phase boundary. Analogously, the changes in
the solubility of A in B with increasing temperature is represented
by the right phase boundary. The two-phase liquid—liquid region
falls between these boundaries. The phase diagram provides considerable information about any sample that is mixed within the conditions represented by that diagram. First, for any binary mixture
of components A and B, it can be discerned whether the sample
will be one or two liquid phases. Second, if the sample is two
phases, it can be discerned what the composition of each phase will
be and how much each phase will weigh. For example, consider the
point circled in Figure 1. Overall, this sample contains 40% A and
60% B. We can immediately see that this sample will separate into

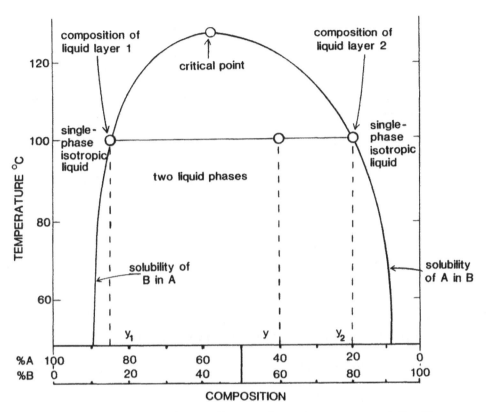

Figure 9.1 Idealized binary diagram for a two-component liquid system having an upper critical point.

two phases at 100°C, with one of the liquid layers having the composition 85% A and 15% B, and the other liquid layer having the composition 20% A and 80% B. The constant-temperature line that connects these three colinear composition points is called a tie line. For binary systems, all of the tie lines are understood to be parallel (i.e., constant temperature) and need not be drawn. Thus, for any composition at 100°C within the two-phase region, the composition of one liquid layer will be 85% A and 15% B, whereas the other liquid layer will have the composition 20% A and 80% B. The only difference will be the relative amount of each of these phases as one moves along the tie line.

To calculate the amount of each layer, remember that the weights of liquid layer 1 (m_1) and liquid layer 2 (m_2) must add up to the total weight of the sample, thus

$$m_{tot} = m_1 + m_2 \tag{1}$$

The sample also has a gross weight percentage of component A that is designated by y in Figure 1. However, as stated before this sample breaks into two phases that have weight percentages of component A equal to y_1 and y_2. Thus, the conservation of component A allows us to write

$$ym_{tot} = y_1 m_1 + y_2 m_2 \tag{2}$$

Substitution of the expression for m_{tot} from Eq. 1 into Eq. 2 gives

$$y(m_1 + m_2) = y_1 m_1 + y_2 m_2 \tag{3}$$

which rearranges to

$$m_1/m_2 = (y_2 - y)/(y - y_1) \tag{4}$$

As seen, the weights of the two phases are in the proportion of the lengths of the tie line segments extending from each side of the sample point. Another way to write this relation is

$$m_1 = m_{tot}[b/(a + b)] \tag{5}$$

$$m_2 = m_{tot}[a/(a + b)] \tag{6}$$

where a and b are the lengths, as shown in Figure 1 [i.e., $(y - y_1)$ and $(y_2 - y)$], respectively. To continue with the example for the point circled in Figure 1, if the total weight of the sample was 5 g, then liquid layer 1 (85% A, 15% B) would weigh 1.54 g, whereas liquid layer 2 (20% A, 80% B) would weigh 3.46 g.

For the diagram shown in Figure 1, as the temperature is raised, the tie lines become shorter because the solubility of each component in the other is increasing. Ultimately, the tie line becomes a point as it crosses the upper portion of the phase boundary. This point could also be considered the point at which the A-in-B solubility boundary joins the B-in-A solubility boundary. A sample having this composition will consist of two roughly equal-volume phases below the upper consult boundary. When heated sufficiently to reach the "critical" point, the interface between the two liquid phases will disappear, and a single-phase liquid will result. Upper critical points, lower critical points, or both upper and lower critical points can occur for binary liquid systems. In systems in

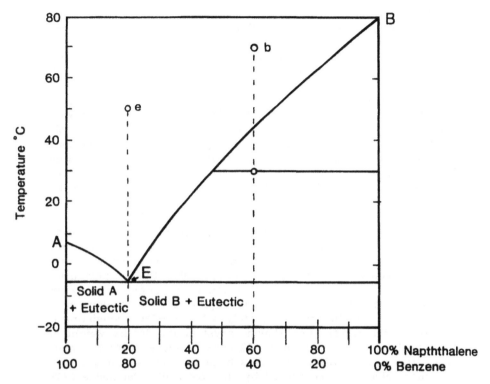

Figure 9.2 Freezing-point diagram for the binary system benzene—naphthalene at 1 atm pressure (After Ref. 9, p. 324).

which both upper and lower critical points exist, the phase boundary (also called the coexistence curve) forms a closed loop. Thus, as the temperature of a miscible mixture is lowered, the mixture first separates into two phases and, then, reforms as a miscible mixture. A very readable article describing these reappearing phases recently appeared in *Scientific American* (8).

If the temperature range of interest for the two-component system is such that one or more solid phases now exist, the binary diagram has a considerably different appearance. As seen in Figure 2, two components that are miscible when liquids, and in which only pure crystalline forms of the two components form as solids, produce distinctive phase behavior when combined. Again, it is understood that the pressure for the system is held constant. Above the curved lines AE and BE, the components are mutually soluble, forming a single liquid phase. Below the horizontal straight line, the temperature is sufficiently low that no liquid phase exits. Be-

tween the curved lines AE and BE is the two-phase region in which
the pure components are in equilibrium with the single-phase liquid.
Just as before, any point in either of these two-phase regions will
have a horizontal tie line passing through it. The location at which
this tie line intersects the phase boundary will give the composition
of the liquid phase, whereas the relative amounts of the liquid phase
and solid pure component can be calculated by Eqs. 4 through 6.

As outlined by Barrow (9), it is instructive to consider what
happens when solutions of two particular concentrations are cooled.
Consider first the composition labeled b (60% naphthalene in benzene)
at 70°C. As the sample is cooled, it will remain a liquid until crossing the phase boundary labeled as curve BE. Upon crossing this
boundary the first solid naphthalene crystals will become detectable
in the sample. Note that for this diagram the length of the tie line
segment to the right of this composition will not change, whereas
the length of the left tie line segment will increase with decreasing
temperature. Thus, the amount of solid naphthalene will continue
to increase in the sample until the horizontal line at approximately
−4°C is reached. As stated previously, below this temperature no
liquid exists; thus, the sample that contained roughly equal amounts
of primary naphthalene crystals and liquid (20% naphthalene and 80%
benzene) at −3°C becomes completely solid at −5°C, consisting of
40% by weight primary naphthalene crystals and 60% by weight of a
fine crystal-grained mixture of 20% naphthalene and 80% benzene
(i.e., the eutectic mixture). This behavior can be compared with
the phase changes encountered by the composition labeled e (20%
naphthalene) at 50°C. As the sample is cooled it remains a single
liquid phase until reaching the −4°C line, at which time the solution
becomes in equilibrium with pure solid benzene and pure solid naphthalene. The sample remains at this temperature until completely
converted into a fine-grained crystal mixture of the two solids, having the composition 20% naphthalene and 80% benzene. The point
labeled E is called the *eutectic point*. Note the eutectic mixture has
a melting point lower than either of the pure components.

Although the benzene—naphthalene system is useful in understanding eutectics and identifying eutectics on a binary diagram,
most systems of interest contain components that interact with one
another in the solid state. Such interactions result in solids or
semisolids that are mixtures of the two components. Interactions of
this type can also result in the formation of solid-state compounds
consisting of simple mole ratios of the two components. The effect
that formation of compounds and mutual component solubility has on
the appearance of the binary phase diagram is shown in Figure 3.
The mutual solid-state solubility of A with B can be seen along the
vertical axis of the plot. When B begins to solidify at temperature

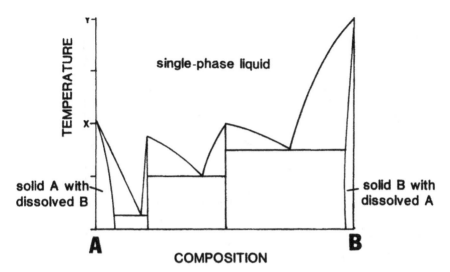

Figure 9.3 Idealized freezing point diagram showing multiple compound formation in the solid state.

y, a certain amount of A is accommodated within the solid to form a solid-state mixture. Analogously, as A is cooled below temperature x, the resulting solid is a mixture of A and B. The vertical lines at 80:20 A/B and 50:50 A/B correspond to the formation of compounds. Although this phase behavior may seem complex at first, the formation of two compounds merely divides the diagram into three simple eutectic diagrams. Each of these eutectics can be evaluated in the same manner as Figure 2 was.

The phase behavior characteristic of miscible solids is considerably different from the eutectic diagrams that result when the solid phases are only partially soluble in each other. Figure 4 shows idealized phase behavior for two miscible solids. The uppermost of the two curves is the freezing point curve and represents the temperature at which the various compositions begin to freeze. The lower curve gives the composition of the solid that separates out at that freezing point (9). Obviously, the solid will always contain more of the higher-melting component than the solution from which it separates. The practical implication of this phase behavior for formulation of semisolids is that adequate mixing must be maintained as the temperature is lowered through the region in which solid is in equilibrium with liquid.

The binary diagrams described thus far have been limited to two components that are chemically pure, and that have sharp melting points. For most "real-life" formulation efforts, especially for semi-

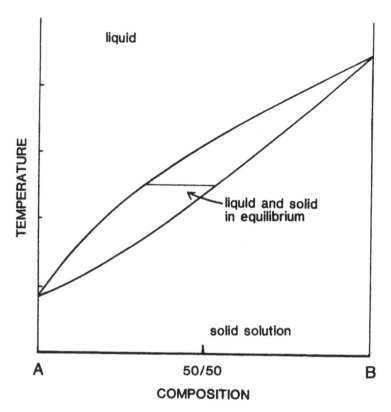

Figure 9.4 Idealized binary diagram for a two-component system in which the components are miscible in both the liquid and solid phases but have different melting points.

solids, the components will be mixtures characterized by melting ranges, rather than exact melting points. Although the phase equilibria of these practical systems are less elegant than the previously described academic systems, the phase diagram representation is still very useful. The primary difference between the systems is that both the single- and multiple-phase regions must be labeled descriptively. Phase equilibrium for complex multicomponent systems often becomes a mapping of characteristics on a composition/temperature grid. For example, the phase equilibria for the white petrolatum—anhydrous lanolin mixture is shown in Figure 5. Whereas the mutual semisolid solubility regions are definitive, the presence of a eutectic is not certain. The phase behavior at high petrolatum concentrations seems to have the appearance of a eutectic;

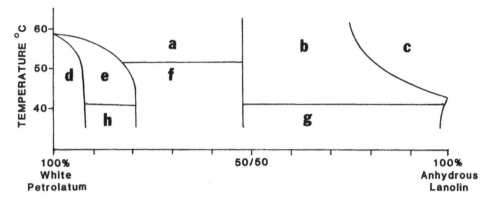

Figure 9.5 Freezing point diagram for the semisolid system white petrolatum—anhydrous lanolin. Component impurity and wide melting ranges result in an appearance significantly different than for idealized diagrams. The phase regions were observed to have the following appearance: (a) single-phase, clear, colorless liquid; (b) two clear liquid phases, top colorless, bottom yellow; (c) single-phase clear yellow liquid; (d) white petrolatum with dissolved lanolin; (e) cloudy semisolid top phase, colorless clear liquid bottom phase; (f) homogeneous semisolid; (g) two semisolid phases; top, white, bottom, yellow; and (h) two semisolid phases.

however, the single phase becomes a semisolid at temperatures above the temperature at which the lanolin is completely liquefied. This raises an important consideration when working with semisolids whose components are mixtures. Solidification is a very nebulous term. It could mean uniform clouding with an increase in viscosity, or solidification could mean hardening into an amorphous solid. For these systems, it is usually sufficient to characterize the temperature effects upon phase equilibria and utilize this information as necessary. For example, the diagram in Figure 5 indicates that any combination from 20% through 45% lanolin in petrolatum will result in a homogeneous cloudy, slightly yellow semisolid upon cooling below 50°C. To obtain this semisolid, no mixing is required, the liquefied, miscible system above 60°C will cool to form a homogeneous semi-solid. If greater than 50% lanolin is used, then the system will separate upon cooling. Although vigorous mixing of the system until reaching temperatures below 40°C may result in an acceptably homogeneous system, this composition range should be avoided. Mixtures containing less than 20% lanolin should be avoided for the same reasons, unless 5% or less lanolin is to be solubilized into the white petrolatum.

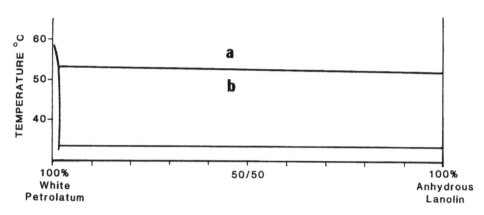

Figure 9.6 Freezing point diagram for the semisolid system white petrolatum—anhydrous lanolin. Comparison with Figure 4 emphasizes the phase behavior changes that can result from lot-to-lot variability of a natural product. The phase regions were observed to have the following appearance: (a) two clear liquid phases; top, colorless, bottom, yellow: (b) multiple phases both liquid and semisolid.

When considering the phase behavior of natural products, such as lanolin, the lot-to-lot chemical variability of the material must be considered. Phase behavior can often be very sensitive to changes in the chemical composition of the component, especially if the impurities are surface active. To illustrate the dramatic changes that can occur, compare the phase behavior of Figure 5 with the behavior shown in Figure 6. Both diagrams are for the same white petrolatum—anhydrous lanolin systems, using the same lot of white petrolatum. The only difference in the systems is the lot from which the lanolin was used. Although both lots of lanolin were from the same supplier, meeting all of the specifications set for the material, the phase behavior is in no manner similar. The region in Figure 5 that provided the homogeneous semisolid is completely absent in Figure 6. To further emphasize the need for consistent raw material quality when formulating components that are mixtures of components having wide melting ranges, Figure 7 has been included. This diagram shows the phase behavior for a third different lot of lanolin from the same supplier, and white petrolatum from the same lot as for Figures 5 and 6. For this system, the phase behavior is characteristic of two components miscible in both the liquid and solid states, but having different melting temperatures. Needless to say, this third lot of lanolin provided a superior ointment after manufacturing.

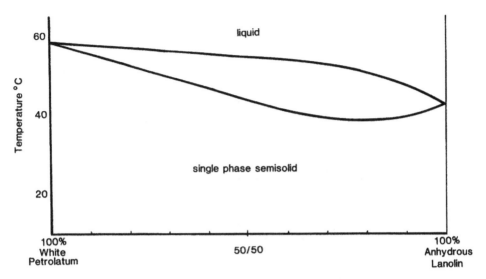

Figure 9.7 Freezing point diagram for the semisolid system white petrolatum—anhydrous lanolin. For this particular lot of lanolin, the phase behavior is similar to the idealized phase behavior for two miscible solids as shown in Figure 4.

2. *Ternary, Pseudoternary, and Quaternary Systems*

The use of ternary plots is a convenient representation of phase behavior when three or more components are involved. Because ternary plots are frequently encountered in the pharmaceutical literature (10), a description of how to fully utilize this type of diagram will follow.

The two-dimensional representation of a three-component system is possible only if the temperature and pressure are fixed. The phase behavior, as a function of composition, can then be represented on a triangular plot. The triangular plot is merely a composition grid in which any point of the triangular plot represents the relative amount of each of the three components. When the amount of each of these components is added up for any single point on the plot, the total composition will always be 100%. The geometric result is that the sum of the three perpendicular distances from any point to the three sides of the triangle is equal to the height of the triangle (i.e., 100%).

First consider the three corners of the ternary plot. These corners represent the pure component, that is, 100% of either A, B, or C. The lines connecting the 100% corners constitute the three

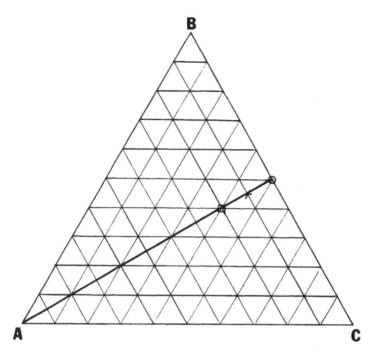

Figure 9.8 The 1:1 constant weight ratio line of B/C. Circle is the composition before addition of A; X represents addition of 10% A, and the square is positioned at 20% A, 40% B, and 40% C.

different binary systems possible, AB, BC, and CA at a given tem-perature and pressure. Therefore, although the corners define the components used in the study, the perimeter of the triangular plot shows the mutual solubilities of the binary systems. This is impor-tant when constructing a diagram, because any solubility extrapolat-ed to the diagram perimeter must correspond to the solubility limits of the binary system.

Every point contained within the interior of the triangle corre-sponds to a mixture of three components. Consider a 50:50 mix of component B and component C denoted by the circle on Figure 8. If enough of component A is added so that it is 10% by weight of the now three-component mixture ABC, then the mixture contains 10% A, 45% B, and 45% C, and the position on the diagram is marked by x in Figure 8. If more of component A is added to the sample, bringing the composition of A to 20% (wt/wt), then the sample would contain 40% B and 40% C, and be denoted by the square in Figure 8. Note that, although the percentage of B and C change with the addition of A, the ratio of these two materials remains the same. This, of course, must be true as long as neither component B nor component C is allowed to enter or leave the sample. From these few examples, it is quickly realized that the three colinear points on Figure 8 fall on a line that passes directly through the 100% A corner, and that this line is actually the 50:50 of B/C constant ratio line.

Now that it is apparent how the addition of a third component to a binary system moves a point from the perimeter of the diagram to the interior of the diagram, it is time to describe the most reliable way to read a composition from a ternary plot. Figure 9 has five compositions denoted by circles of different shading patterns. Consider the open circle. What percentage of A, percentage of B, and percentage of C does this sample contain? First, find the percentage of A by moving to the side of the triangle opposite corner A (side BC). Move perpendicular to the plot lines, parallel to side BC, until the point in question is reached and record the percentage of A. For the sample plotted as the open circle, 20% of the sample is component A. Second, move to the side opposite of corner B (side AC) and follow the same procedure to match the point with the plot percentage line of the diagram. The open-circle sample contains 30% B. Finally, and very important, repeat the procedure for component C as a check, rather than calculating the percentage of C. The summation of the percentages of the components must always equal 100%. If they do not, then an error in reading the diagram has occurred and the researcher should start from the beginning to read the composition from the plot. Remember that the point will be nearest the corner of the component with the largest percentage. One of the more common mistakes in reading ternary plots is to interchange two of the components. This error can be minimized by always beginning at the side opposite the corner of the component whose percentage is being determined. The compositions of the samples denoted by the other shaded circles are listed with Figure 9 for use as examples.

To gain greater dexterity in reading ternary diagrams, consider the following scenario. After combining and mixing numerous two- and three-component samples, it is found that upon standing, some of the samples maintain a homogeneous appearance, whereas other samples separate with either two or three distinct layers. In these two- or three-phase samples, each homogeneous layer is separated by a distinct, unchanging interface, indicating an equilibrium between the phases. The samples can be grouped into three different categories based on their unchanging homogeneous appearance at equilibrium. Likewise, it is noticed that the samples in each of these categories fall within a distinct region on the ternary plot. Thus, boundaries can be drawn on the ternary plot separating the

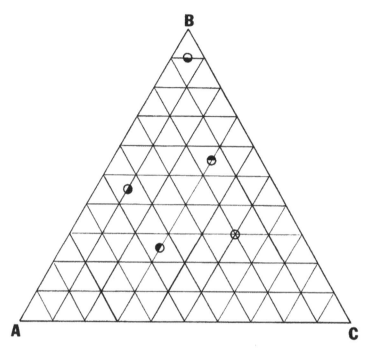

Figure 9.9 Each point of the triangular grid denotes a unique mixture of components A, B, and C. Circles denote 20% A, 30% B, 50% C (○); 15% A, 55% B, 30% C (◖); 5% A, 90% B, 5% C (◕); 45% A, 25% B, 30% C (◐); and 45% A, 45% B, 10% C (◑).

composition forming the single-phase compositions from the compositions forming two- or three-phase regions. Tie lines can also be drawn that will indicate which phases are in equilibrium when the composition of the samples falls outside of the homogeneous, single-phase region. The phase boundaries for this hypothetical system are shown in Figure 10 and the unique categories or diagram regions are labeled α, β, and γ.

From the foregoing discussion, it is evident that we can answer a variety of questions using the phase diagram in Figure 10. For example:

1. How much C can be added to A while maintaining the physical properties characteristic of phase α?
2. If 40% A, 20% B, and 40% C are mixed, will the sample exhibit only those properties characteristic of the β phase?
3. What is the maximum amount of B that can be accommodated within the β phase?

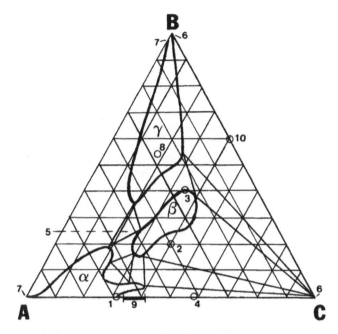

Figure 9.10 Ternary representation of the phase behavior resulting from mixing components A, B, and C. The bold curved lines are the phase boundaries. All compositions within the phase boundary labeled α will form a single phase system (it will not separate after mixing) that is uniquely different from either compositions mixed within the β-phase boundary or the γ-phase boundary. Likewise, single-phase β samples are different from single-phase γ samples. Compositions that are mixed that fall outside of the phase boundaries will be either two-phase systems (separate in two layers) or three-phase systems (separate into three layers). The numbered points or regions on this diagram refer to the questions asked in the text.

4. To reach the maximum amount of B in phase β, to what A/C ratio should B be added?

5. Is it possible to solubilize 25% of B into phase α?

6. Can B and C be mixed to form a single phase?

7. Is A more soluble in B, or is B more soluble in A?

8. If 1 g of A is mixed with 9 g of a 20% A, 60% B, 20% C mixture, will the sample still be single-phase γ?

9. Each of the phases α, β, and γ can be encountered at some point by adding one of the components to a mixture of the other two. What is the ratio of the two components, and which single component must be added?

10. What B/C ratio will neither pass through phase α nor phase β, nor phase γ, even up to the addition of 99% component A?

The answers are

1. Approximately 32% of C can be added to A before crossing the phase boundary.
2. No, the sample will also exhibit characteristics of component C.
3. Slightly more than 40% B can be accommodated within the β phase.
4. The 42:58 ratio of A/C will accommodate the maximum amount of B possible while remaining within the β phase.
5. No, the 25% B line does not pass through phase α.
6. No, the BC side of the triangular plot shows no mutual solubility areas.
7. Less than 1% of B is soluble in A, whereas slightly over 2% of A is soluble in B.
8. Yes, the final composition 28% A, 54% B, 18% C falls within the γ phase boundary.
9. Addition of B to a 67:33 through 59:41 ratio of A/C will first pass through the α phase, then the β phase, and, finally, the γ phase.
10. The B/C ratio of 60:40 will not pass through a single phase region up to 99+% of A.

Understanding how to read the compositions from the single-phase regions of the diagram is obviously an important tool. Because a complete ternary diagram represents the complete characterization of a three–pure-component system, all questions of mutual solubility are answered merely by determining whether the composition is inside or outside of the phase boundary. However, even more information can be retrieved from a completed diagram. Not only can questions of mutual solubility be answered, but also questions of equilibrium. By effectively utilizing the experimentally determined tie lines, multiple-phase equilibrium problems can also be solved.

Just as with binary diagrams, a tie line drawn between two phases in equilibrium can be used to determine the composition of each phase and the amount of each phase. In this manner, they tie the phases together. When two lines intersect at the same location on the single-phase boundary, a three-phase triangle will be formed, mapping out the composition region of the diagram at which three phases are in equilibrium with one another. Consider the diagram in Figure 11. It is seen that the tie lines denoting equilibrium between α and β move along the α-phase boundary until intersecting with the tie line denoting equilibrium between α and γ.

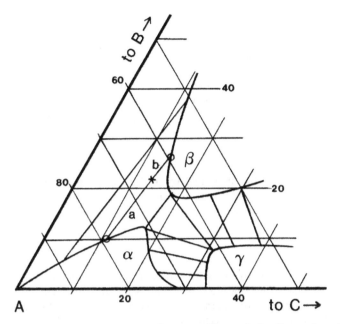

Figure 9.11 Example of the utility of tie lines for determining the amount and composition of each phase in a two-phase system. The composition marked X will split into two phases upon reaching equilibrium. The composition of each phase will be the same as the composition at the phase boundary—tie line intersection (circled points), whereas the amount of each phase is proportional to the geometric distance of the phase boundary—tie line intersection points (circles) and the sample point (X).

Analogous events occur along the β- and γ-phase boundaries, resulting in the tie lines forming the triangular tie line boundary of the three-phase region.

The tie lines can be used to read both the composition of each phase and the amount of each phase. For samples that split into two phases, the composition of the separated phases will be the same as the compositions at which the tie line intersects the phase boundaries. For example, in Figure 11, the sample containing 65% A, 22% B, and 13% C is mixed. Upon standing, the sample splits into an α phase on bottom and a β phase on top, with a distinct interface between the two phases. The total composition of the sample falls on the tie line shown in Figure 11. The tie line that passes through the sample composition intersects the α phase at 79% A, 10% B, 11% C, and intersects the β phase at 60% A, 26% B, 14% C. Thus, a

sample containing 65% A, 22% B, and 13% C will split into a lower
phase with the composition 79% A, 10% B, and 11% C and an upper
phase containing 65% A, 22% B, and 13% C. The amount of each
phase is proportional to the geometric distance of each phase boundary from the sample point. By measuring the distances denoted on
the tie line as a and b and knowing the total amount (mass) of the
sample (m_{tot}), the equation for the amount (mass) of the β phase
(i.e., top phase) would be

$$m_\beta = m_{tot}[a/(a + b)]$$ [7]

whereas the amount of α can be obtained by either having calculated
the amount of phase β and subtracting it from the total amount of
sample, or by use of the equation

$$m_\alpha = m_{tot}[b/(a + b)]$$ [8]

As seen, Eqs. 7 and 8 are identical with Eqs. 6 and 7 and can be
derived analogously once it is realized that the sum of the three
perpendicular distances from any point to the three sides of the triangle is equal to the height of the triangle.

For the samples that are contained within the three-phase triangle, the composition of each of the phases is given by the point
of intersection between the phase boundaries and the three-phase
triangle. Figure 12 shows an expanded three-phase triangle of the
system shown in Figure 11. Any sample that is located within this
triangle of tie lines will separate into three phases with the α-phase
component having the composition 71% A, 12% B, and 17% C, whereas the β and γ phases will have the compositions of 63% A, 19% B,
18% C, and 61% A, 7% B, and 32% C, respectively. Thus, once
within the triangular three-phase region, only the amounts of each
phase will change, whereas the compositions of the phases will remain the same. The amounts are determined in a fashion analogous
to the method used for two-phase samples. Equations for the
amounts of α phase, β phase, and γ phase for the sample denoted
by the open circle in Figure 12 are

$$m_\alpha = m_{tot}[a/(a + b)]$$ [9]

$$m_\beta = m_{tot}[c/(c + d)]$$ [10]

$$m_\gamma = m_{tot}[e/(e + f)]$$ [11]

Thus, a 10 g sample containing 66% A, 12% B, and 22% C will separate into three phases: 4.45 g of phase α with a composition of

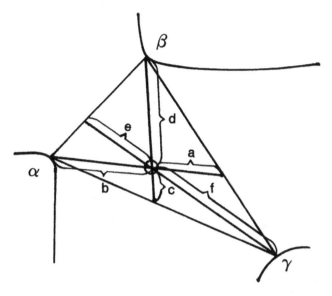

Figure 9.12 Detail of the three-phase region shown in Figure 4. Composition of each phase is given by the phase boundary—tie line intersection, and the amount of each phase is proportional to the distances shown. See text for explanation.

71% A, 12% B, and 17% C; 2.23 g of phase β with a composition of 63% A, 19% B, and 18% C; and 3.32 g of phase γ with a composition of 61% A, 7% B, and 32% C.

Ternary diagrams can also be used to identify critical solution behavior. Critical states of mixtures are considered to be the condition at which the properties of coexisting liquid phases become indistinguishable, that is, the point at which the interface between the phases disappears. In simplest terms consider the phase behavior in Figure 13. The initial composition of 50% A and 50% B will result in a sample with roughly equivalent volumes (depending on density) of the immiscible liquids A and B. As the third liquid component, C, is added, the location of the tie lines show that C will be equally distributed between the immiscible liquids A and B. The volumes of each phase will increase with the addition of C until the point at which the single-phase boundary is crossed. At this point, the *critical point*, the interface disappears between the two phases of roughly equivalent volume.

With this understanding of how to read triangular plots, a few examples of how these diagrams have been utilized for pharmaceutical formulations is warranted. The biologically compatible system,

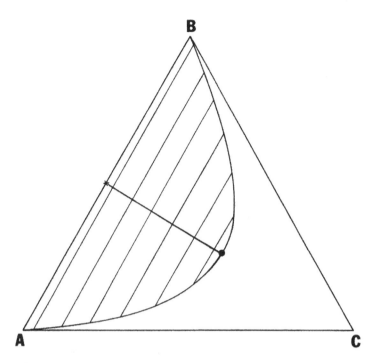

Figure 9.13 Idealized phase behavior representing how addition of
component C to an equal mixture of A to B (point x) will cause an
equal increase in the volumes of phase A and phase B until a critical point is reached (●), and the interface between the two approximately equal-volume phases disappears, resulting in a single-phase
sample.

water—sodium cholate—lecithin is shown in Figure 14 (11). Although
the structure and characteristics of surfactant liquid crystalline
phases are outside the scope of this chapter, it is sufficient to
state that liquid crystals are thermodynamically stable regions, single-phase, and reversible in their formation relative to temperature.
Various texts describe the surfactant systems in detail (7,10). For
the water—sodium cholate—lecithin system, the mixed micellar region
has been used to solubilize vitamin E to form an injectable sterile
solution (12). The liquid crystalline regions, especially the cubic
phase, are stiff phases that are characterized by slow diffusion of
incorporated materials. Similar cubic liquid crystals have been patented for use as sustained-release vehicles (13).
      When considering three-component systems not consisting of pure
materials, tie lines are no longer reliable, and considerably less in-

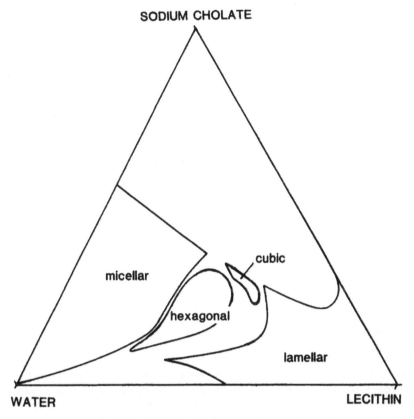

Figure 9.14 Phase diagram for the ternary system egg yolk lecithin—sodium cholate—water at 22°C. (Modified from Ref. 11, with permission from Plenum Press).

formation can be obtained from the multiple-phase regions of the diagram. Triangular plots of systems having more than three pure components are called pseudoternary diagrams. The propylene glycol—petrolatum—emulsifier system is an example of pseudoternary-phase behavior being used to solve a manufacturing problem. As shown in Figure 15, the base having the composition 7% propylene glycol, 88% white petrolatum, 5% glycerol monosterate separates into a petrolatum-rich phase and a propylene glycol-rich phase at 70°C. The physical stability of this base is completely dependent upon kinetic stabilization caused by the increase in viscosity of the vehicle upon solidification of the white petrolatum—glycerol monosterate (melting point, 60°C) mixture. Although the semisolid state appears

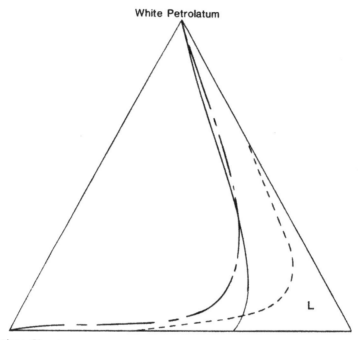

Figure 9.15 Pseudoternary diagram for the propylene glycol—white petrolatum—surfactant systems at 70°C. The mutually miscible solubility area is denoted by L for the surfactants ——— glycerol monostearate, ——— – ——— – ——— sodium stearoyl lactylate, and ------ sorbitan monolaurate.

to be able to accommodate propylene glycol at ambient conditions for an indefinite period, the vehicle will spontaneously separate into two phases if not vigorously mixed when heated. Any separation of a propylene glycol-rich phase will contain high concentrations of the drug, resulting in content uniformity problems for the product. The replacement of glycerol monostearate with a less lipophilic surfactant, sodium stearoyl lactylate (SSL), gives phase behavior more suitable for the stabilization of a propylene glycol—white petrolatum ointment base. Although incorporation of 7% propylene glycol is not possible at higher than 65% white petrolatum, SSL is miscible with both propylene glycol and white petrolatum at 70°C. This implies that the surfactant is appropriately balanced between the polar propylene glycol and the nonpolar petrolatum. Although the desired composition still falls outside the single-phase boundary, compositions above 80% white petrolatum form emulsions that require a few hours to separate at 60°C. Samples in this two-phase region have an additional stabilizing effect because of the ability of SSL to gel propylene glycol. Thus, for these two-phase systems, both the white petrolatum and the propylene glycol solidify upon cooling. Use of the nonionic emulsifier sorbitan monolaurate results in decreasing the single-phase region of the system.

Just as with the binary systems discussed earlier, complex systems can be represented using pseudoternary plots. Each of the three corners should be single phase to maintain ease in sample preparation, and descriptive terms must be used to describe each region within the diagram. Because each of the corners does not represent pure components, tie lines are no longer straight and cannot be assumed. An example of a pseudoternary diagram for an 11-component system is given in Figure 16 (14). Each of the three corners of this diagram were single-phase liquids. The left corner was distilled water in this example, but could have been water—salt or water—preservative combinations. The top corner consists of oleic acid (16.5%), myristic acid (1.9%), triolein (41.8%), palmitic oleate (20.3%), cholesteryl oleate (3.0%), pristane (2.8%), squalene (12.2%), and lecithin (1.5%). The right corner is a mixture of oleic acid neutralized with excess triethanolamine (TEA/OL, 1:1 weight ratio). Although the phase behavior is complex, a large emulsion region (labeled d) that cannot be broken by centrifugation (3600 rpm, 33° fixed-angle rotor, 24 hr) is present from 8% to 64% water. Likewise, if sufficient water is added to region k to break the emulsion, a clear colorless liquid layer will separate from the emulsion. If water were allowed to evaporate from region d, then, an anisotropic liquid crystalline phase will be present in the sample, that is, either region e or f will be crossed provided TEA—OL/SSL ratio is greater than unity.

From the foregoing discussions, the utility of representing phase equilibrium by use of either binary or ternary representations is clear. Because the information contained on these two types of plots is often interrelated, the transfer of information from binary to ternary plots should be briefly described before considering quaternary phase diagrams. In Figure 17, the binary diagram for the surfactant system water—sodium oleate is shown. Because two-dimensional representation of three components is only possible at fixed temperature, the temperature/concentration plot provides phase data for the base of the triangular grid for any ternary plot within the determined temperature range. As seen in Figure 17, the phase behavior shown for the base of the water—oil—sodium ole-

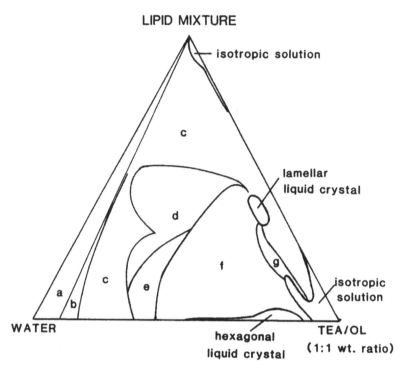

Figure 9.16 Diagram showing the complex phase behavior for the multiple-component water—lipid mixture—TEA:OL system. The multiple-phase regions were observed to have the following appearance after centrifugation: (a) cloudy white and clear gray isotropic layers; (b) cloudy white, clear gray, and anisotropic layers; (c) cloudy white and clear colorless isotropic layers; (d) cloudy white—stable to centrifuging; (e) cloudy white, clear colorless isotropic and anisotropic layers; (f) lamellar and hexagonal liquid crystal mixture; (g) isotropic solution and lamellar liquid crystal mixture.

ate ternary system at 50°C must correspond to the phase encountered by the 50°C line as it extends across the binary composition/ temperature plot. Likewise, the 100°C line must correspond with the phase behavior for the higher temperature ternary plots, because these are identical systems under identical conditions (ambient pressure). If the phases do not correspond, then differences in component purity or investigator interpretation may have occurred.

Often information plotted in one representation is difficult to interpret, but becomes more apparent when plotted in the comple-

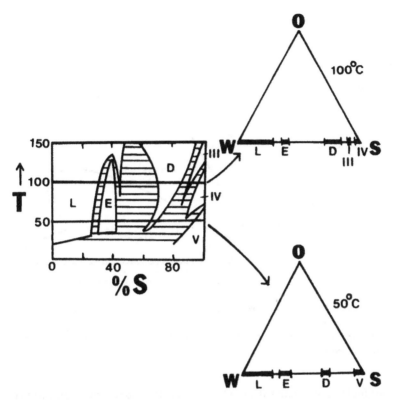

Figure 9.17 Relationship between a temperature/concentration diagram and a ternary diagram. The water—surfactant base of the ternary diagram can be taken from the temperature/concentration diagram for a given temperature.

mentary form. For example a special type of binary temperature/ concentration plot is the cloud point or cloud temperature diagram. These diagrams are plotted for water—nonionic surfactant mixtures containing less than 20% surfactant against a temperature range not exceeding 100°C. As a fluid, isotropic nonionic surfactant solution is heated, the water hydrogen bonded to the ether oxygens decreases. Because this hydration of the ether oxygens is the primary interaction keeping the nonionic surfactant in solution, the micellar weight of the polyoxyethylene-type nonionic surfactant solution increases with increasing temperature (i.e., decreasing hydration). This increase in size ultimately causes the solution to become turbid (*cloud point temperature*), followed by separation into two phases. For an idealized system in which cloud point temperature is plotted

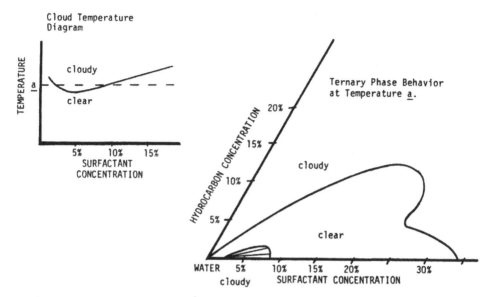

Figure 9.18 Relationship between a cloud temperature diagram for a nonionic surfactant system and the corresponding idealized ternary diagram can be taken from the temperature/concentration diagram for a given temperature.

against surfactant concentration (Fig. 18), the corresponding water—hydrocarbon—nonionic surfactant ternary phase behavior can be estimated. As seen, the minimum in the cloud point temperature versus surfactant concentration curve corresponds to a two-phase region within the normal micelle area on the ternary diagram. As temperature is increased, the ternary phase behavior will change as outlined by Friberg (15). For added nonelectrolytes that are structurally similar to hydrocarbons, the cloud point temperature elevating or depressing capabilities of the added nonelectrolyte can be understood by examining these ternary diagrams. Cloudiness results when the addition of 5% nonelectrolyte causes the final composition to fall outside the micellar single-phase region. If the added nonelectrolyte is more soluble in the micellar region, or if it alters the phase behavior by changing the phase transition temperatures of the surfactant, then the final composition will be a clear solution. If a series of binary diagrams containing progressively larger amounts of nonelectrolyte had been plotted, rather than a ternary plot constructed, then the pseudobinary set of curves would be difficult to interpret. However, the ternary representation shows that a characteristic progression of phase equilibrium exists for these systems.

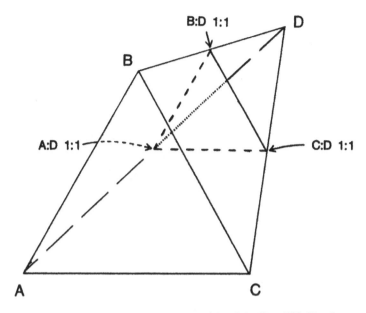

Figure 9.19   Quaternary pyramid with the 50% D plane emphasized.

To represent a system that contains more than three pure components, pseudoternary diagrams can be used, or, if only four pure
components are involved, quaternary (four-component, three dimen-
sional) pyramidal volume diagrams can be constructed or projected
into two dimensions. Consider the four-component diagram of Figure 19. As with the ternary diagrams, the addition of one of the
four components to any mixture of the three other components will
cause the total composition to move in a straight line toward the
added components corner. Also notice that any three-component
mixture will fall on one of the four surface triangles that form the
boundaries of the pyramid. It is immediately obvious that such a
four-component system becomes very complicated, very quickly. To
simplify this representation, slices through the pyramidal diagram
can be taken. For instance, rather than determining the phase behavior for the system A—B—C; if A and D were mixed 1:1 by
weight, B and D were mixed 1:1 by weight, and C and D were
mixed 1:1 by weight, then a pseudoternary diagram for this system
could be determined using the same techniques as used for a true
three-component system. Determination of the mutual solubility areas would be completed exactly the same way; however, what were
tie lines defined by two points for a ternary system are now tie
surfaces that may be curved. Thus, an arbitrary slice through
this surface will produce tie lines that are no longer necessarily linear.

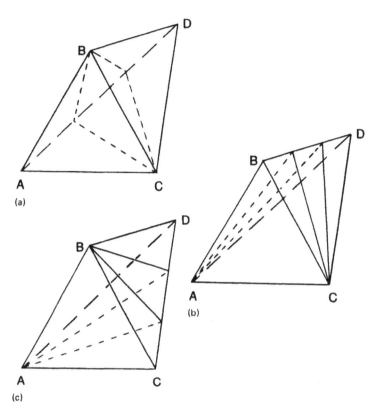

Figure 9.20 Quaternary pyramides in which the ratio of two of the components is varied to form pseudoternary planes. Varying ratio of A/D (a), B/D (b), and C/D (c). Representations resulting from the varying the ratio of A/B, A/C, and B/C for the A, B, C, D system are not shown.

Seldom will component D be soluble up to 50% in A, B, and C. Thus, often the fourth component will be mixed with one of the other three in varying ratios, producing a series of pseudoternary diagrams that will characterize the system. Figure 20a shows how varying the A/D ratio will alter the slices taken through the pyramid. Variations of the B/D and C/D ratios are shown in Figures 20b and 20c, respectively.

Another way to represent a four-variable system is to have the base formed from a triangular grid, while increases in the magnitude of the fourth variable move perpendicular from the base, thus, form-

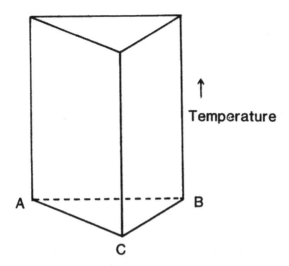

Figure 9.21 Example of a three-component phase diagram with the temperature variable as the vertical direction.

ing a triangular prism. This representation is useful when the fourth variable is temperature, or the fourth component is added to only a few percent (Fig. 21). For these triangular prism, representations for which temperature is the long axis, slices through the prism parallel to the base result in true ternary diagrams, rather than pseudoternary diagrams, whereas parallel slices through a four-component systems result in pseudoternary diagrams.

As seen, quaternary diagrams are actually "built" by constructing a series of pseudoternary diagrams. Thus, all of the concepts described earlier for ternary systems are applicable, with the minor changes that now one or more of the corners is a miscible mixture, the tie lines are no longer necessarily straight, and critical points are found on a surface rather than a curve (Fig. 22; 16).

In summary, this tutorial on binary, ternary, pseudoternary, and quaternary phase diagrams has emphasized some very important facts concerning pharmaceutical fluids and semisolid formulations. First, that homogeneous (i.e., stable) preparations exist only for a discrete set of compositional ranges, which are enclosed by a phase boundary. Preparations whose composition falls outside this phase boundary will be unstable and separate into layers of differing compositions. Within the phase boundary the product characteristics of the preparations (viscosity, drug release, appearance, consumer acceptance) should vary in a continuous and smooth manner through-

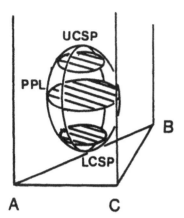

Figure 9.22 Idealized two-phase region in a three-component diagram with an intensive variable as the vertical direction. The upper critical solution point (UCSP) and the lower critical solution
point (LCSP) are connected by a plait-point loop (PPL).

out the range of the component composition. Second, once the sufficient number of samples have been mixed and examined to provide
the information required to construct the diagram, determining the
composition ranges over which to optimize is very straightforward.
Values for the ranges can be read directly from the plots based
upon the phase boundaries. Thus, the determination of the phase
behavior of the fluid or semisolid system is required before the optimization of a topical formulation.

B. Automation of Phase Behavior Determination Using Laboratory Robotics

The introduction of relatively inexpensive, commercially available
laboratory robotic systems provides the formulator with a technique
capable of preparing many samples without the tedium traditionally
associated with phase behavior determination. The laboratory robotic procedure can be conceptually divided into a six-step process:
(1) selection of a preliminary formulation to be studied and determination of the compositions of the components to be mixed; (2) dispense components and check compositions; (3) cap and label sample
tubes; (4) mix samples well; (5) observe the samples for an appropriate length of time to determine when critical product characteristics are established; and (6) record results and repeat steps 1
through 6, as necessary, to complete phase behavior evaluation. In

the following sections, the robotics techniques required to complete each of these steps will be considered in detail. The discussion will be general enough to be applicable to any of the commercially available systems, but specific enough to be used as a guide for automation of the sample preparation required for establishing compositional ranges before formulation optimization.

1. Determination of Compositions to be Mixed

Successful preliminary formulation selection is dependent upon various factors: (1) the robotics system must be able to adequately dispense, process (mix), and evaluate the stability of each sample prepared; (2) the formulation should be as simple (i.e., contain as few components) as possible; (3) the formulation should consist of readily available raw materials that are as chemically pure as possible; (4) the attributes of the formulation that are important for consumer acceptance of the eventual product must be clearly defined and accommodated; and (5) manufacturability of the scaledup formulation must be considered. Once a preliminary formulation is selected, the components can be separated into two groups. The first grouping is for materials that will be used at a preset concentration, regardless of other formulation changes that may occur because of optimization. Concentrations of the active component, preservative, and fragrance may be examples of materials that will be added in small amounts, or over narrow ranges and, thus, are included in this first grouping. The second material group is for those components that may be used over a wide range of concentrations. Examples of this group of materials for topicals would usually include water and other excipients. The concentration ranges forming homogeneous phase regions for this second group of materials will provide the framework of the phase behavior studies. It is hoped that no more than three to five of the components will fall into this second material group.

Next, each of the components falling in this second grouping are divided into polar materials, nonpolar materials, and other materials. This allows pseudoternary diagrams to be constructed. If only two of the foregoing divisions of polarity exists, then a binary diagram may be more appropriate. Five weight percent increments of the component compositions is usually sufficient to obtain a rough diagram that can then be refined along the phase boundaries.

An example of this procedure for a hypothetical system should clarify the steps in determining the compositions to be mixed. The preliminary formulation selected is a cream consisting of 5% to 15% active (oil soluble), 6% emulsifier, 11% coemulsifier, 15% oil, 8% glycerin, and q.s. with water containing the preservative. Glycerin and water are the polar materials, the oil is nonpolar, and the emul-

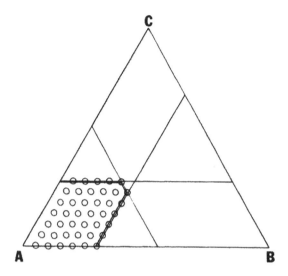

Figure 9.23 Points on the ternary grid at 5% intervals that are ≥ 45% A, and ≤ 30% C, and between 5% and 30% B inclusive.

sifier and coemulsifier are grouped as "other" materials. Preservative challenge testing of the preliminary formulation indicates that 0.1% preservative concentrations adequately protects the formulation. Thus, this level will be added to each sample prepared by the robotic system.   Three discrete levels of the active agent will be evaluated (i.e., 5%, 10%, and 15% by weight).   Because this is a complex seven-component system (five components unconstrained, two components discrete), it is useful to limit the ranges of some of the materials on the basis of previous formulating experience or raw material costs.   For this preliminary formulation we will not evaluate final formulations containing (1) less than 45% water—glycerin, (2) greater than 30% oil, or (3) greater than 30% emulsifier—coemulsifier. With these constraints, the region on the ternary diagram has been limited as shown in Figure 23.   Note, that the polar materials will be plotted at the left corner of the triangle, the oil will be plotted at the top of the triangle, and the emulsifier—coemulsifier mixture plotted at the right corner, in keeping with convention for surfactant systems (7).   For each concentration level of active agent, the 40:60, 35:65, and 30:70 ratios of emulsifier/coemulsifier will be evaluated.   Likewise, two water/glycerin ratios will be evaluated. All combined, 18 pseudoternary diagrams, consisting of 41 sample points each, will be constructed as outlined in Table 1.   This trans-

Table 9.1 A Listing of the Phase Diagrams to be Completed to Establish Smoothness and Continuity of the Hypothetical Cream System Example

Diagram number	A	B	C
5% Active, 0.1% preservative			
1	92% Water/8% glycerin	40% Emulsifier/60% coemulsifier	Oil
2	92% Water/8% glycerin	35% Emulsifier/65% coemulsifier	Oil
3	92% Water/8% glycerin	30% Emulsifier/70% coemulsifier	Oil
4	86% Water/14% glycerin	40% Emulsifier/60% coemulsifier	Oil
5	86% Water/14% glycerin	35% Emulsifier/65% coemulsifier	Oil
6	86% Water/14% glycerin	30% Emulsifier/70% coemulsifier	Oil
10% Active, 0.1% preservative			
7	92% Water/8% glycerin	40% Emulsifier/60% coemulsifier	Oil
8	92% Water/8% glycerin	35% Emulsifier/65% coemulsifier	Oil
9	92% Water/8% glycerin	30% Emulsifier/70% coemulsifier	Oil
10	86% Water/14% glycerin	40% Emulsifier/60% coemulsifier	Oil
11	86% Water/14% glycerin	35% Emulsifier/65% coemulsifier	Oil
12	86% Water/14% glycerin	30% Emulsifier/70% coemulsifier	Oil
15% Active, 0.1% preservative			
13	92% Water/8% glycerin	40% Emulsifier/60% coemulsifier	Oil
14	92% Water/8% glycerin	35% Emulsifier/65% coemulsifier	Oil
15	92% Water/8% glycerin	30% Emulsifier/70% coemulsifier	Oil
16	86% Water/14% glycerin	40% Emulsifier/60% coemulsifier	Oil
17	86% Water/14% glycerin	35% Emulsifier/65% coemulsifier	Oil
18	86% Water/14% glycerin	30% Emulsifier/70% coemulsifier	Oil

lates into the preparation of 738 samples. Dependent on the complexity of the dispensing and mixing steps and whether or not the
samples will be centrifuged, each sample will require approximately
10 min to complete, which leads to an estimate of 123 hr (5-1/8
days) of robotic time. Approximately 2 hr of technician time is required for each day (24 hr of continuous, unattended operation) of
robot time. Thus, the initial samples for this evaluation will require
approximately 1 technician-day and 1 robot-week to complete. Additional samples may be required to pinpoint critical regions of the
phase boundary, as described in the following discussion.

2. Weighing Out Components

Pharmaceutical topicals may contain fluid liquids, semisolids, amorphous solids, crystalline solids, or any combination thereof. Because the raw materials for topicals vary greatly in their physical
properties, a number of different dispensing techniques must be employed to evaluate topical formulations. Solids are generally dispensed by a vibrating robotic hand, whereas liquids and semisolids
are dispensed from automatic syringes. Materials that are not readily dispensed by either of these methods can be added by hand.
Conceptual details of each of these dispensing methods are described below.

Solids. To deliver powders using robotics, a tube filled with
solid is tilted and vibrated in discrete pulses. The vibration
causes the solid to flow from the tube into an appropriate, tared
container before being weighed. Although, in theory, this may
seem a simple task, in practice, it can be very demanding. Accurate delivery of solids depends on a number of factors, including
vibration intensity, grip location, vibration duration, tube diameter,
and tilt angle. The vibration intensity is usually varied during the
course of powder addition. At the beginning of powder addition,
the intensity is at a preset value near, but not necessarily at, the
maximum possible. However, as the target weight is approached
the intensity is reduced, allowing more control over the amount of
solid added. For best results, the tube containing the powder
should be gripped near the base, resulting in maximum transfer of
energy from the hand to the powder in the tube. This ensures that
the solid is added continuously. Vibration duration is also varied
over the course of addition, decreasing as one reaches the target
weight. The amount of variation is dependent on the nature of the
powder being added. For crystalline solids, the duration is reduced, whereas for amorphous solids, a longer vibration time is required. The diameter of the tube being used to dispense the powder should be smaller than the opening of the container receiving

the powder to avoid spillage. A lip on the mouth of the tube containing the powder has proved useful, but requires that the dispensing tube is returned to an exact orientation. This can be accomplished by notching both tube and rack sufficiently to guarantee
return to a constant orientation. The tilt angle at which the tube
is held also determines the rate at which the solid is dispensed.
Accurate addition of solid depends on an appropriate combination of
tilt angle, vibration intensity, vibration duration, and a knowledge
of the physical properties of the solid.

To pour powder, the robot grips the powder tube and tilts it
to an initial angle over the container. A vibration pulse follows,
after which the tube is weighed. A lag time usually occurs, representing the time required for the solid to reach the tip of the tube.
If only a small fraction of the target weight has been reached, the
tilt angle is increased. The hand vibrates and powder is added.
A comparison between the target weight and the amount added is
made. As the target weight is approached, both the vibration intensity and duration are decreased and the pouring routine continued. When the target weight is obtained, the pouring routine is
stopped, and the net weight of solid calculated. By using this
technique, we have obtained good reproducibility with target weights
as low as 0.05 g.

Liquids. Liquids can be dispensed using automatic, gas-tight
syringes that are commercially available and capable of dispensing
accurate volumes. Typically, these syringes are computer-controlled
to position a valve so that the desired volume of liquid can be
pulled from a reservoir into the barrel of the syringe. The valve
is then switched and the liquid dispensed through Teflon tubing into a robot hand-held sample tube. Because the syringe drive is
subdivided into approximately 1000 positions, the reproducibility of
the volume delivered will be 0.1% of the total syringe volume. Thus,
if a 10-ml syringe is used, the volume delivered will be ± 0.01 ml;
5.0-, 2.5-, and 1.0-ml syringes can readily be substituted to provide greater accuracy. Although volumes can be accurately dispensed, it is best to weigh each sample tube after each liquid addition. This allows an exact accounting of what is added into (or
adhered onto) the tube by eliminating volume errors that may result
from air in the syringe lines or unpurged syringes. Because all of
the input and output compositions are recorded in terms of weight
percents, a numerical conversion from weight to volume is necessary
before each liquid addition. This requires knowledge of the density
of the liquid being dispensed. Approximate density values, such
as those given in Table 2, are usually sufficient for this conversion.
The dispensing of nonviscous liquids (viscosity less than 1 cp) is

Table 9.2 Approximate Densities for Liquids

Liquid	Density (chain length)	
n-Alkanols	0.79 (n = 2)	0.84 (n = 17)
n-Fatty acids (saturated)	1.05 (n = 2)	0.91 (n = 8)
n-Fatty acids (monounsaturated)	1.03 (n = 4)	0.85 (n = 22)
n-Alkanes	0.66 (n = 6)	0.79 (n = 20)
Alkanes		
pristane	0.785	
squalane	0.810	
squalene	0.858	
petrolatum	0.82−0.88	
Glyerides		
glycerol trioleate	0.91	
glycerol monooleate	0.94	
Esters		
propyl palmitate	0.84	
cetyl palmitate	0.83 (at 50°C)	
propyl stearate	0.84	
ethyl oleate	0.87	
Surfactants	approx. 1.0	

readily accomplished using virtually any syringe configuration.
However, more viscous materials present a greater challenge. To
accurately dispense more viscous materials (viscosity between 1 and
8000 cp), various changes can be made to the dispensing system.
The diameter of the tubing, dispensing probe (if present), and
bore diameter of the valve can be increased. Tubing lengths can
be minimized, and a smaller syringe volume can be used when possible. The time required for the syringe to travel full stroke can
also be increased. Table 3 gives a series of syringe configurations
for the Hamilton Microlab 900 series dispenser that accurately dispensed liquids of known viscosity without stalling the syringe.

Table 9.3 Filling and Dispensing Configurations for the Hamilton Microlab 900 Series Dispenser

Viscosity (cP @ 25°C)	Syringe volume (ml)	Tube diameter (in.)		Syringe speed (0–15)[a]	
		Inlet	Outlet	Filling	Dispensing
1–100	10	1/8	1/8	4–15	4–15
	2.5	1/8	1/8	4–15	4–15
	1.0	1/8	1/8	2–15	4–15
	2.5	1/8	1/16	2–15	8–15
100–500	10	1/8	1/8	15	15
	2.5	1/8	1/8	4–15	4–15
	1.0	1/8	1/8	4–15	4–15
500–1,000	10	1/8	1/8	15	15
	2.5	1/8	1/8	10–15	8–15
	1.0	1/8	1/8	10–15	8–15
1,000–5,000	10	1/8	1/8		
	2.5	1/8	1/8	10–15	10–15
	1.0	1/8	1/8	10–15	10–15
5,000–8,000	2.5	1/8	1/8	15	15
	1.0	1/8	1/8	15	15

[a]Syringe speed of 4 is equivalent to 1.4-sec full stroke, whereas a syringe speed of 15 is equivalent to 15-sec full stroke.

Semisolids. Viscous materials can be dispensed by use of commercially available liquid application systems that are designed for dispensing adhesives, solder pastes, and other similar materials (EFD, Inc. 977 Waterman Avenue, East Providence, RI 02914). The dispenser is connected to filtered compressed plant air. Output air pressures up to 100 psi provide controlled application of even thick materials that will not pour. An adjustable vacuum system prevents dripping between dispensing cycles for lower-viscosity liquids. These dispensers can readily be controlled by interfacing to the robotic control computer. To dispense the desired amount of material, the syringe barrel, with appropriate dispensing tip, is placed above the sample tube placed on the balance. The viscous material is dispensed into the sample tube for a preset time (e.g., 50 cycles through a timing loop). Dispensing is halted, and the amount of added material weighed. Based upon the amount added for the first 50 cycles of the timing loop, the number of timing-loop cycles required to complete the viscous material addition is calculated, and material is then dispensed into the sample tube for that period. Greater accuracy can be gained by increasing the number of addition iterations.

Hand additions. Some materials, such as stiff amorphous semisolids, are unreasonably difficult to dispense with any of the foregoing methods. Hand addition of these troublesome materials is the most practical method. The robot is programmed to tare each sample tube, after which the researcher adds a quantity of the material to the tared tube. The robot then weighs the sample tube containing the hand addition, calculates the weight of the added material, and calculates the amounts of the remaining liquid or semisolid components that must be added to obtain the originally desired composition. In this way, hand addition can be completed rapidly, provided the total volume of the sample is not required to be constant. The researcher must be careful that the amount of material added by hand is of an appropriate magnitude so that neither overfilling nor underfilling of the sample tube results upon subsequent component additions. Obviously, this procedure allows for only the first material to be added by hand.

3. Containers, Capping, and Labeling

Virtually any container is suitable for a laboratory robotic system, provided the weight of the filled sample container does not exceed the lifting capacity of the robot. Otherwise, containers should be selected based upon the ease of completing proper mixing within the container and the washability or disposability of the container. We have found 16 × 125-mm disposable culture tubes with caps to be ideal for our phase behavior investigations.

Capping stations function by either having a device that will grip and rotate the cap while the tube is held stationary by the robotic hand, or by having a device that will grip and rotate the tube, while the cap is held stationary by the robotic hand. The former system holds the cap until the tube is repositioned to the capping station, and the cap is torqued onto the tube. Thus, only one tube is typically uncapped at any given time. The latter system allows more than one tube to be uncapped during any single command cycle.

The labeling of the numerous samples can be readily accomplished by printing the sample name, composition, and preparation data on peelable adhesive labels that are applied to the sample tube by the researcher. A dot matrix printer can be set up in a manner analogous to that required for printing address labels and interfaced to accept output data from the robotic control computer.

4. Mixing Samples

After component addition and weighing, the next stage is to effectively mix each sample. For fluid samples, an automated vortex mixer can be installed with the robotics system to operate in either a single-action or pump action manner. The length that the sample is vortexed can easily be adjusted to meet the requirements of the system studied. To facilitate mixing of more viscous samples, or samples that contain components that melt at slightly higher than ambient temperatures, a heating block can be positioned within the work envelope of the robot, and samples may be heated for a preset length of time. The method of centrifuging viscous samples through a constricted glass tube can also be useful for mixing samples by use of laboratory robotics. A soft glass tube 5 to 6 in. long is sealed at one end, and a constriction is placed roughly in the middle of the tube. After addition of the components, the mouth of the tube is then flame-sealed, and the matrix is repeatedly centrifuged from end to end of the tube through the matrix. Slight annealing may be required after mixing by this process (17). The construction and sealing of the constricted tubes will need to be completed manually, although use of robotics is ideally suited for the repeated inverting and centrifuging of the constricted tube.

5. Observation Of Samples

After a week of robotic effort, the researcher is faced with evaluating 738 samples and plotting the phase behavior of each sample on one of 18 pseudoternary plots. It quickly becomes apparent why it is necessary that all of the critical processing steps be completed using the laboratory robotics configuration. If each sample required a manual processing step (e.g., mixing with a spatula) the amount of human effort would be unacceptable. Sample evaluation should consist of a simple observation to establish the presence or absence of a homogeneous phase. Thus, each point on each diagram can be quickly plotted as a circle if one phase, and as an x if two or more phases are present. The best smooth curve passing between the xs and os is considered the phase boundary. If a best curve cannot be drawn because of lack of data, then additional

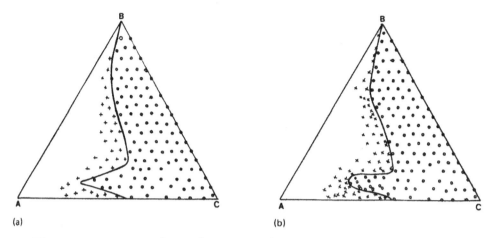

Figure 9.24 Comparison of a rough diagram (a) drawn from an insufficient number of data points and a finalized diagram (b) that has a well-defined phase boundary.

samples can be prepared to better characterize that region of the diagram. Figure 24a shows a rough diagram drawn from a limited number of available points, and Figure 24b shows the refined diagram.

Thus far, we have discussed single-phase regions and homogeneous phases. It may be necessary to clarify what is meant by a phase. A general definition quoted from Ref. 8 is: *A sample of matter is said to be in a certain phase when it has a certain well defined set of macroscopically observable properties. The phase of a sample is really an indication of the degree of order or disorder inherent in the molecules of which the sample is composed.* Because the researcher can establish the set of macroscopically observable properties, a phase can be defined in various ways. For our cream example from the first paragraphs of this section, the phase could be defined as any mixture that remains homogeneous after being heated to 40°C for 10 min and then immediately centrifuged at 4000 rpm for 1 hr. Note, that this was not referred to as a single phase. Usually the term *single phase* is reserved for a mixture of molecules forming a phase that is thermodynamically stable. Such systems include simple solutions at equilibrium, lyotropic liquid crystals, and other surfactant association structures. Emulsions are kinetically stabilized by the presence of the surfactant and, thus, cannot be considered "single-phase" systems. Distinct phase boundaries will enclose the stable emulsion regions on the di-

Table 9.4 Component Ranges of the
Formulation that Provide a Smooth and
Continuous Phase Suitable for Optimization

Range (%)	Component
5 — 15	Active
0.1	Preservative
3.5— 10	Emulsifier
6.0— 16.25	Coemulsifier
3.6— 11.9	Glycerin
38.7— 78.2	Water
5 — 20	Oil

agram as seen in Figure 14 and in Ref. 18.   Thus, for most commercial applications, the researcher defines the phase according to
the stability requirements for the product.

Before the experimental design, modeling, and optimization steps
can be established, the researcher must guarantee that a phase
boundary is not crossed as the composition of each component is
varied.   Pseudoternary diagrams have now been constructed to
bracket the constrained ranges of each of the component variables.
If each of the 41 samples on each of the 18 diagrams was within the
desired phase boundary, then the compositional ranges to be optimized would be identical with the values of the constraints we
placed on each component.   However, if the phase behavior was as
shown in Figure 25, then the compositional ranges would be more
limited.   Note that for 5%, 10%, and 15% active compound, the 30%:
70% emulsifier/coemulsifier provide a stable emulsion only at higher
than 20% to 25% by weight of the emulsifier—coemulsifier mixture.
Thus, the emulsifier/coemulsifier ratio must be greater than 30:70.
For the other two emulsifier/coemulsifier ratios, the amount of oil
can range from 5 to 20 wt%, whereas the amount of emulsifier—coemulsifier can range from 10% to 25%.   Within the above constraints,
the water—glycerin mixture can range from 45% to 85% (remember,
the formulation contains a minimum of 5% oil and 10% emulsifier—coemulsifier).   The component ranges based upon the phase behavior
results shown in Figure 25 are given in Table 4.

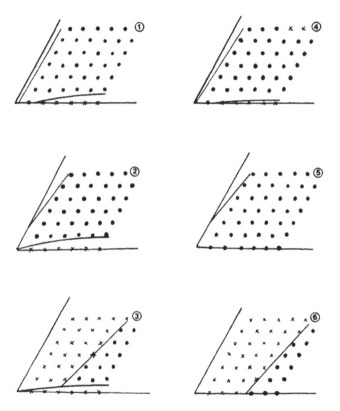

Figure 9.25 Phase behavior results for the ternary systems listed in Table 1 for the hypothetical cream system example. Xs denote either clear single-phase regions, two liquid phases (i.e., an unstable emulsion) or stable emulsion plus another phase.

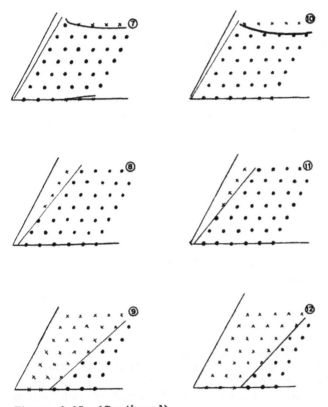

Figure 9.25 (Continued)

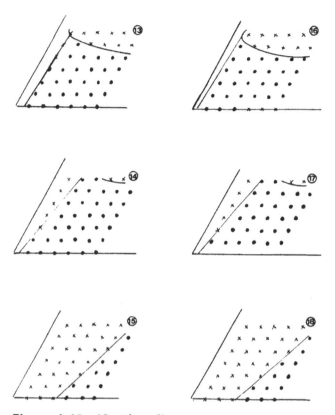

Figure 9.25 (Continued)

III. OPTIMIZATION TECHNIQUES

Before discussing how to design an experiment to optimize a formulation, some terminology should be clarified. First, the difference
between trial and experiment, as used in this secion, should be described. When considering formulations, the result of each *trial* is
a finished product possessing a set of properties (e.g., potency,
viscosity, density, appearance). After completing a sufficient number of trials, an optimized formulation can be selected to satisfy
the critical product characteristics. A set of trials resulting in an
optimized formulation is called an *experiment*. Thus, to design an
experiment is to select a series of unique or replicate trials that
will ultimately provide sufficient information to select an optimized
formulation. Each trial will differ in the conditions, that is, in the
variables used to formulate the product (variables might include

composition, mixing times, shear rates, pH). The variables that
are chosen for inclusion in the study should be those that have the
most dramatic effects upon the critical product characteristics.
Likewise, variables that are unlikely to affect final product characteristics, or variables that are constrained because of equipment or
procedural limitations, should be held constant throughout the trials. Finally, the terms *response* and *level* should be discussed. As
stated earlier, each trial will result in a finished formulation possessing a set of properties. The resultant characteristic properties
of this formulation are called the *responses*. The variable levels
and corresponding responses are combined by use of an appropriate
regression analysis to empirically derive a function. For two variables and one response, an easily visualized three-dimensional response surface results when this function is plotted. As the number of variables or responses increase, a multidimensional response
surface will result that cannot be easily visualized. However, any
three-dimensional "slice" of the multidimensional surface can be plotted. For each variable, a range of values exist that could be tested. In this section, the *level* of the variable will refer to the value
of a variable selected for a particular trial. If the variable range
is viewd in terms of highest, middle, and lowest values, then the
phrases *level of the variable* and *variables at three levels* can be
understood. Similarly, one might be required to evaluate 10 levels
of an excipient between 10 and 20 wt% in a formulation. In summary, an experiment is a carefully designed set of trials in which variable levels have been selected. Each response can then be empirically modeled to provide a response surface. The optimum value
of the response surface is located, and the corresponding variable
values are obtained, thus providing an *optimized* formulation (i.e.,
optimized critical product characteristics).

The quality of an optimized formulation is only as good as the
design of the experiment used for optimization. The techniques to
be described minimize the number of trials in a manner to maximize
the amount of information obtained by setting up a statistically valid experimental design. These techniques require that the response
surface is smooth and continuous and, thus, for disperse systems
in which a phase boundary might be crossed, it is necessary to
have established the phase behavior as detailed earlier. Because
the application of statistical methods for optimization of disperse
dosage forms (e.g., suspension, emulsions) has recently been reviewed by Franz and co-workers (3), these topics will be only
briefly described to serve as an introduction to the reader who is
unfamiliar with experimental design, modeling, and optimization
strategies.

A. Experimental Design

An empirical response model surface can be created by conducting a series of experimental trials having selected variable levels. Each variable level for the trials is carefully chosen such that a maximum amount of information on the relationship between the variables and responses is obtained with a minimum number of trials. A number of assumptions are made about the data to be obtained from the statistically designed experiment. It is assumed that observations are representative of the population, that replicate observations for various combinations of variables have the same variance, and that the random errors are independent and normally distributed. In addition, it is assumed that outliers and missing values will not be obtained. Each of these assumptions can be statistically validated, either by random sampling of a proper size from the appropriate population or by use of standard statistical methods of testing (19). The most commonly used analysis within the pharmaceutical industry is the full factorial design in which all the variables are at two levels. For this design the number of trials will equal 2^k, where k is the number of variables. Because of the exponential increase in the number of trials for this design, it is generally used when five or fewer variables will be investigated. The advantages of this design are that (1) a low number of trials per independent variable are required; (2) both quantitative and qualitative variables can be examined; (3) the results are easily interpreted; and (4) this design forms the basis for several other designs such as the fractional factorial, face centered, and central composite experimental designs. When there are five or more variables, a fraction of the full factorial can be used. All essential information on variable main effects and two-way interactions are still obtained, but with less experimental effort.

The design layout or *matrix* for a 2^k factorial design for k = 2 through 5 is shown in Table 5. This table provides information on the number of trials required for a statistically valid experimental design (2^k) and the value or level that each variable must be evaluated for a given trial. Remember that we have limited ourselves to a two-level design; thus, the maximum and minimum value of the variable range are the only two levels of the variables to be evaluated. The (+) notation means that the maximum value within the variable range will be tested, whereas the (−) notation represents the minimum value of the variable range. Therefore, from the table it is apparent that experimental trial number 25 will use the minimum values for variables X_1, X_2, and X_3, whereas the maximum values of variables X_4 and X_5 are implemented. Similar information can be readily obtained from the table for any trial. Note, these trials should not be performed in the order listed in Table 5, rather the order should be randomized. This will help to ensure that erroneous conclusions will not be drawn from the effects of any systemic or environmental variables not included in the study.

Once the experimental design has been selected, decisions must be made concerning the proper response attributes for the formulation. Remember, when the topic "determination of compositions to be mixed" was considered earlier, it was stated that *successful preliminary formulation selection is dependent upon* ... (4) *the attributes of the formulation that are important for consumer acceptance of the eventual product must be clearly defined and accommodated.* Because the attributes of the formulation were clearly defined before phase behavior characterization, the attributes (or responses) tested for each formulation have already been thought through. For a topical agent, the attribute viscosity will likely be considered. Cost of the formulation and purchase intent, as judged by a consumer panel, may also be possible responses selected for each trial. Potential attribute responses to be tested may include bioavailability (as estimated by in vitro percutaneous absorption studies) or irritation potential (as shown by animal testing). An excellent example of optimizing a topical product for consumer acceptance is given by Moskowitz (20). Because generation of a reliable response surface is the goal of this entire process, the researcher must discern from the very beginning what product characteristics or response attributes are important to the commercial success of the product. If the researcher is interested in optimizing only one response attribute, then the complete characterization of the phase behavior of the system and subsequent optimization techniques described in this chapter are probably not worth the effort required for completion.

B. Modeling and Optimization

We now know the concentration ranges over which the formulation provides a smooth and continuous phase region. The experimental design has been selected to include both compositional and process variables. If the full factorial experimental design at two levels was selected for five variables total, then the 32 formulations or trials (2^5) have been manufactured. Six tests have been selected to evaluate the product attributes, and results from these have been obtained for each of the 32 formulations. These experimental observations provide the response attributes we consider most important for the commercial success of the product. All of this observed data are then used to construct a model of the response surface. Optimums (either maximums or minimums) can be determined by using standard techniques (3) or by using statistical software packages that offer multiple experimental designs and perform

Table 9.5 Two-Level Factorial Design Patterns

Trial	Variable				
	x_1	x_2	x_3	x_4	x_5
1	-	-	-	-	-
2	+	-	-	-	-
3	-	+	-	-	-
$4 = 2^2$	+	+	-	-	-
5	-	-	+	-	-
6	+	-	+	-	-
7	-	+	+	-	-
$8 = 2^3$	+	+	+	-	-
9	-	-	-	+	-
10	+	-	-	+	-
11	-	+	-	+	-
12	+	+	-	+	-
13	-	-	+	+	-
14	+	-	+	+	-
15	-	+	+	+	-
$16 = 2^4$	+	+	+	+	-
17	-	-	-	-	+
18	+	-	-	-	+
19	-	+	-	-	+
20	+	+	-	-	+
21	-	-	+	-	+
22	+	-	+	-	+
23	-	+	+	-	+
24	+	+	+	-	+
25	-	-	-	+	+
26	+	-	-	+	+

Table 9.5 (Continued)

Trial	Variable				
27	−	+	−	+	+
28	+	+	−	+	+
29	−	−	+	+	+
30	+	−	+	+	+
31	−	+	+	+	+
$32 = 2^5$	+	+	+	+	+

both modeling (by use of multivariant curve-fitting routines) and optimization (21). Predicted results can then be confirmed with additional experiments. Although these last steps of modeling and optimization are truly the "guts" of optimization techniques, the thorough reviews and articles dealing with this subject (1−3,20,21) should be consulted for further information.

IV. CLOSING REMARKS

The task of designing an experiment, preforming the trials, developing a model based upon observed data, and confirming the predicted results with additional experiments seems in itself a herculean effort that may appear justifiable only for the most important of projects. Now, added to these efforts, is the burden of proving that the phase behavior of the disperse system is both smooth and continuous before even starting the foregoing procedures. Although this chain of studies may seem too involved to be practical, remember that (1) the final product will be of the highest quality possible with the greatest likelihood of commercial success, and (2) that the steps are well defined and readily predictable in terms of time and cost required for completion. It should also be noted that compared with solid-dosage forms, disperse systems can be formulated in small quantities (2−200 ml) to check the stability and response attributes, thereby minimizing the drug and raw material costs during optimization.

AKNOWLEDGMENTS

Special thanks to Roger Klassen for review of the section on optimization and to Carolyn Pesheck and Kilian O'Neill for review of the section on laboratory robotics. Prof. Stig Friberg's method of teaching surfactant ternary phase behavior was thoroughly utilized in writing this chapter. His mentorship during my graduate studies was, and still is, greatly appreciated.

REFERENCES

1. D. E. Fonner, Jr., J. R. Buck, and G. S. Bander, *J. Pharm. Sci.*, *59*:1587, 1970.
2. M. R. Harris, J. B. Schwartz, and J. W. McGinity, *Drug. Dev. Ind. Pharm.*, *11*:1089, 1985.
3. R. M. Franz, J. E. Browne, and A. R. Lewis, in *Pharmaceutical Dosage Forms: Disperse Systems*, Vol. 1 (H. A. Lieberman, M. M. Rieger, and G. S. Banker, eds.). Marcel Dekker, New York, pp. 427–514, 1988.
4. S. E. Friberg, and C. Solans, *Langmuir*, *2*:121, 1986.
5. G. M. Eccleston, *Cosmet. Toiletries*, *101*(11):73, 1986.
6. S. E. Friberg, I. Blute, H. Kunieda, and P. Stenius, *Langmuir*, *2*:659, 1986.
7. P. Ekwall, in *Advances in Liquid Crystals*, Vol. 1 (G. H. Brown, ed.). Academic Press, New York, 1975.
8. J. S. Walker, and C. A. Vause, *Sci. Am.*, *256*(5):98–105, 1987.
9. G. M. Barrow, *Physical Chemistry*, 4th ed., McGraw-Hill, New York, pp. 321–329, 1979.
10. D. Attwood, and A. T. Florence, *Surfactant Systems: Their Chemistry, Pharmacy and Biology*. Chapman and Hall, London, 1983.
11. D. M. Small, in *The Bile Acids*, Vol. 1 (P. P. Nair, and D. Kritchevsky, eds.). Plenum Press, New York, 1971.
12. European Patent 133-258-A, Vitamin/E mixed micelle injection solns. contg. vitamin/E, phospholipid, e.g., lecithin and bile acid, e.g., glyco:cholic acid, 1983.
13. World Patent 84/02076, method of preparing controlled release preparations for biologically active materials and resulting compositions, 1984.
14. S. E. Friberg, and D. W. Osborne, *J. Am. Oil Chem. Soc.*, *63*:123, 1986.
15. K. Shinoda, and S. E. Friberg, *Emulsions and Solubilization*, John Wiley & Sons, New York, 1986.

16. B. M. Knickerbocker, C. V. Pesheck, H. T. Davis, and L. E. Scriven, *J. Phys. Chem.*, *86*:393, 1982.
17. F. D. Blum, E. I. Franses, K. D. Rose, R. G. Bryant, and W. G. Miller, *Langmuir*, *3*:448, 1987.
18. S. M. Ng, and S. G. Frank, *J. Dispersion Sci. Technol.*, *3*: 217, 1982.
19. G. E. P. Box, and N. R. Draper, *Empirical Model Building and Response Surfaces*. John Wiley & Sons, New York, 1987.
20. H. R. Moskowitz, in *Cosmetic Product Testing: A Modern Psychophysical Approach*, Chap. 10. Marcel Dekker, New York, 1984.
21. J. Dayhoff, *J. Eng. Comput. Appl.*, fall, pp. 101, 1986.

10

General Considerations for Stability Testing of Topical Pharmaceutical Formulations

GARY R. DUKES *The Upjohn Company, Kalamazoo, Michigan*

I. INTRODUCTION

Stability testing is a necessary and vital means to help assure that
a formulation will maintain its integrity (i.e., strength, quality, and
purity) during its assigned shelf life. It would, however, be futile
to attempt to propose a simple, universal stability-testing protocol,
which could reasonably be expected to assure the integrity of every
topical formulation of every pharmaceutical manufacturer, because
there are necessary differences in specific stability protocols between different formulations of the same manufacturer and between
similar formulations of different manufacturers.

Although a number of factors may influence the determination of
an expiry date for a given product, this discussion will be confined
to testing designed to determine the chemical and physical properties critical to the stability of the product, with the assumption
that one, or a combination, of these is the limiting factor. Although
much of the discussion may be applicable to the testing and dating
of nonprescription products, it will be confined to prescription
products that progress from investigational new drug (IND) status,
through the new drug application (NDA) stage, and on to marketing.

The theme of this chapter is based on the premise that despite
the differences in specific stability-testing protocols among products
and manufacturers, there are activities that are common to any stability-testing program. The first set of stability-related activities
in the development of a new active ingredient into a marketed
product generally involves the *profiling* of its physical and chemical

properties. This is followed by a period in which the stability of the bulk active ingredient and its associated formulations is *assessed* and an expiry period established. Subsequently, stability studies are initiated and continued to *confirm and expand* the results of the preceding studies. Finally, the stability of the marketed product is *periodically reassessed*.

Thus, the development of an understanding of the stability performance of a given formulation is basically an evolutionary process. For convenience of discussion, this evolutionary progression may be divided into six stages (1):

Preformulation stage
Formulation development stage
Proposed product stage
New product stage
Established product stage
Revised product stage

Within this progression from profile to assessment, through confirmation and expansion, and finally to periodic reassessment of product stability, a range of testing protocols, methods, and mathematical models may be utilized, with each successive stage designed to augment the data base of the product and, thereby, strengthen and expand the conclusions reached during each preceding stage. At each stage differing types and amounts of information are sought that are important to the assignment of an expiry period, thus, giving rise to widely differing concerns and objectives for each phase of the overall stability-testing program.

It is also important to understand the relationship between these stages and the progression of the product through clinical development to marketing (Table 1). The evolution of the product stability performance profile begins with the preformulation stage, which is associated with the pre-IND phase, then proceeds to formulation development which is generally associated with Phase I and II clinical trials. From there it proceeds to the proposed product stage, which is temporally associated with Phase III clinical trials. The new product stage is associated with scaleup, process validation, and approval and is followed by the established and revised product stages, which are associated with the marketing of the product. Note that the timing relationships given in Table 1 represent a general approach and are not intended to be rigidly interpreted.

As the product progresses through the various stages, different kinds of studies are carried out to meet a variety of objectives. There are a number of elements that are common to all of these stability studies and that must be included in the study protocol.

Table 10.1 Evolution of a Product Stability Performance Profile

Study stage	Timing
Preformulation	Pre-IND
Formulation development	Phase I-II
Proposed product	Phase III
New product	Scaleup and approval
Established product	Postapproval
Revised product	

The design of each study must begin with a clear objective. With the objective in mind, the attributes to be measured are defined, and appropriate test methods are defined or developed; the storage conditions, assay schedules, and types of samples are selected; and the timing and method for data evaluation are determined.

The importance of a clearly stated objective cannot be overemphasized. For example, a study to determine the effects of metal ions on the stability of a new active ingredient in aqueous formulations will be quite different from a study on an established product to confirm its stability performance profile.

II. PREFORMULATION STAGE

The first assessment of the stability of a potential new product begins in the preformulation stage. During this phase the first requirement is the development of analytical methods that are specific for the drug substance in the presence of process impurities and degradation products. Two types of stability studies are conducted during this stage.

A. Profile Studies

The first type is to determine the profile of physical and chemical properties of the new drug substance. To identify probable routes of degradation, the active ingredient should be subjected to short-term accelerated studies to produce obvious changes (Table 2). The storage conditions should be geared to possible degradative conditions, such as heat, light, moisture, and oxidants. Testing schedules are chosen to allow estimates of stability parameters. The sam-

Table 10.2   Profile Studies

Protocol elements	Activities or results
Objectives	Determine any reactivities of the new drug substance
	Establish storage conditions and packaging requirements for the new drug substance
Study design	Short-term studies under accelerated conditions, looking for gross changes
Tests	Potency, appearance, physical properties, degradation products, and the like
Storage conditions	Geared to possible degradative conditions of heat, light, moisture, oxidants, catalysts, pH, and others
Schedules	As appropriate to allow estimates of stability parameters
Samples	Representative of the batch
Evaluation	Review of data to estimate relevant stability parameters

ples should be representative of the batch of material used, and the evaluation will consist of a review of the data to estimate relevant stability parameters. The results of these studies will provide important information about the molecular reactivity of the new drug substance, an indication of the types of formulations that may be successful, and information on the storage requirements of the bulk drug.

An example of the type of study and the results that can be obtained are shown in Figure 1. This study involved the effect of metal ions on the stability of a drug substance in solution at 70°C. It is apparent that zinc (II), even at micromolar concentration, as well as nickel (II), and chromium (VI), to a lesser extent, exert a very deleterious effect on the solution stability of the drug. (The solution with no added metal ions retained 93% of its initial potency after 2 months at 70°C.) Because these metals may well be present in packaging materials, excipients, or even the synthetic process for the bulk drug, this knowledge was of great value in the development of formulations, the selection of packaging materials, and the monitoring of the bulk-drug synthetic process.

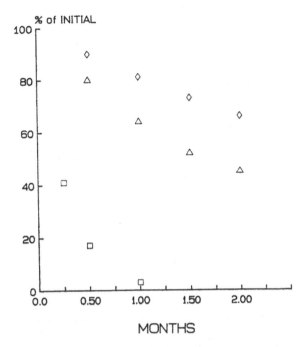

Figure 10.1 Effect of metal ions in solution at 70°C. Squares, 3.0 µM Zn (II); triangles, 1.0 mM Ni (II); diamonds, 1.0 mM Cr (VI).

B. Toxicology Supplies

A second useful source of stability information during the preformulation stage comes from the monitoring of toxicology supplies. These represent the first rudimentary formulations of the new drug substance. The objectives are to assure that the potency and level of significant degradation products are documented during the use of the supplies. An important added benefit is the development of supporting data for use during the formulation development stage.

III. FORMULATION DEVELOPMENT STAGE

The emphasis in the formulation development stage shifts from profiling the physical and chemical properties of the new drug substance to the assessment of the stability of both the bulk drug and its associated dosage forms as they are developed. During this stage, preliminary formulations are manufactured in small-scale lots for use in clinical studies, and their stability is monitored. Early in this stage, additional studies specific to the type of formulation chosen are conducted on the bulk drug to complete the profile of its physical and chemical properties. The effects of pH upon solution stability (2), for example, would be critical if the active ingredient were to be formulated as an aqueous solution, cream, or lotion. If the product is to be formulated as a topical suspension, studies to determine changes in particle size with changes in temperature or over time would be very useful. A knowledge of the various crystal forms in which the active ingredient can exist, their interconversion pathways, and the solubilities of each could prove important to the manufacturing process for the formulations and for their resultant chemical and physical stability (3). Finally, studies should be conducted to determine the compatibility of the new active ingredient with probable formulation excipients. These are short-term studies under accelerated conditions of binary mixtures of the drug substance with potential excipients. Comparison of results provides for the early elimination of potentially deleterious excipients. If the drug substance is sensitive to the presence of moisture, hydrated excipients should be avoided. Sometimes, the studies may indicate the types of controls necessary to ensure the suitability of the excipient for use. An example is shown in Figure 2. A cream that was formulated with two different grades of the same excipient was stored at 40°C. The results indicated the necessity to control the pH of the excipient to avoid the basic pH range.

Later in this stage, studies are conducted to determine the optimum formulation. A number of different formulations should be prepared and compared with each other.

A. Formulation Comparison

A convenient means for conducting formulation comparison studies is to first screen the trial formulations for undesirable physical changes under accelerated conditions. For example, ointments and creams may be subjected to temperature cycling to determine the relative tendency to separate. Topical suspensions should be screened for changes resulting from particle growth by Ostwald ripening, for changes in polymorphic forms, or for a tendency to cake (4). Next, the formulations should be screened for chemical stability. By using the assay for the drug substance from the previous stage as a starting point, a stability-indicating assay can be developed for the most promising formulations. Specificity may be determined by spiking the assay preparation with like compounds or authentic samples of known or suspected degradation products.

Extensive use can be made of short-term studies at high-temperature and high-humidity conditions (5). For example, the ex-

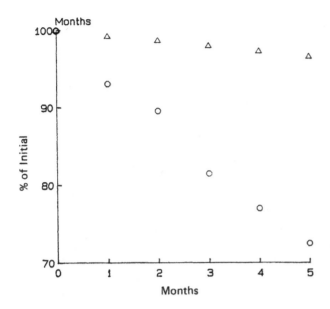

Figure 10.2 Effect of excipient pH at 40°C. Diamonds, pH 5; circles, pH 8.

perimental formulations may be stored at 40°, 47°, 56°, and 70°C and assayed for potency at several intervals. If there does not appear to be a change in the mechanism of the degradation reaction, Arrhenius plots may be utilized to calculate energies of activation, and the shelf life of the proposed formulations at other temperatures may be estimated. Regression analyses may also provide a crude estimate of the shelf life of each formulation. The shelf lives estimated from the use of the Arrhenius plots and the regression analyses can be used to compare the formulations with each other to determine the one that is most stable. The data for the selected formulation are also useful in providing the first indications of a suitable storage condition for the finished product (Table 3).

B. Container and Closure System Selection

The next component of this stage involves studying the selected formulation in several different container and closure systems. Again, short-term testing under conditions of elevated temperature and humidity are used to determine the stability characteristics of

Table 10.3   Formulation Comparison

Protocol elements	Activities or results
Objectives	Determine any formulation reactivities and significant drug substance degradation products
	Compare stability characteristics of the experimental formulations
	Collect preliminary information on possible stability-limiting factors
Study design	Short-term studies under accelerated conditions, looking for gross changes
Tests	Potency, significant degradation products, physical properties, plus individual tests relevant to specific dosage forms
Storage conditions	Geared to possible degradative conditions of heat, light, moisture, and the like
Schedules	As appropriate to allow estimation of stability parameters
Samples	Representative of the batch
Evaluation	Direct comparison of raw data, comparison of rates, and so forth, to estimate relative stability performance

the formulation in the various container—closure systems. Test and assay data are obtained and regression analyses are used where applicable to obtain crude estimates of shelf life for comparison purposes. Any packages that are found to adversely affect product stability are eliminated.

C. Clinical Supplies

An additional component of this stage involves the monitoring of clinical supplies (Table 4).

Here, the objectives are to determine the values of the critical quality parameters of the formulations used to establish safety and efficacy as a function of time and to ensure that only satisfactory material is used in the clinic.

Table 10.4 Clinical Supplies

Protocol elements	Activities or results
Objectives	Determine the values of the critical quality parameters of formulations utilized to establish safety and efficacy
	Assure that only satisfactory material is used in the clinic
Study design	Medium-term studies looking for changes in critical quality parameters
Tests	Basic attributes, such as potency, significant degradation products, physical properties, plus individual tests relevant to specific dosage forms
Storage conditions	Compatible with label storage conditions
Schedules	More frequent for less stable formulations
Samples	Representative of each formulation and container—closure system in use in the clinic
Evaluation	Periodic evaluation to assure continued suitability of supplies by direct comparison to established specifications or statistical evaluation

Generally, the study design for clinical supplies is a medium-term study (covering at least the period for which the supplies are in use) looking for changes in critical quality parameters. The tests should include potency, significant degradation products, physical properties (e.g., appearance, odor, consistency, and weight loss), plus individual tests relevant to the specific dosage form [e.g., particle size (suspensions), pH (aqueous formulations), drug release (transdermal patches), resuspendability (lotions), and (water content (nonaqueous formulations)]. In addition to measuring the potency of bulk ointments and creams, the potency of samples from several locations inside the container should also be measured to determine the tendency of the active ingredient to concentrate. The storage conditions are those that correspond to the label storage conditions. The samples are representative of at least each formulation and container closure system in use in the clinic.

Sometimes, every lot of clinical supplies may be on stability. There is a periodic evaluation of data by direct comparison of the data to established specifications or by statistical evaluation (e.g., regression analysis), to assure the continued suitability of the clinical supplies in the field.

IV. PROPOSED PRODUCT

The proposed product stage covers the period from selection of the final formulation to NDA filing; it roughly corresponds to the period during which Phase III clinical studies are conducted. At this point in the development process, a substantial amount of fundamental stability information will be available. Any inherent sensitivities of the active ingredient have been determined, significant degradation products identified, and major stability-limiting characteristics of the various experimental formulations have been detailed.

The objectives of this stage are (1) to assess the stability performance of the selected formulation for the purpose of defining storage conditions and stability-limiting factors; (2) to confirm degradation pathways in the selected formulation; and (3) to establish the initial expiration dating period for the container—closure system in which the product is to be marketed.

A greater variety of test data can be collected during this phase to establish the stability-limiting factor(s). Both short-term accelerated studies and long-term studies at proposed label storage conditions are conducted. The tests and assays to be conducted should be essentially the same as those given in Section III.C. Many of the tests and assays included in the stability protocols utilized during this stage will not be included in the protocols in later stages because they will be deemed redundant or unnecessary. In general, at least three lots of the proposed formulations are placed on test in the proposed container—closure systems. These lots are usually produced in production-scale equipment (Table 5).

There are cases when scale testing may not be scientifically necessary. If a stable, noncomplex formulation (formulation defined as a fixed ratio of active ingredient(s)/excipients) is being developed in several strengths or concentrations, full testing should be conducted on *the highest and lowest strength or concentration*. The data from this testing regimen will provide information to fully characterize the stability performance profile of formulations *at all intermediate strengths*. For example, the lowest concentration would test the case for which the package surface area per milligram of active ingredient is at a maximum, whereas the highest concentration would test the case for which the probability of nucleation and crystal growth (solutions) or agglomeration (suspensions) is maximized.

Table 10.5 Proposed Product

Protocol elements	Activities or results
Objectives	Assess stability performance to define stability-limiting factors and storage conditions; confirm degradation pathways in the market formulation; establish initial expiration-dating period
Study design	Both short-term studies at accelerated conditions and long-term studies at proposed label storage conditions
Tests	Potency, significant degradation products, physical properties, plus individual tests relevant to specific dosage forms
Storage conditions	Accelerated: geared to known reactivities
Schedules	More frequent testing for less stable formulations
Samples	Representative of the formulation and the container—closure systems; the least protective container—closure should be included
Evaluation	Direct comparison of raw data with proposed standards or statistical analysis with a determined confidence level

Another instance in which full-scale testing may not be scientifically necessary involves bracketing of package sizes within a container—closure system. (A *container—closure system* is defined as a family of packages made with the same materials.) Again, for a stable, noncomplex formulation packaged in several sizes, within a given container—closure system, full testing should be conducted on product packaged *in the smallest and largest sizes*. (The smallest package size provides the greatest possibility for product—package interaction.) Therefore, comparison of data for product packaged in the smallest and largest sizes (as well as that from other container—closure systems) will provide an additional assessment of the effect of the package upon the stability of the product and will support packaging of the product in any *intermediate sizes* of the container—closure system.

V. NEW PRODUCT STAGE

At the time of NDA approval, development enters the new product stage during which the objectives shift from assessment of stability performance and establishment of the initial expiration-dating period to confirmation of the stability performance and expansion of the stability data base. The new product stage is a time of transition to the established product stage and usually lasts from 1 to 2 years. The studies initiated during the proposed product stage are continued and, in general, at least three marketed lots will be placed on stability. These lots are placed on long-term study at label storage conditions. Testing schedules should be selected according to the stability performance of the product.

As the data base for the product matures, the protocol will evolve such that the studies in this stage, although generally less intensive than those of the preceding stage, will be more intensive than those of the succeeding stage. When warranted, any revisions of the expiration dating period for the product are supported by a mature data base.

VI. ESTABLISHED PRODUCT STAGE

The last stage is the established product stage, which continues for as long as the firm manufactures and markets the product. During this stage, concern again shifts. At this point, a comprehensive data base is in place, and the objective becomes one of periodically reassessing the stability of the product.

Because much stability data has been generated on the product during the preceding stages, generally fewer data per product are required in any given time span during this continuation of the stability testing program.

In general, one lot per year should be placed on test. Lot selection should be designed to rotate among the various marketed container—closure systems. The protocol should include tests and assays for significant quality attributes, as established in previous stages. The study design is for full-term studies at label storage conditions. The schedules take into account the relative stability of the product as well as the relative sensitivities of the various stability parameters determined during the previous stages (Table 6).

Table 10.6   Established Product

Protocol elements	Activities or results
Objectives	Confirmation of stability performance determined in preceding stages
Study design	Long-term studies
Tests	Significant quality attributes as determined in preceding stages
Storage conditions	Consistent with label storage conditions
Schedules	Based on relative stability of the product and the relative sensitivities of the various stability parameters
Samples	Representative of the product; lot selection rotated among the various marketed container—closure systems
Evaluation	Periodic evaluations to allow timely decision-making; evaluation by comparison of results with the established stability profile

VII.   PRODUCT REVISION STAGE

The expiration dating periods of established products are assigned or revised based upon a great deal of experience and upon a mature data base. When products are revised, however, there are many instances in which the number of months of stability data are significantly fewer than the number of months of the proposed dating, an event that experience indicates would be entirely reasonable. If the programs previously discussed were conscientiously implemented, treating revised products as entirely new products requiring full, directly supportive data bases would be neither advisable nor necessary. Changes in products occur in several ways and for various reasons. In general, they may be categorized as formulation changes, such as the addition or deletion of excipients; processing changes; changes in the source of active ingredients; or changes in packaging material that comes into direct contact with the dosage form. When such changes are anticipated, a scientific judgment should be made concerning their probable effects on product stability. Such judgments would be based on a sound, technically based understanding of the product's decomposition pathways and physical stability properties that were developed during the new active ingredient, formulation development, proposed product, new product, and established product stages.

If it is determined that the changes will have no effect on stability, then no additional stability testing is necessary. Many changes, such as a change in the supplier of a raw material or deletion of minor excipients, would fall into this category. Here, stability monitoring would be the same as that for established products.

If it is judged that the change is likely to affect the stability performance profile of the product, short-term studies under accelerated conditions should be carried out to compare performance of the revised product with the original product's performance. Comparisons of the data obtained from these accelerated studies are the key to rapidly determining the effect of a change on the product's stability. The conclusions drawn should be confirmed with full-term studies under label storage conditions.

VIII. CONCLUSION

The foregoing discussion reflects the current stability-testing philosophy of the Upjohn Company. It has been general in nature because the finer details of stability testing programs of different manufacturers will almost certainly be different. Indeed, the finer details of the stability programs at the Upjohn Company are dynamic (6,7). However, it is hoped that it has stimulated some thoughts on the basic underlying concepts that should be common to all stability-testing programs.

In particular, the concept of a progression of stability testing with product maturation—a system guaranteeing a certain continuity of data augmentation—should be applicable to any stability program (8).

REFERENCES

1. Stability Concepts, PMA Joint QC-PDS Stability Committee, *Pharm. Tech.*, June 1984.
2. J. J. Winheuser, and T. Higuchi, *J. Pharm. Sci.*, *51*:354, 1962.
3. A. J. Aguair, J. Krc, Jr., A. W. Kinkel, and J. Samyn, *J. Pharm. Sci.*, *56*:847, 1967.
4. C. T. Rhodes, *Modern Pharmaceutics*, (G. S. Banker, and C. T. Rhodes, eds.). Marcel Dekker, New York, p. 329, 1979.

5. J. T. Carstensen: *Solid Pharmaceutics: Mechanical Properties and Rate Phenomena.* Academic Press, New York, pp. 243–244, 1981.
6. G. R. Dukes, *Pharm. Tech.*, January 1982.
7. G. R. Dukes, *Drug Dev. Ind. Pharm.*, 10:1413, 1984.
8. C. H. Bibart, *Drug Dev. Ind. Pharm.*, 5:349, 1979.

11

In Vitro Testing of Topical Pharmaceutical Formulations

RONALD C. WESTER and HOWARD I. MAIBACH *University of California School of Medicine, San Francisco, California*

I. INTRODUCTION

The development of dermatological and transdermal drugs requires knowledge of the percutaneous permeation (local skin) and absorption (systemic) of the drug. The ultimate in both relevance and cost is in vivo human testing. To retain relevance, but compromise on cost, alternate test procedures are desired. One of these procedures is in vitro skin absorption using diffusion cells. A recent report (1) attempted to define principles and practices related to in vitro percutaneous penetration studies. This chapter will discuss and elaborate on those procedures and principles as they apply to drug testing in our laboratory.

We hold only one principle that relates to in vitro skin absorption studies: the system is artificial and subject to all variabilities of its parts. Relevance must be reserved until in vivo (preferably human) testing is done. On this principle, we build our in vitro studies.

We also offer an alternative study design for in vitro skin absorption drug screening that has proved valuable, especially for hydrophobic compounds that will not partition into reservoir fluid.

The information that follows is a small part of in vitro percutaneous absorption. The reader is referred to the fine work of Robert Bronaugh on in vitro percutaneous absorption (2).

II. BASIC STUDY DESIGN

Table 1 gives the basic study design that we approach with in vitro absorption studies. We believe that

Table 11.1 In Vitro Study Design

Human skin source	Number of replicates	
	Formulation 1	Formulation 2
A	4	4
B	4	4
C	4	4
Total	12	12

Assay: Reservoir fluid
        Skin
        Skin surface wash
        Apparatus wash
        Material balance

1. All formulations should be compared within the same skin samples.
2. Variability exists between skin sources, and thus a minimum of three skin sources is desired.
3. Reservoir fluid, skin, skin surface wash, and apparatus wash should be assayed for material balance.

Obviously, the abbreviated study design depicted in Table 1 will not meet all needs. However, it will avoid pitfalls of human skin variability, and the material balance is a check on scientific quality. In addition, as will be discussed later in this chapter, assay of chemicals within skin may be a viable alternative when solubility in reservoir fluid is a problem.

III. FACTORS

A. Membranes

The literature is full of reports showing that skin absorption is different in most animals than in humans (Table 2; 3). Human skin is available, so it should be the membrane of choice.

Human skin should be used as soon as possible. Obviously, freshness is best. Reports suggest that the freezing of skin changes its permeability characteristics. If skin metabolism data are warranted, then appropriate procedures and verifications are warranted.

Table 11.2 Absorption of Paraquat and Water Through Human
and Laboratory Skin

In vivo/in vitro	Paraquat permeability rate ($\mu g/cm^2$)	
Human (in vivo)	0.03	
Human (in vitro)	0.5	

	Permeability constant (cm/hr $\times 10^5$)	
	Water	Paraquat
Human	93	0.7
Rat	103	27.2
Hairless rat	130	35.3
Nude rat	152	35.5
Mouse	164	97.2
Hairless mouse	254	1065.0
Rabbit	253	92.9
Guinea pig	442	196.0

B. Cell Design

The diffusion cells, connecting lines, and collection chambers should
be made from inert, nonreactive material. The drug can disappear
through volatility and apparatus adsorption. Material balance in
the study design is a check on this. Table 3 shows this difference
(4). If solubility in reservoir fluid is a problem, then the larger
volume from a flow system may result in relevant data (Table 4; 5).

C. Temperature

It is fortunate that circulating bath water of 37°C results in skin
surface temperature of 32°C, just as in humans in vivo. Obviously,
temperatures should be verified.

D. Dose

This is easily determined by taking a known volume of formulation,
spreading it over human skin, and measuring the skin area covered

Table 11.3 In Vitro Percutaneous Absorption and Skin Distribution
of Chemicals in Water Solution

	Chemical (%)		
Parameter	Benzene	54% PCBs	Nitroaniline
Percutaneous absorption (systemic)	0.045 ± 0.037	0.03 ± 0.00	5.2 ± 1.6
Surface bound/stratum corneum	0.036 ± 0.005	6.8 ± 1.0	0.2 ± 0.17
Epidermis and dermis	0.065 ± 0.057	5.5 ± 0.7	0.61 ± 0.25
Total (skin/systemic)	0.15	12.3	6.1
Skin wash/residual	2.51 ± 0.94	71.2 ± 5.5	92.4 ± 0.8
Apparatus wash	0.006 ± 0.005	0.25 ± 0.2	0.47 ± 0.03
Total (accountability)	2.67	83.3	99.0

Table 11.4 In Vitro Percutaneous Absorption of Triclocarban in
Adult and Newborn Abdominal and Foreskin Epidermis

Type	Dose absorbed (% ± SD)
Static system, 37°C	
human adult abdominal (n = 14)	0.23 ± 0.15
human newborn abdominal (n = 6)	0.26 ± 0.28
human infant abdominal (n = 4)	0.29 ± 0.09
monkey adult abdominal (n = 6)	0.25 ± 0.09
human adult foreskin (n = 4)	0.60 ± 0.25
human newborn foreskin (n = 7)	2.5 ± 1.6
Static system, 23°C	
human adult abdominal (n = 8)	0.13 ± 0.05
Continuous flow system, 23°C	
human adult abdominal (n = 12)	6.0 ± 2.0
Man in vivo (n = 5)	7.0 ± 2.8

by the formulation (spreadability). Generally, 1 to 5 µl of formulation will cover 1-cm^2 skin surface area. The drug mass within that amount will be determined by the concentration of drug within the formulation.

E. Open Versus Occluded

The use of the product under normal circumstances will determine its application.

F. Finite Versus Infinite Dose

For dermatological products, a dose relevant to actual use will usually suggest a finite dose. A delivery system that employs constant delivery may justify infinite dose. Common sense says to follow how the product will be used.

G. Equilibration

Physiologically (except for some psoriasis treatment) a person does not sit submerged in a tub before applying a dermatological or transdermal drug. Therefore, it seems irrational to use equilibration beyond 30 min to determine if the system is working (bubbles). Because human skin tends to fall apart after 24 to 48 hr in an *in vitro* run, adding more water time generally seems irrelevant.

H. Time

The relevant in vivo drug use should determine in vitro run time. Most absorption is determined over a 24 to 48 hr period because of the tendency of human skin to fall apart after this.

I. Receptor Fluid

The previous published recommendation is as follows (1): For most studies, an isotonic solution buffered to pH 7.4 is a suitable and preferred receptor fluid; a different pH can be used if it can be justified. In all instances, the thermodynamic activity of drug in the receptor fluid should not exceed 10% of its thermodynamic activity in the donor medium to maintain a favorable driving force for permeation and to assure reasonable and efficient collection of permeant. The receptor medium may need alteration from a strictly aqueous medium to attain this endpoint for hydrophobic compounds. This factor supercedes concern for maintaining a minimal receiver volume.

Hydrophobic compounds may be defined as compounds that have uniquely low solubilities in water (less than 10 mg/L), and solubilities in both water-miscible (alcohol, propylene glycol) and water-immiscible (ether, octanol, chloroform, hexane) solvents that are orders of magnitude larger. These compounds have large octanol/water partition coefficients that are about 10^3 or higher. Such compounds have a low tendency to partition into the receiver medium beneath the skin. It may be necessary to use a nonphysiological medium in which the drug is more soluble to efficiently elute such substances. When this is a concern, studies should be performed either by using a lipophilic receptor fluid that is without effect on the skin membrane, as demonstrated experimentally, or by using an isotonic solution containing an appropriate concentration of solubilizer for the hydrophobic compound. Flow-through cell designs also tend to minimize collection problems.

The practical problem is with hydrophobic compounds that will not partition into water receptor fluid. The alternative is to go to a nonwater medium. This creates a nonphysiological situation. The data can then reflect an extraction of compound from skin, possibly with damage to the skin membrane. A pseudoextraction system without in vivo human verification does not appear to be a strong position.

J. Alternative Approach for Drug Research

Historical precedent states that with the in vitro system, drug content is assayed (only) in reservoir fluid, and these data are subjected to vigorous mathematical treatment for lag time and permeability determinations. Drug development may not be so interested in vigorous lag time as they might be interested in which formulation allows a drug to penetrate skin. We offer this alternative approach.

Table 5 gives the in vitro absorption screening in human skin for spironolactone topical formulations. Spironolactone will not partition into the water-based reservoir fluid. Therefore, formulations were screened for spironolactone content in skin, and it was the skin content that was used to determine differences in formulations. With appropriate control and total recovery of material balance, an effective screening system can be instituted.

We believe in in vivo verification, and the data in Table 6 show the in vitro absorption in human skin and in vivo verification in rhesus monkey and humans for the same formulation. Comparison of human skin content in the in vitro system (with material balance) with in vivo percutaneous absorption in rhesus monkey and humans (by the standard urinary ^{14}C excretion method) shows excellent correlation. The in vitro system is verified.

Table 11.5 Spironolactone Formulation Screening

	Spironolactone content (%)			
Formulation	Reservoir fluid	Skin	Surface wash	Total recovery
Control[a]	0.0	1.0	95.6	96.6
1	0.0	2.3	94.1	96.4
2	0.0	21.4	74.4	95.8
3	0.0	3.5	96.1	99.6
4	0.0	3.1	93.0	96.1

[a]control = 0-hr application time (system verification); 1–4 = 48-hr application time.
HLPC assay for spironolactone only.

Table 11.6 Percutaneous Absorption of Spironolactone

	Total percent dose (n = 11)			
In Vitro (human skin)	Skin	Receptor fluid	Surface wash	Total
	2.4 ± 2.0	0.06 ± 0.07	89.7 ± 14.5	92.2 ± 14.5
In Vivo				
Rhesus monkey	2.9 ± 3.2	(24-hr skin application time)		
Man	0.5 ± 0.4	(8-hr skin application time)		

IV. CONCLUSIONS

In vitro percutaneous absorption is best done with human skin. The
    skin should be used as soon as possible, and stored in the re-
    frigerator no longer than 7 to 10 days.
In vitro penetration into skin gives results suitable to distinguish
    among drug formulations, especially when the drug will not par-
    tition into reservoir fluid.
Material balance in an in vitro study design adds to the overall
    data presentation.
In vivo verification of skin absorption, preferably in humans, adds
    relevance to the in vitro data.

REFERENCES

1. J. P. Skelly, V. P. Shah, H. I. Maibach, R. H. Guy, R. C.
    Wester, G. Flynn, and A. Yacobi, *Pharm. Res.*, 4:265, 1987.
2. R. L. Bronaugh, in *Percutaneous Absorption*, (R. Bronaugh,
    and H. Maibach, eds.). Marcel Dekker, New York, p. 267,
    1985.
3. R. C. Wester, and H. I. Maibach, in *Percutaneous Absorption*
    (R. Bronaugh, and H. Maibach, eds.). Marcel Dekker, New
    York, p. 251, 1985.
4. R. C. Wester, M. Mobayen, and H. I. Maibach, *J. Toxicol.
    Environ. Health*, 21:367, 1987.
5. R. C. Wester, H. I. Maibach, J. Surinchak, and D. A. W.
    Bucks, in *Percutaneous Absorption*. (R. Bronaugh, and H.
    Maibach, eds.). Marcel Dekker, New York, p. 223, 1985.

12

Drug Delivery from Topical Formulations: Theoretical Prediction and Experimental Assessment

WILLIAM J. ADDICKS* and NORMAN D. WEINER   *University of Michigan College of Pharmacy, Ann Arbor, Michigan*

RANE L. CURL   *University of Michigan College of Engineering, Ann Arbor, Michigan*

GORDON L. FLYNN   *Cygnus Research Corporation, Redwood City, California*

I. INTRODUCTION

Drug delivery from topical formulations for local or systemic effects essentially involves passive diffusion of the drug through the skin. The release of a drug molecule from a vehicle into the skin and diffusion across the skin are controlled by physicochemical factors sensitive to the molecular properties of the permeant, the vehicle, and the skin (1). Katz and Poulsen (2) have identified four categories of interactions with bearing on the delivery process. Rate-influencing interactions are possible between the (1) drug and skin, (2) vehicle and skin, (3) drug and vehicle, and (4) drug, vehicle, and skin. Examples of drug—skin interactions include alteration of the surface structure of the skin by the drug components of formulations (e.g., hydration of the stratum corneum by sodium pyrrolidine carboxylate) as well as the binding of drugs to constituents of the skin as they diffuse through the tissue field. A vehicle—skin interaction occurs when one or another of the vehicle's main components, here meaning those components that principally determine the delivery matrix of the formula, effects a change in the physical state of the skin, in turn, affecting the skin's permeability. Vehicle—skin interactions can be divided into three categories: (1) penetration enhancement effects brought about by direct solvent action on the skin, (2) the vehicle's general influence on the state of

---

Current affiliation: E. I. du Pont de Nemours and Company, Wilmington, Delaware

the skin's hydration, and (3) vehicle effects on the local vasculature, which alter drug clearance and skin temperature, in either
case, subtly influencing permeability. Drug—vehicle interactions
are those in which physicochemical interactions between the drug
and the vehicle kinetically or thermodynamically govern the release
of the drug into the skin. Such interactions can become the rate-
controlling factors and be clinically highly important when the stratum corneum is impaired as the consequence of disease or injury,
for, when the skin is damaged, solution and diffusion in the vehicle
may be relatively slow relative to skin permeation. Here, one is
concerned about the solubility of the drug and its diffusive mobility
within the vehicle, as each is a factor influencing the rate of presentation of the drug to the vehicle—skin interface. For the more
usual case, in which the permeability of the intact stratum corneum
is low, partitioning of the drug into the skin can still exert a profound, if not dominant, influence on the rate of delivery.

As one critically reviews the literature dealing with vehicle effects in transdermal drug delivery, a confused state of affairs
about the theory of topical drug delivery is apparent. In a review
of the subject, Idson (3) has cited several reasons for this confusion: (1) Experiments have been performed on a large variety of
animals; (2) different methods have been used to estimate skin penetration; (3) there has been a lack of awareness of possible drug—
vehicle interactions; and (4) there has been a lack of consideration
of thermodynamic factors in the interpretation of results. Although
all the possible types of interactions between drug, vehicle, and
skin mentioned here are important to topical bioavailability, this
paper mainly focuses on drug—vehicle and drug—vehicle—skin (membrane) interactions.

II. IN VITRO RELEASE FROM VEHICLES

A. Theoretical Background

As part of the assessment of a vehicle's potential for delivering a
drug to the skin for topical absorption, researchers have long
sought methods to determine a drug's release rate from its vehicle.
It is generally assumed that the results obtained in such experiments at least qualitatively relate to the release of the drug to the
skin in a clinical situation. Failure to properly interpret and in-
tegrate results from such studies could lead to a substandard drug
product. Thus, by necessity knowledge of the thermodynamics and
kinetics involved in the in vitro situation is important if one wishes
to successfully project the results to the in vivo circumstance (4).

There are several identifiable processes that, theoretically, can
establish the rate at which a drug is delivered from a film into an
absorbing membrane. Minimally, these include diffusion through

the vehicle to its interface with the skin and diffusion across the outer skin. Even in the absence of specific drug–vehicle–skin interactions, there are special circumstances in which dissolution of drug suspended in the vehicle or diffusion through the inner living strata of the skin may become rate-determining.

A theoretical basis for the study of the release kinetics of drugs from both suspension and solution ointments for the case in which release from the vehicle is rate-limiting was established by Higuchi (5). Higuchi first depicted the situation in which the ointment vehicle is initially saturated with solute, with excess solute uniformly suspended as tiny particles. The exact assumptions for the derivation of the time dependency of release are as follows:

1. The particles are present in a fine enough state so that dissolution of the particles is not rate-limiting.
2. Q (the total concentration (mass/volume) of dissolved and undissolved drug) is much greater than C_S (the solubility (mass/volume) of the drug in the ointment).
3. A sink condition prevails at the ointment–receiver phase interface.
4. Release occurs through a planar surface.
5. There is no significant boundary layer adjacent to the ointment (assumed implicitly).
6. Quasi–steady-state diffusion exists between the dissolution interface at the edge of the particle field and the interface with the sink.
7. Although not explicitly stated, the model is semi-infinite, as in the original derivation no limit was placed on how far the boundary could recede.

Assumption (2) is in effect a directive that dissolution takes place rapidly at the edge of the receding particle field in the course of drug release. Thus, unit activity (activity of the crystalline state) is continuously experienced along this front, and, with the passage of time, a well demarcated zone lying between the releasing surface and the suspension phase develops that is spanned by a linear concentration gradient given by the solubility divided by the thickness of the layer, that is, (C_S/h). The release of drug through the surface is controlled by the momentary steepness of this gradient. As time progresses, the distance between the edge of the saturated suspension phase within the ointment and the ointment's sink-side surface widens, resulting in a decrease in the rate of drug delivery. The following equation describing the release of solute was derived (5).

$$M_t = \sqrt{2 D C_s \left(Q - \frac{C_s}{2} t\right)} \qquad [1]$$

where D is the diffusivity (cm^2/sec) of the drug in the ointment. After differentiation for time, an expression for the instantaneous rate of release is obtained.

$$\frac{dM_t}{dt} = \frac{1}{2} \sqrt{\frac{D(2Q - C_s)C_s}{t}} \qquad [2]$$

When $Q >> C_s$, the amount of drug released into a sink bears the following relationship to time:

$$M_t = \sqrt{2QDC_st} \qquad [3]$$

and the rate becomes

$$\frac{dM_t}{dt} = \sqrt{\frac{QDC_s}{2t}} \qquad [4]$$

Equation 3 predicts that a plot of the amount of drug released (per unit area) versus the square root of time should be linear, whereas Eq. 4 predicts that the rate of drug release is proportional to the reciprocal of the square root of time.

Higuchi also deduced a relationship characterizing the release of drug from an ointment with its drug totally in solution ("solution ointment"), also from a planar surface directly into a diffusional sink (6). This equation is a solution to Fick's second law, and can be found in several standard texts concerning diffusion processes (7,8). As in the previous case, uptake into a sink is assumed, with diffusion to the releasing interface being the rate-limiting step in the overall process. The following mathematical description of the process was presented:

$$M_t = hC_0 \left\{ 1 - \frac{8}{\pi^2} \sum_{m=o}^{\infty} \frac{\left[\exp\left[-\frac{D(2m+1)^2\pi^2t}{4h^2}\right]\right]}{(2m+1)^2} \right\} \qquad [5]$$

In this expression, h is the thickness of the ointment phase and C_0 is the initial drug concentration in the ointment. The following simplified equation closely describes diffusion for the first 30% of release (9):

$$M_t = 2C_0 \sqrt{\frac{Dt}{\pi}} \qquad [6]$$

The solutions to the solution release case embodied in Eqs. 5 and 6 assume a semi-infinite geometry for the ointment phase. As in the suspension case, it is once again predicted that the amount of drug released will be proportional to the square root of time, but in this instance, as a practical matter, this is so only up to 30% of the total release.

Paul and McSpadden (10) have pointed out that although the Higuchi approximation works well in instances in which the total concentration of drug in the ointment is much larger than the solubility of the drug ($Q >> C_S$), appreciable departures from actual theoretical behavior occur in the limit $Q \to C_S$. In this case, Eq. 1 reduces to

$$M_t = C_s \sqrt{D t} \tag{7}$$

In the limit of $Q \to C_S$, Eq. 7 predicts 11.3% less drug will be released than does Eq. 6. Paul and McSpadden have eliminated this discrepancy by deriving a solution for this case that can be applied over the complete spectrum of drug concentrations, including the suspension situations. In this derivation, boundary conditions have been applied that remove the pseudo—steady-state approximation of Eq. 1. In addition, a semi-infinite geometry is assumed. The resulting relationship for the mass released at a given time is

$$M_t = \left[\frac{2C_s}{\text{erf}(\beta)} \right] \sqrt{\frac{D t}{\pi}} \tag{8}$$

In this equation, $\text{erf}(\beta)$ is defined by the following transcendental expression:

$$\sqrt{\pi} \, (\beta) \exp[(\beta)^2] \, \text{erf}(\beta) = \frac{C_s}{(Q - C_s)} \tag{9}$$

As in Eqs. 1 and 6, Eq. 8 indicates that release should follow a square root of time dependence. However, the dependence on Q in Eq. 8 is different than in the previous treatments. To illustrate the properties of this exact solution, Paul and McSpadden have examined some limiting cases. When $Q \to C_S$, Eq. 9 indicates that β becomes very large and, thus, $\text{erf}(\beta)$ approaches unity. As a result, Eq. 8 reduces to

$$M_t = 2C_s \sqrt{\frac{D t}{\pi}} \tag{10}$$

This equation is the same as Eq. 6, for which all of the drug was assumed to be in solution. For the case of $Q >> C_S$, Eq. 9 predicts that β is small. A series expansion of Eq. 9 permits an explicit calculation of $erf(\beta)$ as follows:

$$erf(\beta) = \sqrt{\frac{2}{\pi} \frac{C_s}{(Q - C_s)}} \tag{11}$$

Upon combining this with Eq. 8, the following result is obtained:

$$M_t = \sqrt{2 D C_s (Q - C_s) t} \tag{12}$$

Equation 12 is identical in form with Eq. 1 except that C_S appears in lieu of $C_S/2$, which is a result of the limit process. Paul and McSpadden point out that no relevant difference exists in the theoretical treatments when $Q >> C_S$. In effect, they conclude that the pseudo—steady-state assumption that resulted in Eq. 1 is adequate when $Q >> C_S$. To test the utility of Eq. 8, Paul and McSpadden (10) performed a series of experiments for which the release rates of an organic dye, diffusing out of a silicone polymer and into an acetone sink, were measured over a wide range of solute loadings ($Q/C_S < 1$ to $Q/C_S > 1$). The theoretical predictions were experimentally borne out over the entire range of solute loadings.

Yet another approach to the problem of drug release from a nonerodible solution—suspension matrix has been developed by Lee (11). This approach involves the application of a refined heat balance integral method to the moving boundary problem first solved by Higuchi and then more rigorously addressed by Paul and McSpadden. As before, Lee assumed that diffusion occurs from a planar surface, and that a sink condition existed at the matrix—receiver interface. The result of Lee's derivation is

$$M_t = \frac{1 + H}{\sqrt{3H}} [C_s \sqrt{D t}] \tag{13}$$

where H is defined as

$$H = 5\left(\frac{Q}{C_s}\right) - 4 + \sqrt{\left(\frac{Q}{C_s}\right)^2 - 1} \tag{14}$$

Like the Higuchi treatment, Eq. 13 is an approximate solution to the problem. Lee compared the predictions of Eq. 13 and the Higuchi

solution (Eq. 1) against the exact solution of Paul and McSpadden
(Eq. 8) over a wide range of Q/C_S values (11). The deviations of
Lee's solution from the exact solution were consistently one order
of magnitude smaller than the deviations obtained upon using
Higuchi's equation. Thus, it was concluded that the Lee equation
could be applied safely over a wider range of Q/C_S ratios.

The models discussed thus far have described drug release from
topical formulations directly into a sink. Although these models are
useful in terms of their ability to characterize the diffusional behavior of a drug within a formulation, they do not describe the re-
alistic situation in which a drug diffuses from the vehicle, partitions
into, and subsequently diffuses across a resistant membrane (i.e.,
the skin). To treat situations such as these, Paul and McSpadden
also derived an exact solution for drug release from a suspension
matrix when a finite mass transfer resistance is encountered between the edge of the releasing matrix and the sink (10). The following equations to describe this case were put forth:

$$M_t = \frac{2 C_s}{erf(\beta)} \sqrt{\frac{D t}{\pi}} - Q\left(\frac{D}{\omega}\right) \tag{15}$$

and

$$M_t = \frac{2 C_s}{erf(\beta)} \sqrt{\frac{D}{\pi}} (\sqrt{t} - \sqrt{t_0}) \tag{16}$$

where

$$\sqrt{t_0} = \frac{1}{\omega}\left(\frac{Q}{C_s}\right) \sqrt{\frac{D}{2\left(\frac{Q}{C_s} - \frac{1}{2}\right)}} \tag{17}$$

Here, ω is the mass transfer coefficient, and $erf(\beta)$ takes the identical form as in Eq. 9. Notably, Eqs. 15 and 16 are asymptotic relationships that are only applicable at long time. The term $\sqrt{t_0}$ is
the finite intercept on the $\sqrt{t}$ axis.

Roseman and Higuchi (12) also developed equations for the situation in which finite mass transfer resistance is present; here, in
the form of a hydrodynamic boundary layer. In this model, release
from a suspension matrix depends upon two distinct processes.
The first of these involves an initial zero-order phase of release in
which the drug partitions from the matrix into the boundary layer.
During this period, diffusion through the boundary layer determines the overall release rate. The duration of this period depends
mostly upon the drug's external phase/matrix partition coefficient

as diffusion in the liquid film at the surface of the matrix is pre-
dictably fast and should be reasonably similar from liquid to liquid.
In a more general case in which the film is any type of material,
the duration would be comparably dependent on the permeant's diffusivity through it. Regardless and eventually, a transition period
is reached at which the matrix gradually assumes control of the
process, ultimately with the release rate being proportional to the
square root of time. In this model, it is assumed that $Q >> C_s$,
where Q is once again the total drug concentration and C_s is the
drug's solubility in the matrix. The expression for the quantity of
drug released within an area A is

$$M_t = Q A h_m \qquad\qquad [18]$$

where h_m is the thickness of the zone of depletion within the matrix.
This quantity is related to time through the following expression:

$$h_m{}^2 + 2 \frac{D_m h_m}{\omega K} = 2 \frac{D_m C_s t}{Q} \qquad\qquad [19]$$

In this equation, D_m is the drug's diffusion coefficient in the matrix,
ω is the mass transfer coefficient, and K is the external phase/matrix
partition coefficient. (In the original paper, Roseman and Higuchi
attributed the mass transfer resistance to an aqueous boundary layer
and thus assumed ω to be D_{bl}/h_{bl}. Clearly, another value could be
given to this resistance, justifying the generalizations made in the
text.) Two limiting cases were presented. During the initial, usually
fleeting, period the receding boundary within the matrix, characterized by h_m, is sufficiently thin that

$$2 \frac{D_m}{K \omega} >> h_m$$

In this circumstance the amount of drug released is

$$M_t = A C_s K \omega t \qquad\qquad [20]$$

This equation discloses that the release of the drug is directly proportional to the drug's solubility in the matrix and the mass transfer coefficient. Zero-order release is predicted, as the excess of
solid drug ensures the maintenance of a saturated solution at the
releasing surface. When differentiated for time, an expression for
the instantaneous rate of release is obtained.

$$\frac{dM_t}{dt} = A K C_s \omega \qquad [21]$$

However, as time elapses, the receding boundary thickens, with a commensurate increase in its resistance. Eventually, the following condition is met:

$$h_m >> \frac{2 D_m}{K \omega}$$

and, as a result, Eq. 19 simplifies to

$$M_t = A \sqrt{2 Q D_m C_s t} \qquad [22]$$

which will be recognized as the Higuchi equation presented earlier (Eq. 3). The shift to square root of time behavior indicates the release kinetics have come under matrix control. Although Roseman and Higuchi applied these equations to a situation in which a drug was suspended in a silicone rubber matrix and released into an aqueous sink, it has been pointed out (4) that they could conceivably be applied to other situations, such as drug release from an ointment across a dialysis membrane or weeping wound, the latter as might occur when a topical drug is spread over badly damaged skin.

An elaborate solution to the problem of release of a drug from a suspension ointment through a resistant membrane has been presented by Ayres and Lindstrom (13). In their model, terms are even included to account for a finite rate of dissolution of any suspended drug particles. Mathematical solutions for the possible situations were derived with the aid of Laplace transformation techniques. The arrived at equations describe the loss of a drug from an ointment into the bloodstream as a function of time under many scenarios. As such, the Ayres and Lindstrom equations represent the most rigorous theoretical characterization of drug delivery from thin-film applications into the skin yet attempted. Unfortunately and unavoidably, the resulting equations are unwieldy. When interpreted by example, one does get an intuitive feel for the release and permeation processes. However, numerical analysis is required to implement this method when actually analyzing data and, at least up to the present time, there are no real data precise enough to take advantage of the power of the theory.

Guy and Hadgraft (14) have also evaluated the theoretical release from ointments, particularly focusing on delivery of the drugs as a function of the thickness of the applied film. They provide

theoretical curves in which optimum dosage regimens may be
achieved. Another treatment, admittedly a much more approximate
solution to the problem of drug permeation of the skin from a suspension ointment, was presented by Flynn et al. (15). This solution is an adaptation of a standard solution to Fick's second law for
diffusion through a resistant membrane under conditions such that
drug is maintained at a high, steady concentration on one side of
the membrane and diffuses into a sink that is maintained at the
other side. These seem to be reasonable assertions covering many
real instances, as the skin's stratum corneum is a thin, but highly
resistant, membrane and the body of the patient to which the topical system is applied is overwhelmingly large relative to the volume
of the applied film. The equation in terms of the fraction of total
drug initially present from a system in which the drug is finely
suspended is

$$
\frac{M_t}{M_\infty} = \left(\frac{KC_s}{Qh_v}\right)\left[\frac{D_{sc}t}{h_{sc}} - \frac{h_{sc}}{6} - \frac{2h_{sc}}{\pi^2} \sum_{i=1}^{\infty} \frac{(-1)^n}{n^2} \right.
$$
$$
\left. \times \exp\left(\frac{-D_{sc}n^2\pi^2 t}{h_{sc}^2}\right)\right]
\tag{23}
$$

where K is the stratum corneum/vehicle partition coefficient and Q
is the initial loading of drug in the vehicle. Other terms are as
defined in previous derivations, with the subscripts sc and v re-
ferring to the stratum corneum and vehicle, respectively. An im-
plicit assumption is that the drug remains diffusionally well mixed
within the vehicle phase, even as it is permeating through the skin,
because the later process is slow. Equation 23 correctly indicates
that the fractional drug delivery through the skin is a function of
the reciprocal of the product of Q and h_v, which gives the total
drug content in the application (M_∞). It also suggests that as long
as one has a suspension, and providing the vehicle has no influence
on the skin membrane, the release rates from different vehicles will
be equal, because the product of K and C_s will be constant (15).

The situations discussed to this point have mostly dealt with
drug release from topical formulations in which the drug is present
as a suspension. For such situations, as long as the skin acts as
a highly resistant membrane and there is residual suspended drug,
an invariant drug concentration (thermodynamic activity) within the
formulation is expected. Drug diffusion through the skin from a
topical solution can be expected to be more complex, as the concentration (activity) of the drug decreases continuously irrespective of

whether the solution vehicle is functionally well mixed or is marked with steep gradients. It must be realized that the typical application thickness of a topical product is on the order of 20 μm (13). Diffusion of even a small amount of drug from such a thin application results in a rapidly decreasing drug concentration in the donor phase, resulting in non—steady-state kinetics. There is scant information in the literature about topical drug delivery under these circumstances. Flynn et al. (15) have deduced a solution to the problem of diffusion from a solution ointment through a resistant membrane into a sink. The boundary conditions are as follows:

$$C = 0, \quad x = 0$$

$$K = \frac{C_{sc}}{C_v}, \quad x = h_{sc}$$

$$\frac{C_v}{\partial t} = D_{sc}\left[\frac{\partial C_{sc}}{\partial x}\right], \quad x = h_{sc}$$

The initial conditions to the problem are

$$C = C_0, \quad h_v > x > h_{sc}$$

$$C = 0, \quad h_{sc} > x > 0$$

where $x = 0$ is defined as the membrane-sink interface, h_v and h_{sc} are the thicknesses of the vehicle and stratum corneum (membrane), respectively, and D_{sc} is the drug's diffusivity within the stratum corneum. It is assumed that the drug is diffusionally well mixed within the vehicle. The following concentration—distance—time relationship was deduced:

$$\frac{M_t}{M_\infty} = \sum_{n=1}^{\infty}\left[\frac{2K}{h_v}\right] \frac{\left[1 - \exp^{(-D_{sc}\beta_n^2 t)}\right]}{\cos\beta_n h_{sc}\left[h_{sc}\left\{\beta_n^2 + \left[\frac{K}{h_v}\right]^2\right\} + \frac{K}{h_v}\right]} \quad [24]$$

where β_n are the roots of $\beta_n h_{sc}\tan(\beta_n h_{sc}) = Kh_{sc}/h_v$. This equation is an adaptation of the solution to an analogous heat transfer problem (16). Computer simulations have been presented that show the effects of varying the parameters in the equation (15). The ratio K/h_{sc} was shown to affect fractional drug release with time.

Increasing K, with all other parameters kept constant, increases the absolute rate of delivery, whereas increasing the thickness of application slows the *relative* clearance of the drug from the film. (Relative clearance is in terms of the fraction of the total drug present. This should not be confused with the absolute rate of release, which is always equal to, if not greater than, that of a thinner film.) Experimental verification of some of the basic dependencies has been provided by Addicks and co-workers (17).

III.   APPLICATIONS OF THEORY: IN VITRO
   RELEASE STUDIES

A.   Experimental Designs

According to Gemmell and Morrison (18), ...*in vitro methods may be of limited predictive value but they are the means of assessing the ability of a vehicle or base to liberate medicament under the conditions of the test.* This statement very accurately describes the usefulness of in vitro release studies. In vitro studies have invariably involved release from applications that are much thicker than those seen clinically. Additionally, the chemical nature of the receptor phase used in such studies often bears no resemblance to the skin, and is usually configured as a thick slab. Overall, the conditions are such that the release kinetics reflect features of diffusion from a semi-infinite medium into a semi-infinite medium. If the receptor phase is rate-controlling, and if this phase bears little chemical resemblance to the skin, then it is foolishly unproductive to make extrapolations to in vivo situations (4).

Among the first experiments performed on the release of drugs from ointments were those developed 60 years ago by Reddish (19, 20). The procedure involved the measurement of the antiseptic properties of ointments containing various drugs. The experiments were carried out by applying ointment to agar that had been inoculated with *Staphylococcus aureus*. After incubation, the agar plates were assessed to determine the presence of a zone of inhibition. Thus, a crude measure of the ointment's ability to liberate a drug was achieved. A variation of this method was introduced by Clark and Davies (21) in which the minimum time required for an ointment to cause inhibition of bacterial growth was used as an index of ointment efficacy.

The practice of gauging the bioavailability of a drug from an ointment from the drug's ability to diffuse from the ointment into agar was extended by Lockie and Sprowls (22). The drugs used in their experiments were sulfathiazole and iodine, both of which can be detected colorimetrically. The procedure consisted of placing a tube filled with agar (containing an appropriate colorimetric indicator) in contact with a tube filled with a drug-containing ointment. The diffusion of the drug (as evidenced by a color change in the agar) was monitored, and the results were plotted as diffusion distance versus time. This method was later used by Billups and Sager (23) to measure the antiseptic properties of various ointments. In these experiments, diffusion of drugs from ointments into bacterially inoculated agar was assessed by the measurement of the length of a resulting zone of inhibition.

The method that has been used most ·frequently to assess the release of a drug from a vehicle involves the release of the drug from the vehicle directly into a stirred solvent. Most often the receptor solvent is present in a volume that allows sink conditions to be approximated throughout the course of an experiment. In this configuration, the vehicle slab is in direct contact with the receptor phase and, therefore, a solvent is chosen that does not dissolve the vehicle. This system was introduced by Poulsen et al. (24) to study the release of fluocinolone acetonide and its acetate ester from various propylene glycol—water gels into isopropyl myristate. With use of similar experimental designs, this group of researchers have systematically investigated the release of other topical steroids from various gelled solvent systems (25,26). Rank-order correlation between in vivo physiological responses and in vitro release for these drugs was shown. The choice of isopropyl myristate as a receptor phase has been given additional credibility through experiments by Scheuplein (27), who showed that topical corticosteroids are transported through the skin by way of a lipoidal pathway. Other investigators have utilized systems in which the vehicle is in direct contact with an aqueous receptor phase (28,29). In an interesting modification of Poulsen's method, Busse et al. (30) devised a three-layer system that consisted of a vehicle phase, an alcohol—water phase, and a chloroform phase. The alcohol—water phase was intended to represent the skin, and the chloroform phase acted as a sink. The release of betamethasone 17-valerate from two different ointment systems was studied with this experimental configuration.

When performing studies involving release from a topical formulation into a solvent receptor phase, it is preferable to reduce the probability that cross-contamination of the two phases will occur. Jurist (31) introduced a system for the study of the release of water-soluble materials from ointment bases. He described a dialysis cell method for measuring the ion-exchange capacity of a resin incorporated into an ointment base. The method involves the measurement of ion exchange through a semipermeable membrane. Mutimer et al. utilized this system for measuring the drug-release characteristics of several ointment bases (32). In their study, an ion-exchange resin was incorporated into each base, whereupon the formulation's ability to release drugs was judged by its ability to deliver hydrogen ions to an alkaline solution.

On occasion, a dialysis membrane has been used to separate the donor from the receiver phase. Wood and associates (33) introduced a diffusion cell for the study of the transport of a drug within a vehicle. The cell consisted of donor and receptor compartment separated by a cellophane membrane. The donor phase contained drug incorporated into the vehicle components, whereas the receptor phase contained only the components of the vehicle. Both phases were stirred, and the cell was kept at a constant temperature. The major drawback inherent in this experimental configuration was that it allowed for a one-point analysis only, which did not allow the extraction of a diffusion coefficient from the data. A refined version of this method was used by Howze and Billups (34). As in the previous method, a cellophane membrane was used to separate the donor and receiver phases. However, in this method, the receiver compartment was filled with water, and samples were withdrawn at specified time intervals. Thus, plots could be generated showing the amount of drug released versus time. This procedure has been used frequently by others as a means of assessing the relative abilities of ointment bases to release drugs (15,32,35−38). A common feature of the dialysis systems that have been used is that the aqueous phase is usually stirred. The stirring effectively sets a limit on the resistance of the watery regions within and adjacent to the membrane. As pointed out by Poulsen and Flynn (4), the effect of this hydrodynamic resistance is the creation of an initial zero-order release situation, the duration of which is a function of the physicochemical properties of the membrane and ointment, the thickness of the membrane, and the rate of stirring. The initial zero-order release phase is governed by the drug's (vehicle/water) partition coefficient, because the drug's release rate is dependent upon the steepness of the concentration gradient within the aqueous layers associated with the membrane during this time. As diffusion proceeds, a point is reached at which release of the drug becomes proportional with the square root of time.

Although dialysis membranes composed of cellulose esters have generally been chosen for release studies, some investigators have used other membranes. Bottari et al. (39,40) and Di Colo et al. (41) used silicone rubber membranes in release studies involving suspension gels and solution ointments. Similar studies have been described by others (42−44).

A novel technique was employed by Tanaka et al. (45) in which silicone rubber sheeting was used as the sole receptor phase for the release of hydrocortisone butyrate propionate from various creams and gels. In this study, the effect of the evaporation of vehicle components was investigated. When using a nonporous, nonpolar membrane of this type, one must be aware of the relative diffusional resistance offered by the membrane and be prepared to quantitatively account for it. This points to a fundamental flaw in many of the studies that have employed membranes, namely that the resistance offered by the membrane is ignored when analyzing the results. Although the assumption that this resistance is negligible may generally be safe when the formulation is applied as a very thick layer, nonetheless, without specifying values for K, D_m, and h_m, it does not seem possible that the experimental circumstances under which the assumption is valid can be assured.

Although the aforementioned experimental designs have been used frequently, there are several innate drawbacks in the designs that should cause one to exercise caution when attempting to interpret results (4). Although release of a drug from a vehicle occurs from thin vehicle films clinically, the vehicle is presented to the receiver as a relatively thick slab in most in vitro experiments. Release from such thick slabs is expected to assume an overall square root of time dependency, which might not occur if the vehicle were present as a thin film. This problem may be present in any type of release study involving a thick ointment application, regardless of whether or not a membrane is used. Another problem inherent in these studies is that experiments are usually conducted under closed conditions, which prevents the occurrence of compositional changes in the vehicle that might otherwise be manifest if the formulations were to be exposed to the atmosphere. In addition, the absence of a membrane could lead to solubilization of vehicle components by the solvent present in the receptor phase. The choice of a receptor phase itself creates problems, particularly if the phase is purported to be representative of the skin. For instance if one is studying the release characteristics of a steroid, isopropyl myristate might, in fact, be an appropriate choice for a receptor phase. However, for a small polar molecule, a phase with some capacity to dissolve the drug, perhaps water itself, might be a better choice.

B. Measurement of Diffusion Coefficients

Perhaps the first demonstration of the applicability of the theory concerning release of drugs from solution ointments was reported by Higuchi et al. (7). Using data that had been previously generated by Patel and co-workers (46), W. Higuchi was able to successfully apply a simplified version of the theoretical equation derived by T. Higuchi (Eqs. 5 and 6) to show that the kinetics of drug release from ointments do follow a square root of time pattern. The release experiments involved the liberation of radiolabeled iodine from hydrophilic ointments into a stirred aqueous reservoir. Visking membranes were used to separate the two phases, and it was assumed that the membrane contributed negligibly to the release resistance. Higuchi selected a diffusion coefficient (D) that yielded a best-fit curve for each set of data using the following modification of Eq. 6:

$$R = 200 \sqrt{\frac{Dt}{\pi h^2}} \qquad\qquad [25]$$

In this equation, R is the percentage of drug released, and h is the thickness of the ointment slab. The application of these theoretical equations to experimental data as a means of extracting diffusion coefficients from the data was significantly advanced by Spang-Brunner and Speiser (37). With the use of a dialysis cell method employing a cellulose ester membrane, the diffusion coefficients of resorcinol in various hydrogels were found to be on the order of 2.3 to 4.1 × 10^{-6} cm^2s^{-1}. Diffusion from ointments was found to be 50 to 300 times slower than diffusion from hydrogels. In yet another study, Koizumi and Higuchi (47) applied Eq. 25 to the measurement of the diffusion coefficient of pyridine from water—oil emulsions.

Subsets of the theoretical equations have been employed by Bottari et al. (40) and Di Colo et al. (39,41) in studies involving the release of drugs from gels and ointments across silicone rubber membranes. While investigating the release of salicylic acid as a function of concentration from emulsions (40), Eq. 6 was used to calculate diffusion coefficients in the emulsions. By plotting release rates against drug concentration for each ointment, the authors attempted to determine drug solubility by noting the concentration at which a deviation from linearity occurred. Reasonable agreement was claimed between these values and the independently obtained solubilities as determined visually. Through the use of Eq. 6, it was assumed that the membrane offered negligible resistance to diffusion of the salicylic acid. However, an appreciable lag time was observed in the mass versus square-root of time plots, thus indicating the possibility of membrane involvement in the release process during early times. Similar studies in which benzocaine was suspended in aqueous gels were performed by the same group (39). However, here, the membrane was assumed to offer resistance to diffusion, and Roseman and Higuchi's vehicle—boundary diffusion layer model was employed. In this instance, excellent agreement was found between the experimental data and theory.

Wahlgren et al. (48), in an interesting application of a standard solution to Fick's second law (7), calculated the diffusion coefficient of hydrocortisone in a lecithin—water lyotropic liquid crystalline solution. In this method, the diffusion coefficient was extracted from concentration/distance profiles. A glass cylinder containing a hydrocortisone–liquid crystal solution was brought into contact with a cylinder of plain liquid crystal. Diffusion was allowed to proceed for several days, and the cylinders were then separated. The liquid crystal was pressed from the cylinders, scraped off, and weighed. Each liquid crystal segment was assayed for hydrocortisone, and the data obtained was used to calculate the diffusion coefficient. The following equation was employed in the calculations:

$$D = \frac{x^2}{4t} \left[\frac{1}{\text{erf}^{-1}\left[1 - \frac{2C}{C_0} \right]} \right]^2 \tag{26}$$

where x is the distance along the diffusion axis, t is time, C is the concentration of drug at x, and C_0 is the initial concentration of drug in the gel.

IV. DIFFUSION STUDIES INVOLVING SKIN

A. In Vitro Studies

A frequently employed technique to test the relative permeability of topical drugs involves the in vitro use of excised skin mounted in diffusion chambers. Historically, the method used most often is one in which skin is mounted as a barrier between two stirred, fluid-filled chambers. The donor chamber is filled with a drug solution with samples withdrawn from the receptor as a function of time. Typically, a steady-state or quasi–steady-state condition arises in this diffusional situation, because generally there is an inappreciable alteration of the concentration differential existing between donor and receiver chambers over the course of the experiment. This method has been used frequently to systematically study mechanisms of skin permeability (49–52). Poulsen et al. utilized infinite dosing in the development of several topical formulations (25,26,53). One must continually remember, however, that this type of diffusion cell experiment has little in common with the delivery of drugs from topical applications and is used to get mass transfer coefficients for membranes.

To more adequately depict true clinical situations, so-called finite-dose diffusion cells have been used (54,55–57). A membrane, typically skin, is mounted vertically in the diffusion cell, a small amount of formulation is applied to the membrane, and samples are withdrawn from a stirred receiver compartment beneath the membrane (skin) as a function of time. An advantage of this method is that, in principle, the amounts of formulated drug applied can reasonably reflect the usage situation. In addition, the cell can be left open to the ambient conditions of the laboratory for the formulation to undergo the same compositional changes as occur clinically. Another advantage to the finite dose method is that phenomena of topical drug delivery, such as skin penetration enhancement, can be realistically evaluated.

To test the usefulness of the finite-dose diffusion cell, Franz (54,55) studied the in vitro diffusion of various compounds from thin solid deposits through human skin. Data obtained were compared with previously obtained in vivo data. Good agreement was reported between the two sets of data. Mollgaard and Hoelgaard, with the use of finite-dose diffusion cells, investigated the diffusion of estradiol from 21 different vehicles through excised human abdominal skin (58). It was concluded that, for the vehicles chosen, drug permeability was most influenced by the direct vehicle effect on the barrier. The degree to which the thermodynamic activity of the drug in the vehicle influenced the course of events was relatively minimal. One questionable aspect of this work was that the baseline diffusion rate of estradiol through the skin was measured from a solid drug deposit, a great departure from the situation in which the drug is in solution in the vehicle. This certainly affects the quantitative interpretations of the studies.

Finite-dose diffusion cells have often been used incorrectly to assess drug delivery, in that formulations have been applied so thickly that diffusion occurs from an infinite slab, changing the whole kinetics of the drug transport process. This basic flaw is evident in several studies (59,60—63).

B. Role of Thermodynamic Activity in Topical Drug Delivery

Important conceptual work concerning the role of topical vehicles in the delivery of drugs through the skin was pioneered by Poulsen, Ostrenga, and others (25,26,53,64). These studies focused on the role of the drug's thermodynamic activity in the overall diffusional flux through the skin. The first of these studies dealt with the effect that the drug's thermodynamic activity has on its release rate from the vehicle (24). The work was based on T. Higuchi's theoretical supposition that the steady-state diffusion of a drug dissolved in a vehicle through the skin can be represented by

$$\frac{dM_t}{dt} = A \ K \ Cv \left[\frac{D_{sc}}{h_{sc}} \right] \qquad\qquad [27]$$

and

$$\frac{dM_t}{dt} = A \, \frac{a_v}{\delta_s} \left[\frac{D_{sc}}{h_{sc}} \right] \qquad [28]$$

Where dM_t/dt is the rate of penetration, K is the stratum corneum/
vehicle partition coefficient, C_v is the drug's concentration in the
vehicle, A is the area of the application site, h_{sc} is the effective
thickness of the skin, D_{sc} is the effective stratum corneum diffusivity, a_v is the thermodynamic activity of the drug in the vehicle,
and δ_s is the effective activity coefficient of the drug in the skin
barrier. The latter equation indicates that only certain terms are
available for manipulation by the formulator to increase the rate of
diffusion of a drug. These are K, the partition coefficient, and
C_v, the drug's concentration in the vehicle. K is altered when the
composition of the vehicle is altered, and C_v can be adjusted up to
the level of solubility in the vehicle. Note that, for saturated solutions in different vehicles, K C_v remains constant, and this product
is constant irrespective of the vehicle differences. If one assumes
that no vehicle—skin interactions take place, and if the drug itself
does not alter the skin in any way, then δ_s can be regarded as a
constant. The drug's effective thermodynamic activity is proportional to the product of K and C_v, which, in fact, yields the drug's
concentration at the surface of the skin. Any change in this product leads to a change in the tissue concentration that establishes
the gradient across the skin.

   To test the foregoing concept, two topical steroids were incorporated into a series of propylene glycol—water gels, and drug release into isopropyl myristate was studied (24). Drug solubility determinations were made for each system, and a receptor—vehicle
phase partition coefficient was calculated. It was found that the
maximum drug release occurred from gels containing the minimal
amount of propylene glycol required to exactly dissolve all the drug.
Excess propylene glycol required to exactly dissolve all the drug.
Thus decreasing the thermodynamic activity of the drug and hence
its release rate from the vehicle, is fully consistent with the foregoing argument. At the other extreme, when insufficient propylene
glycol was available to completely dissolve the steroid, dissolution
and intragel diffusion became part of the rate-limiting factors in the
overall diffusion process over the multihour collection periods. The
concept of using the thermodynamic activity of a drug in its vehicle
as an indicator of its bioavailability from that vehicle was solidified
in a subsequent series of studies by the group (25,26,54). For both
fluocinonide and fluocinolone in propylene glycol—water gels, positive
correlations were seen when data were compared from in vitro release,
in vitro permeation, and in vivo vasoconstriction studies (26). The
latter directly relate to concentrations of the steroids that build up

in the local tissues beneath the applications. It was found that maximum bioavailability was obtained from the gel system in which just enough propylene glycol was present to completely solubilize the drug, a result totally consistent with the in vitro release studies. Similar results were obtained when fluocinonide was incorporated into various creams and ointments (54).

Along these same lines, in a study investigating the diffusion of tritiated water from a wide range of polar solvents, the absorption rate of water through human epidermis was proportional to the thermodynamic activity of water in a particular solvent (65). Barry et al. (18) correlated the thermodynamic activity of benzyl alcohol vapor with the diffusional flux of the vapor through human abdominal skin. These experiments were done in such a way that the vehicle was not in contact with the skin, thus ruling out the possibility of any direct vehicle—skin interactions. In an interesting practical application of the finite-dose method, Bronaugh et al. (60) studied the absorption of n-nitrosodiethanolamine through excised human skin. This compound is an impurity, believed to be a carcinogen, found in many cosmetic products. It was found that the permeability of n-nitrosodiethanolamine increased with an increase in the stratum corneum/vehicle partition coefficient of the agent. In another study, Turi et al. studied the permeation of diflorasone diacetate through hairless mouse skin from various vehicle formulations (59). It was observed that the permeation rate increased with the degree of solubilization up to a point, and then decreased. This observation is because, at maximum permeability, the vehicle exactly dissolves all of the drug in the formulation. Any further increase in drug solubilization results in a decrease in the partition coefficient (but no further increase in the drug's vehicle concentration). This decrease in the thermodynamic activity of the drug results in a decrease in its permeability. Sloan et al. (62,63; see Chap. 13) have compared the experimental flux of drugs through skin with theoretical predictions obtained through the use of solubility parameters. Studies of this type are significant in that they illustrate the ways in which theoretical concepts can be applied to the development of topical drug products. However, experimental conditions used in these studies generally fail to approximate the clinical situation.

C. In Vivo Studies

When all is considered, most topical products on the market have been developed using more clinical research strategies than benchtop ones. In vivo assay methods have played an important role in the development of many important topical products. Some of these studies have involved the use of animal models; however, the present discussion is limited to studies involving human subjects. The in vivo study of the penetration kinetics of drugs has often involved the detection of some local biological response that the drug elicits upon penetration into the skin. Examples of responses that have been utilized include vasodilation, vasoconstriction, keritinization, epidermal proliferation, and changes in blood pressure. In an early study that recognized the importance of standardizing the amount of drug placed onto the skin, Barrett et al. (52) studied the relative erythema produced by methyl nicotinate when incorporated into various vehicles. What is interesting about this study is that it involved the application of a discrete disk of vehicle of known thickness to the forearms of human volunteers, perhaps the first time that the importance of applying a standard amount (and thickness) of vehicle was recognized. Other in vivo techniques involve the analysis of blood or serum levels of drugs absorbed through the skin or the amounts of drugs excreted in the urine. Urinary analysis, in particular, often has been used by Feldman and Maibach to study percutaneous absorption in human subjects (66–70).

Perhaps the most well-refined of all biological assays used to develop and assess dermatological formulations are topical corticosteroid bioassays. In some instances, an inflammatory stimulus is applied to the skin, which is immediately followed by the administration of the steroid. The anti-inflammatory response is then graded. Using a method involving the application of croton oil and kerosene to induce pustules or blisters in humans, Kaidby and Kligman (71) obtained a rank order of corticosteroid activity that correlated well with the relative clinical efficacies of the drugs. Similar studies using irritants such as mustard oil and nitric acid (72), tetrahydrofurfuryl alcohol (73), and histamine (74) have been used to test corticosteroid activity. The fact that certain steroids have the ability to constrict local vasculature and turn the site of application pale has often been utilized to study both the relative potencies of steroids, as well as the ability of vehicles to deliver steroids topically. McKenzie and Stoughton (75) discovered that treating psoriasis with steroids under plastic wrap blanched the lesion as well as normal skin. This observation was used to develop a test to rank commercially available steroids in order of potency (76). This method was adopted by Sarany et al. (77) to measure the efficacy of several topical steroids from various vehicles. A method of application that had been previously employed by this group (78) was used in this study. Subsequently, Woodford and Barry (79), among others (16,26,53), have used the blanching test to compare topical steroid formulations.

REFERENCES

1. B. Barry, in *Dermatological Formulations*, Marcel Decker, New York, p. 145, 1983.
2. M. Katz, and B. Poulsen, in *Handbook of Experimental Pharmacology*, New Series, Vol. 28, Springer—Verlag, Heidelberg, 1971.
3. B. Idson, *J. Pharm. Sci.*, *64*:901, 1975.
4. G. Flynn, and B. Poulsen, in *Percutaneous Absorption: Mechanism—Methodology—Drug Delivery*, Marcel Dekker, New York, pp. 431—459, 1985.
5. T. Higuchi, *J. Pharm. Sci.*, *50*:874, 1961.
6. T. Higuchi, *J. Soc. Cosmet. Chem.*, *11*:85, 1960.
7. J. Crank, in *The Mathematics of Diffusion*, 2nd ed. Clarendon Press, Oxford, p. 48, 1975.
8. M. Jacobs, in *Diffusion Processes*, Springer—Verlag, New York, p. 44, 1935.
9. W. Higuchi, *J. Pharm. Sci.*, *51*:802, 1962.
10. D. Paul, and S. McSpadden, *J. Membr. Sci.*, *1*:33, 1976.
11. P. Lee, *J. Membr. Sci.*, *7*:255, 1980.
12. T. Roseman, and W. Higuchi, *J. Pharm. Sci.*, *51*:353, 1970.
13. J. Ayres, and T. Lindstrom, *J. Pharm. Sci.*, *66*:654, 1977.
14. R. Guy, and J. Hadgraft, *Int. J. Pharm.*, *6*:321, 1980.
15. G. Flynn, E. Topp, and G. Amidon, in *Topics in Pharmaceutical Science 1985.* Elsevier Science, New York, pp. 313—328, 1985.
16. J. Crank, in *The Mathematics of Diffusion*, 2nd ed. Clarendon Press, Oxford, pp. 44—68, 1975.
17. W. J. Addicks, G. L. Flynn, and N. Weiner, *Pharm. Res.*, 4: 337, 1987.
18. D. Gemmell, and J. Morrison, *J. Pharm. Pharmacol.* 9:641, 1957.
19. G. Flynn, and S. Yalkowsky, *J. Pharm. Sci.*, *61*:846, 1972.
20. G. Reddish, and H. Wales, *J. Am. Pharm. Assoc.*, *18*:576, 1929.
21. G. Clark, and G. Davies, *Br. J. Dermatol.*, *00*:521, 1949.
22. L. Lockie, and J. Sprowls, *J. Am. Pharm. Assoc.*, *38*:222, 1948.
23. N. Billups, and R. Sager, *Am. J. Pharm.*, *137*:57, 1965.
24. B. Poulsen, E. Young, V. Coquilla, and M. Katz, *J. Pharm. Sci.*, *57*:928, 1968.
25. J. Halblian, B. Poulsen, and K. Burdick, *Curr. Ther. Res.*, *22*:713, 1977.
26. J. Ostrenga, C. Steinmetz, and B. Poulsen, *J. Pharm. Sci.*, *60*:1175, 1971.

27. R. Schueplein, I. Blank, G. Brauner, and D. MacFarlane, *J. Invest. Dermatol.*, *52*:63, 1969.
28. R. Dempski, J. Portnoff, and A. Wase, *J. Pharm. Sci.*, *58*: 579, 1969.
29. Z. Chowhan, and R. Pritchard, *J. Pharm. Sci.*, *64*:754, 1975.
30. M. Busse, P. Hunt, K. Lees, and N. Maggs, *Br. J. Dermatol.*, *81*:103, 1969.
31. A. Jurist, *J. Invest. Dermatol.*, *20*:331, 1953.
32. M. Multimer, G. Riffkin, J. Hill, M. Glickman, and N. Gilman, *J. Am. Pharm. Assoc.*, *45*:212, 1955.
33. J. Wood, L. Rising, and W. Hall, *J. Pharm. Sci.*, *51*:668, 1962.
34. J. Howze, and N. Billups, *Am. J. Pharm.*, *138*:193, 1966.
35. G. Bramanti, G. Disaturo, G. Mazzi, P. Mura, and P. Papini, *Boll. Chim. Farm.*, *119*:738, 1980.
36. J. Ayres, and P. Lasker, *J. Pharm. Sci.*, *63*:1402, 1973.
37. B. Spang-Brunner, and P. Speiser, *J. Pharm. Pharmacol.*, *28*: 23, 1976.
38. P. York, and A. Suleh, *J. Pharm. Sci.*, *65*:493, 1976.
39. F. Bottari, G. Di Colo, E. Nampieri, M. Saettone, and M. Serfini, *J. Pharm. Sci.*, *66*:926, 1977.
40. F. Bottari, G. Di Colo, E. Nannipieri, M. Saettone, and M. Serafini, *J. Pharm. Sci.*, *63*:1779, 1974.
41. G. Di Colo, V. Carelli, B. Giannaccini, M. Serafini, and F. Bottari, *J. Pharm. Sci.*, *69*:387, 1980.
42. F. Broberg, and A. Brodin, *Acta Pharm. Suec.*, *19*:229, 1982.
43. A. Broden, and A. Nygvist-Mayer, *Acta Pharm. Suec.*, *19*: 267, 1982.
44. A. Kahil, S. Davis, and J. Hadgraft, *Pharm. Res.*, *3*:214, 1986.
45. S. Tanaka, Y. Takashima, H. Murayana, and S. Tsuchiya, *Int. J. Pharm.* *87*:29, 1985.
46. K. Patel, G. Banker, and H. DeKay, *J. Pharm. Sci.*, *50*:300, 1961.
47. T. Koizumi, and W. Higuchi, *J. Pharm. Sci.*, *57*:87, 1968.
48. S. Wahlgren, A. Lindstrom, and S. Friberg, *J. Pharm. Sci.*, *73*:1484, 1984.
49. C. Behl, G. Flynn, T. Kurihara, N. Harper, W. Smith, W. Higuchi, N. Ho, and C. Pierson, *J. Invest. Dermatol.*, *75*: 346, 1980.
50. H. Durrheim, G. Flynn, W. Higuchi, and C. Behl, *J. Pharm. Sci.*, *69*:781, 1980.
51. G. Flynn, H. Durrheim, and W. Higuchi, *J. Pharm. Sci.*, *70*: 52, 1981.
52. R. Scheuplein, and I. Blank, *J. Invest. Dermatol.*, *60*:286, 1973.

53. J. Ostrenga, J. Haleblain, B. Poulsen, B. Ferrell, N. Mueller, and S. Shastri, *J. Invest. Dermatol.*, *56*:392, 1971.
54. T. Franz, *Curr. Probl. Dermatol.*, *7*:58, 1978.
55. T. Franz, *J. Invest. Dermatol.*, *64*:190, 1975.
56. R. Bronaugh, and R. Stewart, *J. Pharm. Sci.*, *74*:64, 1985.
57. G. Hawkins, and W. Reifenrath, *J. Pharm. Sci.*, *75*:378, 1986.
58. B. Mollgaard, and A. Hoelgaard, *Int. J. Pharm.*, *15*:185, 1983.
59. J. Turi, D. Danielson, and J. Woltersom, *J. Pharm. Sci.*, *68*: 275, 1979.
60. R. Bronaugh, E. Congdon, and R. Schueplein, *J. Invest. Dermatol.*, *76*:94, 1981.
61. E. Cooper, *J. Pharm. Sci.* *73*:1153, 1983.
62. K. Sloan, K. Siver, and S. Koch, *J. Pharm. Sci.*, *75*:744, 1986.
63. K. Sloan, S. Koch, K. Siver, O. Flowers, and F. Flowers, *J. Invest. Derm.*, *87*:244, 1986.
64. T. Malone, J. Haleblian, B. Poulsen, and K. Burdick, *Br. J. Dermatol.*, *90*:187, 1974.
65. P. Dugard, and R. Scott, *Int. J. Pharm.*, *28*:219, 1986.
66. R. Feldman, and H. Maibach, *Arch. Dermatol.*, *91*:661, 1965.
67. R. Feldman, and H. Maibach, *Arch. Dermatol.*, *94*:649, 1966.
68. R. Feldman, and H. Maibach, *J. Invest. Dermatol.*, *48*:181, 1967.
69. R. Feldman, and H. Maibach, *J. Invest. Dermatol.*, *50*:351, 1968.
70. R. Feldman, and H. Maibach, *J. Invest. Dermatol.*, *52*:891, 1969.
71. K. Kaidby, and A. Kligman, *J. Invest. Dermatol.*, *63*:292, 1974.
72. A. Scott, and F. Kalz, *J. Invest. Dermatol.*, *26*:361, 1956.
73. H. Gray, and R. Wolfe, *Arch. Dermatol.* *84*:18, 1961.
74. B. Reddy, and G. Singh, *Br. J. Dermatol.*, *94*:191, 1976.
75. A. McKenzie, and R. Stoughton, *Arch. Dermatol.*, *86*:608, 1962.
76. A. McKenzie, *Arch. Dermatol.*, *86*:611, 1962.
77. I. Sarkany, J. Hadgraft, G. Caron, and C. Barrett, *Br. J. Dermatol.*, *77*:569, 1965.
78. J. Barrett, J. Hadgraft, J. Sarkay, *J. Pharm. Pharmacol.*, *16*(suppl.):104T, 1964.
79. B. Barry, in *Dermatological Formulations: Percutaneous Absorption*, Marcel Dekker, New York, p. 242, 1983.

13

The Use of Solubility Parameters of Drug and Vehicle to Describe Skin Transport

KENNETH B. SLOAN *University of Florida, Gainesville, Florida*

I. INTRODUCTION

The use of solubility parameters of drugs and vehicles to describe the transport of drugs through skin is based on the efforts of Hildebrand (1) and others (2–4) to account, in a rational way, for solvent–solvent and solute–solvent interactions in the process of dissolution. The results of those efforts to develop a theoretical basis for solubility is represented in simplified form by Eq. 1 where X_i^v is the mole fraction solubility of drug (i) in the solvent or vehicle (v), ΔH_f is the heat of fusion of the drug at its melting point, T_m is the melting point of the drug, T is the temperature at which the solubility is being measured and γ_i^v is the activity coefficient of the drug in the vehicle, which can be calculated from Eq. 2. In Eq. 2, δ_i and δ_v are the solubility parameters of the drug and vehicle, respectively, ϕ_v is the fraction of the solution volume occupied by the vehicle, and V_i is the molar volume of the drug. The calculated mole fraction solubility of the drug, then, is composed of two components; one is a contribution from its ideal solubility, the other is a contribution from its activity coefficient to take into account deviations of the actual solubility from the ideal solubility. The evolution of this approach to predicting solubility strictly from theoretical considerations of the physicochemical properties of solute and solvent (drug and vehicle) has recently been reviewed by Martin and Mauger (5).

$$- \log X_i^v = (\Delta H_f / 2.3\ RT)[(T_m - T)/T_m] + \log \gamma_i^v \qquad [1]$$

$$\log \gamma_i^v = (V_i \phi_v^2 / 2.3\ RT)(\delta_i - \delta_v)^2 \tag{2}$$

Because transport through skin is a diffusion process that depends on a concentration gradient, the application of calculated solubilities, derived in part from δ, to predict the concentration gradients and, hence, diffusion is a logical progression. Fick's first law for diffusion (Eq. 3) contains terms for the concentrations of the drug in the membrane (C_i^{s1} and C_i^{s2}) that constitute the concentration gradient (Fig. 1). They are difficult to determine. Hence, Fick's first law is usually expanded into a form (Eq. 4) that contains contributions from more readily measured terms, where $J_i^{s,v}$ is the flux of drug (i) delivered from vehicle (v) through a membrane that, here, is skin (s), D_i^s is the diffusion coefficient of the drug in the skin, h_s is the thickness of the skin membrane, C_i^{s1} and C_i^{s2} are the concentrations of the drug in that layer of the skin next to the vehicle and receptor phase or plasma (p), respectively, C_i^v is the concentration of the drug in the vehicle, and $K_i^{s,v}$ is the distribution or partition coefficient of the drug between the vehicle and skin equal to C_i^{s1}/C_i^v. In proceeding from Eq. 3 to Eq. 4, $K_i^{s,v} C_i^v$ has been substituted for C_i^{s1} and C_i^{s2} has been assumed to approach zero, because sink conditions are maintained on the plasma side of the membrane, that is, $C_i^p = 0$ (see Fig. 1).

$$J_i^{s,v} = (D_i^s / h_s)(C_i^{s1} - C_i^{s2}) \tag{3}$$

$$J_i^{s,v} = (D_i^s / h_s) K_i^{s,v} C_i^v \tag{4}$$

The partition coefficient for the distribution of the drug between two phases (here, vehicle and skin) to form saturated, nonideal solutions is given by Eq. 5. Because the activity coefficients can be calculated from Eq. 2, the partition coefficient can be calculated by substituting the identities for γ_i^v and γ_i^s from Eq. 2 into Eq. 5 to give Eq. 6. If it is assumed that the drug is not very soluble in either the vehicle or the skin, ϕ_s^2 and ϕ_v^2 approach a value of 1 and Eq. 6 reduces to Eq. 7.

$$\log K_i^{s,v} = \log \gamma_i^v / \gamma_i^s \tag{5}$$

$$\log K_i^{s,v} = [(\delta_i - \delta_v)^2 V_i \phi_v^2 / 2.3\ RT] - [(\delta_i - \delta_s)^2 V_i \phi_s^2 / 2.3\ RT$$

$$\tag{6}$$

$$= [(\delta_i - \delta_v)^2 - (\delta_i - \delta_s)^2] V_i / 2.3\, RT \qquad [7]$$

II. GENERAL APPLICATIONS OF REGULAR SOLUTION THEORY TO PARTITIONING PROCESSES

Various forms of Eq. 6 and of relationships between δ values and bioactivity have been used successfully to describe a number of biological processes besides percutaneous absorption. Ferguson (6) showed that the toxic concentration of a chemical was a function of the intrinsic toxicity of the chemical and its distribution into a target phase. Here, the distribution is given by the difference in the partial molal free energies of the chemical (phase 1) in its standard state in two phases (phases 2 and 3; Eq. 8), where, for example, the molal free energy of the chemical in phase 2 is given by 2.3 RT log $\gamma_{i,2}$ (7).

$$\log K = (\overline{F}_3^0 - \overline{F}_2^0)/2.3\, RT \qquad [8]$$

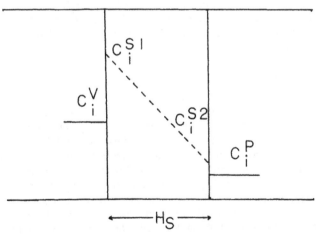

Figure 13.1 Concentration gradient of a solute (i) across a membrane of thickness h_s where C_i^v is the concentration of the solute in the vehicle in the donor phase and C_i^p is the concentration of the solute in the plasma in the receptor phase. C_i^{s1} and C_i^{s2} are the concentrations of the solute in the first and last layers of the membrane, respectively.

Khalil and Martin (8) used solubility parameters in their analysis of the transfer of a solute from one phase to another in partitioning experiments. They measured the rate of transfer of salicylic acid from a pH 2.0 aqueous solution (representing the stomach), through nonpolar immiscible liquids exhibiting solubility parameter values of from 7.1 to 10.7 $(cal/cm^3)^{1/2}$, and into a pH 7.4 aqueous solution (representing plasma). The model was set up to analyze absorption from the gastrointestinal tract, but a similar system could have been used to model transfer through other membranes, such as skin. The results for salicylic acid in this model showed that the closer the δ value of the nonpolar liquid was to that of salicylic acid (δ_i = 10.8), the faster the rate of disappearance of salicylic acid was from the pH 2.0 solution and the slower its rate of appearance in the pH 7.4 phase. It was also observed that the optimum transfer of salicylic acid to the pH 7.4 solution occurred at a δ value of the nonpolar liquid of about 9.7 $(cal/cm^3)^{1/2}$ to balance the rate of transfer from the pH 2 solution into the nonpolar phase with its rate of release into the pH 7.4 solution.

Ostrenga (9) suggested that molar attraction constants (F values) for chemical structures, derived from solubility parameters (δ) where δ = F/V and where V = molar volume, could be used to correlate bioactivity with chemical structure. Because the F values were readily available for most organic functional groups and were additive on a constitutive basis, they could be calculated for most chemical structures without resorting to additional experimentation. This was in contrast with the Hansch approach (10) which used π values derived from the experimental difference in logarithms of the partition coefficients of a model chemical (RH) and its substituted derivative (RX; Eq. 9).

$$\pi_x = \log K_{(RX)} - \log K_{(RH)} \qquad [9]$$

Subsequently, Davis (11) deduced Eq. 6 from the Hildebrand–Scatchard equations for the excess free energy of a regular solution and then used this solubility parameter-based partition coefficient to show that π and F were related to each other through Eqs. 6 and 9. Thus, π values could be calculated for various functional groups (X) from their respective F_x and V_x values and the solubility parameter values for the aqueous (δ_a) and organic phases (δ_o) (Eq. 10). Davis found that the agreement between the experimentally determined π and the π calculated from Eq. 10 was not particularly good in many cases and argued against using calculated π or F values to correlate chemical structure with bioactivity.

$$2.3\ RT\ \pi_x = V_x(\delta_a^2 - \delta_o^2) + 2F_x(\delta_o - \delta_a) \qquad [10]$$

Cohen and co-workers (12,13) have used regular solution theory and a more exact form of Eq. 6 (14) to explain how structurally similar molecules could be either depressors (anesthetics) or stimulants (convulsants) of the nervous system. Their explanation of this difference was based on an extension of the concept of Mullins (15) that linked the biological properties of a chemical with its biophase of action. Thus, drug specificity is a consequence of its preferential solubility in a subregion of the membrane exhibiting a bulk solubility parameter similar to that of the drug. In this case, low δ drugs cause excitation, whereas high δ drugs cause depression.

Martin and co-workers (16) examined the effect of solvent polarity and its ability to solvate theophylline on the in vivo absorption of theophylline from mixtures of polyethylene glycol 400 and water into the rat intestinal membrane. Their results were qualitatively similar to those observed for salicylic acid in a model system (8). The closer the δ value of the polyethylene glycol 400—water mixture was to the δ value of theophylline the slower the rate of disappearance of theophylline from the solvent mixture into the rat intestine. The authors also used Eq. 6 to calculate a δ value for the intestinal membrane of 12.6 $(cal/cm^3)^{1/2}$ from their data.

Finally, Bustamante and Selles (17) have shown that the binding of drugs to plasma protein can be analyzed by Eq. 11 in which δ_E and δ_D are the solubility parameters for albumin (or the drug-binding site in albumin) and drug, respectively, where the activity coefficient (α) is the ratio of maximum binding (B_D^M), to the real binding (B_D) by analogy to the ideal and actual solubility from the Hildebrand equation. Data for the binding of sulfonamides, penicillins, phenothiazines, and barbituric acids to human serum albumin were found to fit Eq. 11, which describes a saturation-type process, better than an Eq. like 6, which describes a distribution-type process. It was also observed that maximum binding to serum albumin occurred when the solubility parameters of the drugs were similar to that of the plasma protein. For the sulfonamides, δ_E was calculated to be 12.33 $(cal/cm^3)^{1/2}$ which closely corresponds to the average solubility parameter of the seven-amino acids sequence in human serum albumin thought to be the primary binding site for sulfonamides.

$$\log \alpha = \log(B_D^M /B_D) = V_D \phi_E^2 /2.3\,RT(\delta_E - \delta_D)^2 \qquad [11]$$

III. APPLICATION OF REGULAR SOLUTION THEORY TO PERCUTANEOUS ABSORPTION PROCESSES

Two groups initiated the application of regular solution theory to the partitioning process in percutaneous absorption. In 1984 Cohen

and co-workers described the absorption of alkanoic acids through
skin in terms of the solubility parameters of the acids (18). Shortly after the publication of the 1984 paper, Sloan, Sherertz, and co-
workers published a series of eight articles (19–26), starting in
1986, that used regular solution theory to explain the effects of vehicles and the effect of changes in the physicochemical properties
of drugs caused by their transient chemical modification (prodrugs)
on the delivery of drugs through skin. The articles by Sloan,
Sherertz, and co-workers will be discussed first, then the initial article by Cohen and co-workers, along with two very recent articles
that have appeared from that group, will be discussed.

The experimental designs used in the work reported by Sloan
et al. (19–26) were somewhat different from the previous designs
used in similar experiments (18). First, only saturated solutions of
drug–vehicle were applied to the membranes in the diffusion cells
as the initial part of each experiment. Second, after the initial part
of each experiment a "second" application of a standard solute–vehicle was made onto the membranes.

This second application part of each experiment enabled the investigators to determine whether or not the membranes were damaged
by the initial treatment of drug–vehicle by comparing the rate of
delivery (flux) of the standard solute through skin after an initial
application (pretreatment) of drug–vehicle, with the rate of delivery
of the standard solute without pretreatment. The higher the ratio
between the two fluxes, the greater the degree of damage to the
skin and the greater the degree of deviations of the experimental
flux of drug–vehicle from the theoretical flux predicted by regular
solution theory. In this way, it was possible to separate the thermodynamic effects (1) of the drug-vehicle interaction or (2) of the
physicochemical properties of the prodrug on the partitioning process from the chemical or physical effects of the vehicles themselves
on the integrity of the membranes.

Saturated solutions of drug–vehicle were used to ensure that
the drugs were being tested at their maximum activity. In addition,
only at saturation, on a mole fraction concentration basis, does $\alpha_i^V = \alpha_i^S$, where α_i^V and α_i^S are the activities of the drug (i) in vehicle
(v) and skin (s), respectively. Hence, $K_i^{s,v}$ from Eq. 5 can be
calculated using Eq. 6, but only if saturated solutions of the solute
in the two phases of interest are being evaluated.

The effect of vehicles on flux of a drug through skin can be
correlated with regular solution theory through the relationship between permeability coefficient ($P_i^{s,v}$) and partition coefficient
($K_i^{s,v}$). Flux ($J_i^{s,v}$) is equal to the product of $P_i^{s,v}$ and C_i^v, so
from Eq. 4, $P_i^{s,v}$ is equal to the product of the partition coefficient
and the diffusion coefficient (D_i^s) divided by the thickness of the

membrane (h_S) (Eq. 12). If $D_i^{\bar{S}}$ is independent of the vehicle used and h_S does not change much within the limits of biological variability, then D_i^S/h_S is a constant and Eq. 12 reduces to Eq. 13 (19). It has already been shown that $K_i^{S,V}$ can be calculated from Eq. 6 or 7. Thus, to determine whether or not variations in the rates of delivery of a drug from a series of vehicles can be explained by regular solution theory, it is merely necessary to plot the log calculated (theoretical) $K_i^{S,V}$ against the solubility parameters of the vehicles (δ_V), and determine if the plot of log experimental $P_i^{S,V}$ versus δ_V approximates the shape of the log theoretical $K_i^{S,V}$ versus δ_V curve and is separated from it by a constant amount (D_i^S/h_S).

$$P_i^{S,V} = K_i^{S,V}(D_i^S/h_s) \qquad [12]$$

$$P_i^{S,V} = K_i^{S,V} \cdot \text{constant} \qquad [13]$$

There are two reasons why it is useful to be able to calculate these theoretical partition coefficients based on regular solution theory to understand the dermal delivery process. First, from the foregoing discussion, it is obvious that if it is possible to predict $K_i^{S,V}$, it is possible to predict permeability coefficients and, hence, flux. The ability to predict the effect of changes in vehicles on the solubility of a drug in a formulation and, hence, its flux is essential for the rational design of topical formulations. Second, it is virtually impossible to obtain an accurate and appropriate experimental partition coefficient for a solute distributing between a vehicle and skin. The concentration of a drug in a vehicle is easily determined, but the concentration in the skin is difficult to quantitate because of the different degrees of changes in hydration and of changes in other physicochemical properties of the skin that would result if samples of skin were allowed to equilibrate with different vehicles for extended periods (27). Thus, a method for determining theoretical $K_i^{S,V}$ would be advantageous because there are serious difficulties in obtaining experimental $K_i^{S,V}$.

Conversely, it is possible to calculate theoretical partition coefficients using other methods. The most popular method uses fragment constants for the various functional groups in a drug to calculate the drug's distribution between water and a water-immiscible solvent that serves as a model for the biological membrane (29,30). Usually, the water-immiscible solvent is octanol, but many other water-immiscible solvents have also been used. The fragment constants for a two-phase partitioning system are derived from experimentally determined partition coefficients for model chemicals in the same two-phase system. However, by using the available data, this approach is useful only if water is the vehicle being considered. In addition, many, if not most, of the vehicle components for which it would be useful to calculate theoretical partition coefficients are either miscible with, or at least partially soluble in, both octanol and water. These include, for example, solvents such as propanol, dimethyl sulfoxide, dimethylformamide, propylene glycol and formamide. Other solvents, such as isopropyl myristate and oleic acid, are soluble in octanol or other solvents that serve as models for biological membranes. Thus, it is virtually impossible to experimentally measure partition coefficients for model chemicals in a two-phase system that would give fragment constants that could be used to calculate theoretical partition coefficients for a drug distributing between the solvents (vehicles) of interest and skin.

For the drugs that were investigated by Sloan and co-workers (5-fluorouracil, 6-mercaptopurine, theophylline, and salicylic acid), the theoretical partition coefficients, $K_i^{s,v}$, were calculated assuming that the solubility parameter of the skin, δ_s, was about 10 (cal/ cm^3)$^{1/2}$ (18). The solubility parameters of the vehicles, δ_v, were literature values (28) except for isopropyl myristate (19) and are listed in Table 1, whereas the solubility parameters of the drugs, δ_i, and the molar volumes of the drugs, V_i, were calculated according to Fedors (29). These theoretical partition coefficients are given in Table 1. The experimental permeability coefficients, $P_i^{s,v}$, for the delivery of these four relatively polar molecules from vehicles exhibiting a broad spectrum of polarity, from water to oleic acid, through hairless mouse skin are also given in Table 1 along with the corresponding fluxes, $J_i^{s,v}$, and solubilities, C_i^v.

The values for log theoretical $K_i^{s,v}$ and log experimental $P_i^{s,v}$ for each drug are plotted against δ_v in Figures 2 through 5. In each case, the plot of the log experimental $P_i^{s,v}$ versus δ_v approximates a parabola that is similar to that of the plot of log theoretical $K_i^{s,v}$ versus δ_v, except for the data obtained when octanol and isopropyl myristate (IPM) were used as vehicles. However, both of those vehicles caused a much higher flux of the standard solute— vehicle in the second application part of the experiment than did the other vehicles; therefore, it has been assumed that the inordinately high flux of drug, hence, $P_i^{s,v}$, caused by those vehicles is due to damage to the skin. If the data for these obvious outliers are disregarded, a plot of log theoretical $P_i^{s,v}$ versus δ_v can be constructed from log theoretical $P_i^{s,v}$ values, each of which can be obtained from the sum of the corresponding log theoretical $K_i^{s,v}$ and the average of the differences between the log theoretical $K_i^{s,v}$ and log

Table 13.1 The Effect of Single Component Vehicles on the Delivery of Drugs through Hairless Mouse Skin: Solubilities (C_i^v), Fluxes ($J_i^{s,v}$)[a], Permeability Coefficients ($P_i^{s,v}$) and log Theoretical Partition Coefficients (log $K_i^{s,v}$)[b]

Solvent[c] [δ_v; (cal/cm^3)1/2]	Solubility (C_i^v), (mg/cm3)	Flux ($J_i^{s,v}$) [mg/cm^{2}h × 10^3 (± SD)]		$P_i^{s,v}$ [cm/h × 10^3 (± SD)]		log $K_i^{s,v}$	
Theophylline[d]							
IPM (8.5)	0.062	41.2	(3.7)	660.0	(61.0)	1.12	
OCT (10.3)	1.70	547.0	(146.0)	320.0	(86.0)	−0.18	
DMF (12.1)	34.6	17.0	(4.0)	0.49	(0.12)	−0.98	
PG (14.8)	8.07	1.54	(0.31)	0.19	(0.04)	−1.22	
EG (16.1)	7.40	1.94	(0.72)	0.26	(0.10)	−0.92	
FOR (17.9)	1.57	3.69	(0.25)	2.35	(0.16)	−0.06	
Salicylic acid[e]							
OA (7.6)	30.3	640	(69)	21.2	(2.3)	1.74	(0.68)
IPM (8.5)	41.4	870	(190)	21.0	(4.7)	0.91	(0.34)
OCT (10.3)	145	1600	(60)	11.2	(0.4)	−0.39	(−0.027)
PRO (12.0)	228	1090	(55)	4.8	(0.25)	−1.04	(0.00)
PG (14.8)	207	430	(41)	2.1	(0.20)	−1.30	(0.66)

Table 13.1 (Continued)

Solvent[c] $[\delta_v; (cal/cm^3)^{1/2}]$	Solubility (C_i^v), (mg/cm^3)	Flux $(J_i^{s,v})$ $[mg/cm^2 h \times 10^3 (\pm SD)]$		$P_i^{s,v}$ $[cm/h \times 10^3$ $(\pm SD)]$		$\log K_i^{s,v}$
FOR (17.9)	145	1150	(46)	7.9	(0.32)	−0.64 (2.55)
6-Mercaptopurine[f]						
OA (7.6)	0.0030	0.043	(0.021)	14.3	(7.0)	1.57
IPM (8.5)	0.0034	0.60	(0.30)	176.0	(88.0)	0.90
DET[g] (10.0)	4.40	3.2	(0.24)	0.72	(0.055)	0.06
OCT (10.3)	0.23	18.6	(1.6)	81.0	(7.0)	−0.15
MEG[g] (12.1)	10.0	0.75	(0.10)	0.075	(0.010)	−0.82
DMF (12.1)	14.5	3.8	(0.71)	0.26	(0.044)	−0.82
DMSO (13.0)	34.8	2.1	(0.16)	0.059	(0.0046)	−1.02
PG (14.8)	6.2	0.093	(0.006)	0.015	(0.00097)	−1.12
EG (16.1)	3.0	0.10	(0.010)	0.033	(0.0033)	−0.96
FOR (17.9)	9.1	1.5	(0.10)	0.16	(0.011)	−0.42

H_2O (23.4)	0.17	0.36 (0.21)	2.1 (1.2)	3.76
5-Fluorouracil[h]				
OA (7.6)	0.74	100 (12)	150 (16)	1.35
IPM (8.5)	0.0051	28 (2)	5400 (390)	0.78
OCT (10.3)	0.60	440 (14)	730 (23)	−0.13
DMF (12.1)	62.3	25 (2)	0.41 (0.03)	−0.75
PG (14.8)	16.5	1.6 (1.0)	0.097 (0.061)	−1.13
EG (16.1)	19.6	4.7 (1.3)	0.24 (0.066)	−1.08
FOR (17.9)	13.7	15.0 (1.4)	1.1 (0.10)	−0.75

[a] Steady-state fluxes, number of cell run = 3.

[b] Calculated from Eq. 6; R = 1.98 cal/degree, T = 305 K, $\delta_s = 10$ $(cal/cm^3)^{1/2}$; for theophylline $\delta_i = 14.0$ $(cal/cm^3)^{1/2}$, $V_i = 110$ cm^3/mol; for salicylic acid $\delta_i = 14.4$ $(cal/cm^3)^{1/2}$ or $\delta_i = 11.0$ $(cal/cm^3)^{1/2}$ for values in parentheses, $V_i = 93.9$ $cm^3/mole$; for 6-mercaptopurine $\delta_i = 14.4$ $(cal/cm^3)^{1/2}$, $V_i = 81.2$ cm^3/mol; for 5-fluorouracil $\delta_i = 15$ $(cal/cm^3)^{1/2}$, $V_i = 70.4$ cm^3/mol.

[c] OA, oleic acid; IPM, isopropyl myristate; DET, diethyltoluamide; OCT, 1-octanol; PRO, 1-propanol; MEG, methoxyethanol; DMF, dimethylformamide; DMSO, dimethyl sulfoxide; PG, propylene glycol; EG, ethylene glycol; FOR, formamide.

Sources: Data from [d] Ref. 19; [e] Ref. 20; [f] Ref. 21; [g] unpublished results; [h] Ref. 22.

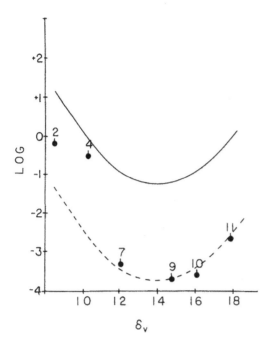

Figure 13.2

Figures 13.2–13.5 The effect of the solubility parameters of vehicles (δ_V) on the log experimental permeability coefficients ($P_i^{s,v}$) of theophylline (Fig. 2), salicylic acid (Fig. 3), 6-mercaptopurine (Fig. 4) and 5-fluorouracil (Fig. 5). In each figure the log theoretical $P_i^{s,v}$ versus δ_V relationship is represented by (---) and the log theoretical partition coefficient ($K_i^{s,v}$) versus δ_V relationship is represented by (—). In each figure the solvents are 1, oleic acid; 2, isopropyl myristate; 3, diethyltoluamide; 4, 1-octanol; 5, 1-propanol; 6, 2-methoxyethanol; 7, dimethylformamide; 8, dimethyl sulfoxide; 9, propylene glycol; 10, ethlyene glycol; 11, formamide; 12, water. In Figure 3 there are two sets of log $P_i^{s,v}$ versus δ_V and $K_i^{s,v}$ versus δ_V curves given. The curves labeled as (o) are for $\delta_i = 14.4$ $(cal/cm^3)^{1/2}$ whereas those labeled as (Δ) are for $\delta_i = 11.0$ $(cal/cm^3)^{1/2}$.

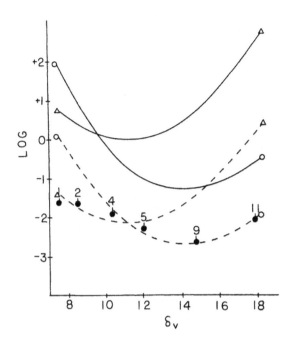

Figure 13.3

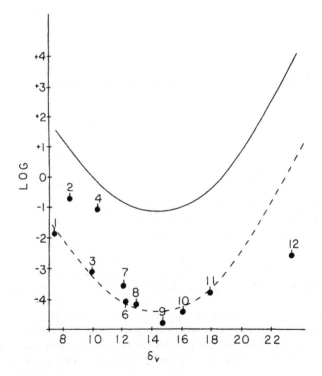

Figure 13.4

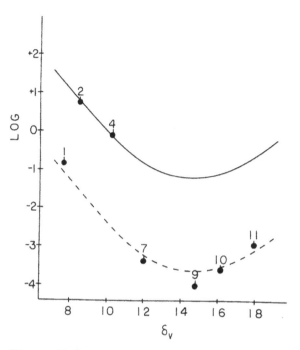

Figure 13.5

experimental $P_i^{s,v}$. This average difference corresponds to the
constant in Eq. 13. If the thickness of the stratum corneum is assumed to be about 10 μm, then D_i^s is about 10^{-10} cm^2/s for most of
the drugs studied, except for D_i^s for salicylic acid, which is one to
two orders of magnitude greater. The calculated D_i^s, then, is of
about the right magnitude for such polar molecules.

There are five conclusions that can be reached on the basis of
the results given in Table 1 and Figures 2 through 5. First, there
is generally a minimum in the log experimental $P_i^{s,v}$ versus δ_V curve
that corresponds to those vehicles exhibiting δ_V similar to that of
δ_i. The drug also usually exhibits its greatest mole fraction solubility in those same vehicles so that, generally, $P_i^{s,v}$ are inversely
related to solubility (30). Second, for single-component vehicles,
those vehicles that exhibit δ_V similar to that of δ_S also cause the most
damage to the skin as assessed by second application studies. Be-
cause those δ_V are similar to δ_S, they may be solubilizing a portion of
the membrane that is responsible for providing the diffusional resistance to drug permeation and, especially, to polar molecules exhibiting
a δ_i quite different from δ_S. Third, from the knowledge of the solu-

bilities of a drug in various vehicles, the relationship between $P_i^{s,v}$ and δ_v, and the flux of that drug from one specific vehicle it should be possible to determine the flux of that drug from other vehicles of interest, assuming that the vehicles do not damage the skin. Fourth, the parabolic nature of the relationship between δ_v and $P_i^{s,v}$ expected from regular solution theory has been confirmed for a number of drugs. Fifth, for "chameleonic" solutes (drugs) such as salicylic acid (see Fig. 3) that apparently can exhibit two different δ_i values, two different log theoretical $K_i^{s,v}$ and, hence, two different log experimental $P_i^{s,v}$ curves are generated.

In addition to the results obtained for single-component vehicles, the rates of delivery of 5-fluorouracil (5-FU) and 6-mercaptopurine (6-MP) from a binary component vehicle (oleic acid—propylene glycol) were also determined (Table 2). Here, although the log experimental $P_i^{s,v}$ values for the delivery of 5-FU and 6-MP from oleic acid and propylene glycol fell on the log theoretical $P_i^{s,v}$ versus δ_v curve, none of the log experimental $P_i^{s,v}$ values for the mixtures did. In each case, starting with mixtures that were rich in oleic acid (21,22), as the solubility of the drug in each mixture increased, its experimental $P_i^{s,v}$ value decreased. Also, in each case, there was considerably more damage to the skin observed after treatment with the binary-component solutions than after treatment with either of the single components (21,23), regardless of the δ_v of the mixture. Thus, enhanced delivery of a drug by a mixture of vehicles may be primarily the result of enhanced damage to the integrity of the skin, rather than to some thermodynamic advantage.

In the investigations of the ability of prodrugs to deliver their respective parent drugs from various vehicles through skin, there was a good correlation between the calculated solubility parameter of a series of homologous prodrugs and their ability to deliver the parent drugs through hairless mouse skin (24—26). In addition, a vehicle effect on rates of delivery was observed that was directly attributable to solubility phenomena. Generally, the results from these articles were qualitatively similar to the results of Ostrenga (9) that showed that there was a good correlation between molar attraction constants (F values) and bioactivity.

The most extensive studies of the effect of relative solubilities of a homologous series of prodrugs on their abilities to deliver a parent drug through skin are the studies of $S^6,9$-bis-acyloxymethyl-6-MP (6,9-bis-6-MP) and S^6-acyloxymethyl-6-MP (6-mono-6-MP) prodrugs. The acetyloxymethyl through octanoyloxymethyl along with the pivaloyloxymethyl derivatives of both the bis- and mono-series were synthesized, characterized, and the rates at which they delivered 6-MP through hairless mouse skins were measured. The solu-

Table 13.2 The Effect of Binary-Component Vehicles on the Delivery of Drugs Through Hairless Mouse Skin: Solubilities (C_i^v), Fluxes ($J_i^{s,v}$), and Permeability Coefficients ($P_i^{s,v}$)

Solvent[a] (mole ratios)	δ_v[b] $(cal/cm^3)^{1/2}$	C_i^v, mg/cm^3	$J_i^{s,v,c}$ $mg/cm^2h \times 10^3$ (± SD)	$P_i^{s,v}$, $cm/h \times 10^3$ (± SD)
6-Mercaptopurine				
OA	7.6	0.0030	0.043 (0.021)	14.3 (7.0)
OA:PG (3:1)	8.1	0.087	0.44 (0.021)	5.1 (0.24)
OA:PG (1:1)	9.3	0.39	0.95 (0.13)	2.4 (0.33)
OA:PG (1:3)	10.6	1.23	1.6 (0.19)	1.3 (0.15)
OA:PG (1:5)	12.0	2.25	1.6 (0.15)	0.71 (0.067)
OA:PG (1:14.5)	13.5	4.35	1.9 (0.17)	0.44 (0.039)
PG	14.8	6.2	0.093 (0.006)	0.015 (0.00097)
5-Fluorouracil				
OA	7.6	0.74	100 (12)	150 (16)
OA:PG (1:1)	9.3	0.98	50 (12)	51 (12)
OA:PG (1:3)	10.6	4.3	50 (2)	12 (0.47)
OA:PG (1:14.5)	13.5	11.1	88 (5)	7.9 (0.45)
PG	14.8	16.5	1.6 (1.0)	0.097 (0.061)

[a]OA, oleic acid; PG, propylene glycol.
[b]Calculated on a volume fraction basis.
[c]Steady-state fluxes, number of cells run = 3.
Sources: Data for 6 = MP from Ref. 21; that for 5 = FU from Ref. 22.

Table 13.3 Solubilities of $S^6,9$-Bis-acyloxymethyl-6-mercaptopurine (6,9-bis-6-MP) Derivatives, the Fluxes ($J_i^{s,v}$) and Log Permeability Coefficients (log $P_i^{s,v}$) for the Delivery of 6-Mercaptopurine (6-MP) by the Derivatives from Isopropyl Myristate (IPM), Propylene Glycol (PG) and Water

$$SCH_2O_2CR$$

$$CH_2O_2CR$$

Derivative, R = /vehicle	Solubility[a]	$J_i^{s,v}$,[b] mg/cm^{2}h × 10^3 (± SD)	log $P_i^{s,v}$, cm/h
1, 6-MP/IPM[c]	0.0034	0.60 (0.30)	− 0.73
2, R = CH$_3$/IPM	0.80	34.5 (2.8)	− 1.36
3, R = C$_2$H$_5$/IPM	5.11	35.2 (9.9)	− 2.16
4, R = C$_3$H$_7$/IPM	13.8	21.5 (5.4)	− 2.81
5, R = C$_4$H$_9$/IPM	26.5	15.5 (7.4)	− 3.23
6, R = C$_5$H$_{11}$/IPM	7.56	1.75 (0.44)	− 3.63
7, R = C$_7$H$_{15}$/IPM	2.60	0.19 (0.014)	− 4.14
8, R = C(CH$_3$)$_3$/IPM	31.4	5.48 (0.55)	− 3.76
1, 6-MP/PG[c]	6.2	0.093 (0.0060)	− 4.82
2, R = CH$_3$/PG	2.79	0.58 (0.04)	− 3.68
3, R = C$_2$H$_5$/PG	8.63	1.95 (0.28)	− 3.65
4, R = C$_3$H$_7$/PG	9.54	1.60 (0.050)	− 3.76
5, R = C$_4$H$_9$/PG	8.32	0.77 (0.060)	− 4.03
6, R = C$_5$H$_{11}$/PG	1.71	0.18 (0.045)	− 3.99
1, 6-MP/H$_2$O[c]	0.17	0.36 (0.21)	− 2.68

Table 13.3 (Continued)

Derivative, R = /vehicle	Solubility[a]	$J_i^{s,v}$,[b] mg/cm^{2}h × 10^3 (± SD)	log $P_i^{s,v}$, cm/h
2, R = CH_3/H_2O	0.44	1.58 (0.88)	− 2.44
3, R = C_2H_5/H_2O	0.25	1.97 (0.16)	− 2.11
4, R = C_3H_7/H_2O	0.030	0.81 (0.081)	− 1.57
5, R = C_4H_9/H_2O	0.0071	0.50 (0.12)	− 1.15

[a]The solubilities are given in mg equivalents of 6-MP per ml of solution.
[b]Steady-state fluxes, number of cells per run = 3.
[c]*Source*: Fluxes of 6-MP from the vehicles from Ref. 21.

bilities of the bis-derivatives in various solvents are given in Table 3 along with their respective $J_i^{s,v}$ and $P_i^{s,v}$ from the same vehicles, and the comparable data for the mono-series are given in Table 4. Plots of the calculated δ_i of the prodrugs versus their corresponding log experimental $P_i^{s,v}$ for the delivery of 6-MP by both series of prodrugs from water, isopropyl myristate (IPM), and propylene glycol (PG) are given in Figure 6.

It is clear from Figure 6 that there is a fairly good correlation between log experimental $P_i^{s,v}$ and δ_i for each series for the delivery of 6-MP from each solvent. However, the form of the relationship depends on the vehicle. For the delivery of 6-MP from IPM, as the prodrug becomes more lipophilic and more like the vehicle (i.e., the value of δ_i decreases), the log experimental $P_i^{s,v}$ decreases. Whereas for the delivery of 6-MP by the prodrugs from water, as the prodrug becomes more lipophilic and less like the vehicle, the log experimental $P_i^{s,v}$ increases. Conversely, there is not much change in the log experimental $P_i^{s,v}$ values for the delivery of 6-MP by the prodrugs from propylene glycol. This last result may be because the δ_i are not that different from that of δ_v.

The other article that has been published in this area describes the effect of the structure of N-Mannich base prodrugs of 5-FU and theophylline on their ability to deliver their respective parent drugs

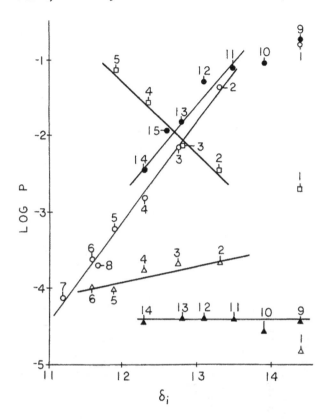

Figure 13.6 A plot of the log experimental permeability coefficient (P) versus the calculated solubility parameter (δ_i) for the S^6-acyloxymethyl-6-mercaptopurine prodrugs delivering 6-mercaptopurine from isopropyl myristate (closed circles) and from propylene glycol (closed triangles) and for S^6,9-bis-acyloxymethyl-6-mercaptopurine prodrugs delivering 6-mercaptopurine from isopropyl myristate (open circles), from propylene glycol (open triangles), and from water (open squares). The numbers refer to the compounds in Tables 3 and 4.

through hairless mouse skin (26). The results are very similar to those found for the acyloxymethyl-type prodrugs of 6-MP. Because of stability consideration, these N-Mannich base-type prodrugs were tested in only one aprotic vehicle—isopropyl myristate. A direct correlation between log experimental $P_i^{s,v}$ and calculated δ_i was observed. Again, the more lipophilic the prodrug (lower δ_i value and

Table 13.4 Calculated Solubility Parameters (δ) of S6-Acyloxymethyl-6-mercaptopurine Derivatives, and the Fluxes ($J_i^{s,v}$) and log Permeability Coefficients (log $P_i^{s,v}$) for the Delivery of 6-Mercaptopurine (6-MP) by the Derivatives from Isopropyl Myristate (IPM) and Propylene Glycol (PG)

$$SCH_2O_2CR$$

Compound, R =	Solubility[a] IPM, PG	Calculated δ_i (cal/cm3)$^{1/2}$	Isopropyl myristate Flux ($J_i^{s,v}$), b mg/cm2h × 10^3 (± SD)	log $P_i^{s,v}$, cm/h	Propylene glycol Flux ($J_i^{s,v}$), b mg/cm2h × 10^3 (± SD)	log $P_i^{s,v}$, cm/h
1, 6-MP[a]		14.4	0.60 (0.30)	−0.73	0.093 (0.006)	−4.82
9, CH$_3$	0.16, 14.2	14.4	30.8 (8.8)	−0.72	0.55 (0.22)	−4.41

10, C_2H_5	0.35, 19.0	13.9	32.5 (14.8)	−1.03	0.54 (0.073)	−4.55
11, C_3H_7	0.50, 22.7	13.5	39.8 (19.7)	−1.10	0.96 (0.17)	−4.37
12, C_4H_9	0.64, 22.1	13.1	33.5 (15.7)	−1.28	0.95 (0.13)	−4.37
13, C_5H_{11}	0.56, 11.2	12.8	8.3 (3.9)	−1.83	0.48 (0.065)	−4.37
14, C_7H_{15}	0.63, 16.7	12.3	2.0 (1.1)	−2.50	0.29 (0.024)	−4.45
15, $C(CH_3)_3$	0.80,	12.6	9.4 (1.4)	−1.94		

[a]The solubilities are given in mg equivalents of 6-MP per ml of solution.
[b]Steady-state fluxes, number of cells per run = 3.
Source: Data for 6-MP from Ref. 21.

more like the vehicle), the lower the value for $P_i^{s,v}$ for the delivery
of the parent drug.

Thus, once the log experimental $P_i^{s,v}$ has been determined for
the delivery of a drug by a particular type of prodrug from a specific
vehicle, it is possible to predict $P_i^{s,v}$ for the delivery of drug by other
prodrugs, in the same homologous series, from that vehicle using the
slope of a previous plot of log experimental $P_i^{s,v}$ versus δ_i. To work
backward from $P_i^{s,v}$ to determine $J_i^{s,v}$, it is only necessary to determine experimentally the solubility of the prodrug in the vehicle of interest. In view of the difficulty in predicting C_i^v, it is almost easier
to predict $J_i^{s,v}$ once C_i^v is known than it is to predict C_i^v. This may
be because $J_i^{s,v}$ is predicted from calculated $K_i^{s,v}$ which represent the
ratio of the solubilities of the prodrug in two phases at equilibrium.
Therefore, any contribution to the irregularity of the solubility of the
solute peculiar to its structure would affect the solubilities in both
phases, and the results would tend to cancel each other (31).

The initial article by Cohen and co-workers (18) studied the effect of the structure of alkanoic acids on their ability to diffuse
through porcine skin and on the effect of the physicochemical properties of various solvents on the delivery of propionic acid from the
solvents through porcine skin. For the first, it was assumed that
when a maximum flux of the alkanoic acids was reached the δ of the
skin matched the δ of the acid. Because the maximum flux was obtained for butyric acid ($\delta = 10.0$), the apparent δ of skin was assumed to be about 10.0 $(cal/cm^3)^{1/2}$. There was an excellent correlation between the permeability coefficients for the delivery of
various straight- and branched-chain alkanoic acids from pentane
with the square root of the volume cohesive energy product. However, the expected parabolic relationship between the permeability
coefficient for the delivery of propionic acid from various solvents
and the δ of the solvents with a minimum in rate for the solvent
exhibiting a δ closest to that of propionic acid was not observed.
Instead, there was essentially a direct relationship with a steady decrease in $P_i^{s,v}$ corresponding to a steady increase in δ of the solvent. This result was attributed to a breakdown in regular solution
behavior beyond a certain polarity. However, this result also may
have been because saturated solutions could not be used in these
experiments.

The focus of the two most recent articles by Cohen and co-
workers (32,33) has been an attempt to separate and to quantify
the two variables responsible for the flux of a solute across the
skin barrier. These two variables are the thermodynamic activity
of the solute in the vehicle, on one hand, and the extent of the alteration of the barrier function of the skin caused by the vehicle,

on the other. The first paper describes the effect of alkanoic acids on the delivery of theophylline through human skin. To separate the purely thermodynamic effect of the alkanoic acid vehicles ("push effect" resulting from excess free energy) from the ability of the alkanoic acids to solubilize the penetrant in the skin ("pull effect"), a ratio of partition coefficients $K_i^{s,v(1)}/K_i^{s,v(2)}$ calculated from regular solution theory (Eq. 6) was determined for the delivery of theophylline from the two alkanoic acids being compared. Because $P_i^{s,v} = K_i^{s,v} \cdot$ constant (from Eq. 13), if the experimentally determined ratio $P_i^{s,v(1)}/P_i^{s,v(2)}$ was equal to the theoretically determined ratio $K_i^{s,v(1)}/K_i^{s,v(2)}$, then only thermodynamically generated push effects were responsible for the flux of theophylline from the two solvents. On the other hand, for example, if the vehicle v(1) affected δ_s by making it more polar, then the experimental ratio would be larger than the theoretical ratio. Conversely, if vehicle v(2) is the one that causes δ_s to become more polar for that diffusion experiment, then the experimental ratio would be smaller than the theoretical ratio.

The experimental results showed that the $P_i^{s,v}$ for the delivery of theophylline from propionic, hexanoic, octanoic, and a mixture of octanoic and hexanoic acids was inversely related to the δ_v for the acid, that is, $P_i^{s,v}$ decreases as δ_v increased and became more like δ_i. However, the $P_i^{s,v}$ for the delivery of theophylline from mixtures of alkanoic acids containing propionic acid was much higher than expected based on considerations of the mixture's solubility parameter. Also, the experimentally determined ratio of $P_i^{s,v(1)}/P_i^{s,v(2)}$ in any comparison of the performance of two alkanoic acid was always smaller than the theoretically determined ratio of $K_i^{s,v(1)}/K_i^{s,v(2)}$ if propionic acid was v(2) and always larger if propionic acid was v(1). Therefore, it was concluded that the flux of theophylline from single- or binary-component mixtures of alkanoic acid vehicles was caused by the thermodynamic push effect, if propionic acid was not included but, if propionic acid was included, then an additional pull effect by the vehicle was observed.

The second of the two most recent articles by Cohen and co-workers describes the effect of alkanoic acids on the delivery of adenosine through human skin. For mixtures of hexanoic and propionic acids a bell-shaped dependence of observed ln $P_i^{s,v}$ on volume fraction of hexanoic acid was observed. This dependence was attributed to the same two opposing variables just identified. Increases in the volume fraction of hexanoic acid in a vehicle caused an increase in the excess free energy of adenosine in that vehicle and, hence, an increased push effect. On the other hand, increases in the volume fraction of propionic acid in a vehicle caused an increase in the flux of propionic acid from that vehicle and, hence, an increase in the pull effect. A similar result was obtained for mixtures of propionic acid in isopropyl myristate.

The enhancement caused by the pull effect of propionic acid in the hexanoic—propionic acid mixtures was determined by first calculating the corresponding ratio of theoretical $K_i^{s,v(1)}/K_i^{s,v(2)}$ and equating it to a ratio of theoretical $P_i^{s,v(1)}/P_i^{s,v(2)}$ (Eq. 13) derived from regular solution theory. The ratios of theoretical $P_i^{s,v(1)}/P_i^{s,v(2)}$ values were normalized by assuming that the flux of adenosine from pure hexanoic acid was caused only by a push effect, therefore, $K_i^{s,v(2)}$ and, hence, $P_i^{s,v(2)}$ in each calculation is for hexanoic acid. The product of the experimental $P_i^{s,v}$ for hexanoic acid and the theoretical $P_i^{s,v(1)}/P_i^{s,v(2)}$ was assumed to give a value for the contribution to the total or experimental permeability coefficient $P_{i,total}^{s,v}$ only because of the push effect, or $P_{i,push}^{s,v}$. The ratio of $P_{i,total}^{s,v}/P_{i,push}^{s,v}$ was taken as the enhancement factor from the pull effect. The largest enhancement factor observed was about 6.4 for a mixture containing 0.4 volume fraction of hexanoic acid.

IV. CONCLUSION

Regular solution theory can be applied to numerous partitioning processes of biological importance to allow investigators to predict the effect of vehicles or solvents and the effect of systematic changes in the physicochemical properties of drugs on the partitioning process. In percutaneous absorption, the effect of vehicles on the thermodynamic driving force for diffusion has been separated from vehicle (solvent) effects on the integrity of the skin itself by using second application experiments or by comparing experimental permeability coefficients with theoretical permeability coefficients calculated from regular solution theory. Analyses of results of the thermodynamic effect of vehicles on percutaneous absorption suggests that such experimental systems can be explained by regular solution theory and that such analyses have predictive value. Similarly, the experimental permeability coefficients of homologous series of prodrugs for the delivery of a parent drug from different vehicles can be correlated with the calculated solubility parameters of the prodrugs. Whether the relationship is direct or inverse depends on the solubility parameter exhibited by the vehicle.

REFERENCES

1. J. H. Hildebrand, *J. Am. Chem. Soc.*, *51*:66, 1929.
2. G. Scatchard, *Chem. Rev.*, *8*:321, 1931.
3. M. J. Chertkoff, and A. Martin, *J. Am. Pharm. Assoc.*, *Sci. Ed.*, *49*:444, 1960.
4. A. Martin, P. L. Wu, Z. Liron, and S. Cohen, *J. Pharm. Sci.*, 74:638, 1985.
5. A. Martin, and J. Mauger, *Am. J. Pharm. Ed.*, *52*:68, 1988.
6. J. Ferguson, *Proc. R. Soc. (Lond.)*, *B127*:387, 1939.
7. J. H. Hildebrand, and R. L. Scott, *The Solubility of Nonelectrolytes*, 3rd ed, Chap. 3. Dover, New York, 1964.
8. S. A. Khalil, and A. N. Martin, *J. Pharm. Sci.*, *56*:1225, 1967.
9. J. A. Ostrenga, *J. Med. Chem.*, *12*:349, 1969.
10. T. Fujita, J. Iwasa, and C. Hansch, *J. Am. Chem. Soc.*, *86*: 5175, 1964.
11. S. S. Davis, *Experimentia*, *26*:671, 1970.
12. S. Cohen, A. Goldschmid, G. Shtacher, S. Srebrenik, and S. Gitter, *Mol. Pharmacol.*, *11*:379, 1975.
13. E. M. Landau, J. Richter, and S. Cohen, *J. Med. Chem.*, *22*: 325, 1979.
14. S. Srebrenik, and S. Cohen, *J. Phys. Chem.*, *80*:998, 1976.
15. L. J. Mullins, *Chem. Rev.*, *54*:289, 1954.
16. A. Adjei, J. Newburger, S. Stavchansky, and A. Martin, *J. Pharm. Sci.*, *73*:742, 1984.
17. P. Bustamante, and E. Selles, *J. Pharm. Sci.*, *75*:639, 1986.
18. Z. Liron, and S. Cohen, *J. Pharm. Sci.*, *73*:538, 1984.
19. K. B. Sloan, S. A. M. Koch, K. G. Siver, and F. P. Flower, *J. Invest. Dermatol.*, *87*:244, 1986.
20. K. B. Sloan, K. G. Siver, and S. A. M. Koch, *J. Pharm. Sci.*, *75*:744, 1986.
21. R. P. Waranis, K. G. Siver, and K. B. Sloan, *Int. J. Pharm.*, *36*:211, 1987.
22. E. F. Sherertz, K. B. Sloan, and R. G. McTiernan, *J. Invest. Dermatol.*, *89*:147, 1987.
23. E. F. Sherertz, K. B. Sloan, and R. G. McTiernan, *J. Invest. Dermatol.*, *89*:249, 1987.
24. R. P. Waranis, and K. B. Sloan, *J. Pharm. Sci.*, *76*:587, 1987.
25. R. P. Waranis, and K. B. Sloan, *J. Pharm. Sci.*, *77*:210, 1988.
26. K. B. Sloan, E. F. Sherertz, and R. G. McTiernan, *Int. J. Pharm.*, *44*:87, 1988.
27. R. J. Scheuplein, *J. Invest. Dermatol.*, *45*:334, 1965.

28. A. F. M. Barton, *Chem. Rev.*, *75*:731, 1975.
29. R. F. Fedors, *Polym. Eng. Sci.*, *14*:147, 1974.
30. P. H. Dugard, and R. C. Scott, *Int. J. Pharm.*, *28*:219, 1986.
31. S. H. Yalkowsky, in *Design of Biopharmaceutical Properties Through Prodrugs and Analogs*, Chap. 13. (E. B. Roche, ed.), American Pharmaceutical Association, Washington, D.C., 1977.
32. R. Kadir, D. Stempler, Z. Liron, and S. Cohen, *J. Pharm. Sci.*, *76*:774, 1987.
33. R. Kadir, D. Stempler, Z. Liron, and S. Cohen, *J. Pharm. Sci.*, *77*:409, 1988.

14

Use of an Epidermal Culture for Cutaneous Toxicity Studies

F. L. VAUGHAN   *University of Michigan School of Public Health, Ann Arbor, Michigan*

I. INTRODUCTION

The skin is one of the organs most frequently exposed to gaseous, liquid, and solid environmental substances that can result in a breakdown of homeostatic mechanisms. Contact with chemicals used in hygiene, medication, cosmetic application, as well as chemicals that result from the development of industrial products, is a constant public health concern. Thus, substances that may cause skin irritation are among those that are constantly under surveillance. Animal assay systems usually involve the application of these substances onto the shaven backs of rabbits and other animals, followed by incubation periods that result in lesions of various severity as quantified by subjective pathological examinations. The Draize skin test (1) is used most frequently by industry and is recommended by regulatory agencies in assaying and classifying chemicals potentially irritating to the human skin. The significant variability of this and similar in vivo animal tests in accurately predicting irritancy to human skin has been observed and reported (2). Studies have also shown that results obtained from such animal experiments are often equivocal and sometimes do not reflect responses that occur when humans are similarly exposed (3). It is also particularly difficult to study mechanisms of toxicity in the whole animal, in part, because of the high levels of exposure necessary to elicit a measurable response and also because of the undescribed interactions of various physiological systems. However, regulatory agencies offer little in the way of alternatives for preevaluating products before they can be released for public use (4). What is needed is an in vitro sys-

tem that will allow the investigator to examine the etiology, at the
molecular level, of the toxicity of certain classes of chemicals and
obtain more information on the predictability of the cutaneous irritancy potential of such chemicals.

Mainly for these reasons, in vitro systems are being developed
to study cutaneous toxicity. Such systems include mammalian cells
and tissues maintained in vitro by using standard tissue culture
techniques, as well as cell-free systems prepared from living tissue
for assay of subcellular changes caused by xenobiotics. Valuable
information is currently being generated with these systems. One
of the more promising contributions of in vitro systems anticipates
the use of human biological material in designing models for studying toxicity.

Correlations between in vitro cytotoxicity and local irritancy in
vivo were described by Schmidt, et al. (5), who used established
cell lines of mouse liver and connective tissue, for a system assaying anesthetics. Another early developed cell culture system was
described by Guess et al. (6) in which cytotoxicity was determined
after exposing agar-overlay monolayer cultures to chemical irritants
of varying potency. The test was reported to correlate well with
implantation, intradermal, dermal, and intraocular tests in rabbits.

Some investigators have used cells derived from the skin in developing cutaneous toxicity in vitro assays. It is felt that the response of cells from this tissue would more closely reflect skin irritation. Early experiments by Hu and co-workers (7) demonstrated
a correlation between in vitro cytotoxicity of topical medicants to
keratinocyte cultures and local irritancy in vivo. Kao and associates (8) employed an organ culture system composed of full-thickness mouse skin to quantitate cutaneous toxicity. The 48-hr organ
culture was treated topically with a xenobiotic and deleterious morphological changes were correlated with the incorporation of radiolabeled DNA and protein precursors, as well as with the leakage of
intracellular enzymes into the culture medium.

As a result of these, and other earlier studies, some desired
characteristics of mammalian cell systems have come to the forefront.
For example, it has been established that, in some instances, cell
types that become permanent lines with indefinite mortality through
genetic alterations, lose much of the metabolic activity expressed by
their progenitors. As a result, many cell culture systems currently
under study use primary cells, which generally have retained many
of the original genetic markers of the parent cell. Another example
of the desirability of specialized in vitro cellular activity is the observation that cells cultivated and maintained in a microenvironment
similar to that in vivo tend to display similar morphological and metabolic activities. Tissue cell types growing as monolayers sub-

merged in growth medium generally do not display the homeostatic interaction seen in cell cultures that have organizational structure similar to that observed in vivo. As a result of these and other observations, some investigators are developing in vitro models using primary cells isolated from living tissue. Attempts are also made to reconstruct the in situ microenvironment of the original tissue as closely as possible. It is hoped that this approach will contribute to obtaining more valid information in assaying the toxicity of xenobiotics.

The skin is divided into two main parts: the dermis or supportive connective tissue containing blood vessels, nerves, etc.; and the epidermis, which is the outermost structure of the skin and which is most responsible for protecting the body from external chemical and physical agents. We feel that an ideal in vitro system for studying the mechanism of action of chemicals, and for screening chemicals to determine their potential toxicity to humans, would be one that mimics the human epidermis both in morphological and in biochemical structure. Such a system, therefore, should meet the following criteria:

1. Morphologically, it must exhibit all of the stages of epidermal differentiation as seen in intact skin. This should result in keratinization of the outer cell layers as a result of ordered, step-by-step differentiation or orthokeratosis. Also, most of the unique morphological markers of the epidermis should be produced.
2. Biochemical markers of epidermal differentiation should be exhibited as it takes place in situ. Among those changes are

   a. Progressive cell surface alterations in the keratinocytes as they differentiate
   b. The production of specific keratin peptides found in terminally differentiated keratinocytes
   c. The synthesis of a specific epidermal histidine-rich protein found only in that tissue

The cell type that is most responsible for the organized epidermis is the keratinocyte. Other cells such as melanocytes, Langerhan cells, and macrophages are also present, but they are not involved in forming the anatomical barrier that is characteristic of the epidermis. The *basal cells*, which are the only keratinocytes capable of proliferation, form a single layer above the dermis. This layer is connected to the underlying dermis by the basement membrane. Basal cells differentiate to *spinous cells* which form multilayers immediately above the basal layer. Keratinocytes in the next stage of differentiation are called *granular cells* because they contain keratohyalin granules, an important structure in the formation of keratin. The cells at the terminal stage of differentiation are the *cornified cells* that form the outermost desquamated layers of the epidermis. It is believed that tonofilaments and keratohyalin granules combine in some way to form the keratin that characterizes the terminally differentiated keratinocyte. Another characteristic unique to the epidermis is the formation of *desmosomes*—structures that form very tenacious bonds between the cells in various layers. *Hemidesmosomes* are bonds between the basal keratinocyte and the basement membrane.

Therefore, any chemical that contacts the epidermis is presented with all of the cellular components of the epidermis. We have developed a differentiated keratinocyte culture system that forms a stratified squamous epithelium with morphological and biochemical markers like those observed in an epidermis formed in situ. Keratinocytes are obtained as primary isolates (to retain as much of the original metabolic activity as possible), seeded on an appropriate substratum, allowed to attach and proliferate, then, raised to the air—medium interface to stimulate further growth and differentiation. This method creates a microenvironment similar to that in which a normal epidermis develops. As a result, the cells differentiate and form a homeostatic relationship with each other that results in a tissue that would be expected to react to a xenobiotic applied to its surface in a manner similar to the epidermis. We have also developed methods of applying test substances to the surface of this keratinocyte culture system and have obtained data concerning the effects that certain classes of chemicals have on its metabolism of macromolecules. Epidermal cultures of this type have been established using both rat and human basal keratinocytes isolated from epidermal fragments. The system currently used for studying the effects of xenobiotics applied to the surface of the culture has as its substratum a commercially available nylon membrane, rather than biological materials such as collagen.

The early, initial molecular events that take place after topical application of bis(β-chloroethyl)sulfide (BCES) to the surface of the culture has been identified using this culture system.

II. DEVELOPING CULTIVATION PROCEDURES

A. Isolation of Rat or Human Keratinocytes for Cultivation

Two types of cells are the major structural components of the normal mammalian full-thickness skin: fibroblasts and keratinocytes. *Fibroblasts*, found in the dermis and responsible for the formation of collagen, can be cultured in vitro and subcultured in a basal medium with serum as the only supplement (9). *Keratinocytes*, which form the epidermal structure, do not proliferate as well in vitro in primary cultivation or after passage when serum is the only supplement to basal medium. Only the basal cells harvested from the living epidermis engage in proliferative activity (10,11). The more differentiated cells appear to have active DNA that engaged in repair but not semiconservative synthesis (12).

Several systems have been developed to promote increased serial cultivation of keratinocytes. Improvements resulted when cells were grown on collagen (13,14), on feeder layers of irradiated fibroblasts (15), or in conditioned medium (16). Later reports indicate that these systems may not be necessary (16–18) and that serial cultivation of keratinocytes can be achieved by supplementing the medium with growth promoters such as epidermal growth factor (19,20) and hormones (17,21). We reported a description of the morphological and cultural characteristics of rat keratinocytes after supplementing the growth medium with various concentrations and combinations of hydrocortisone and insulin (22).

In our investigations into successfully cultivating keratinocytes, full-thickness skins from both newborn rats and human biopsy specimens were obtained from which the desired cells were isolated, using essentially the same procedures. The skins were aseptically removed, placed in a petri dish and loose connective tissue and blood vessels scraped from the underside of the dermis. This scraping also caused the skin to stretch slightly. The resulting tissue was chilled to 4°C and covered with crude trypsin (1:250) prechilled to the same temperature. This combination of stretching and trypsinizing the skin at 4°C caused the dermis and epidermis to separate at the basement membrane and the loosening of desmosomal attachments between basal and lower spinous cells. These cells, mostly in the form of aggregates, were removed from the epidermal fragment by brushing them into growth medium containing fetal bovine serum (FBS). The remaining epidermal fragments containing differentiated cells and stratum corneum were discarded. Brushing the cellular components into growth medium also suspended single basal cells, fibroblasts, and cellular debris. These were removed by layering the resulting suspension on 10% Ficoll followed by centrifugation for 5 min at $13 \times g$ (22). The pellet containing aggregates of basal cells was resuspended in fresh medium, and this washing procedure repeated seven times to remove most of the single cells and debris. Cell aggregate viability was determined using a dye exclusion test with 0.5% erythrosin B (23). The viability of these suspensions was routinely greater than 98% as indicated by this method.

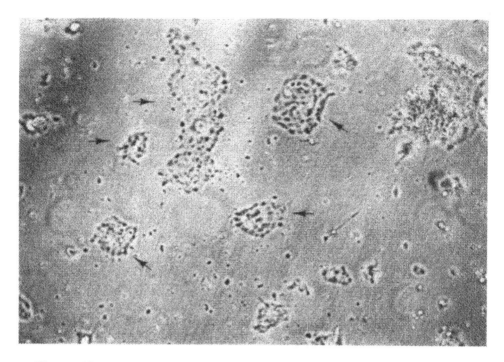

Figure 14.1 Attachment and growth of primary cultures of rat basal keratinocytes 24 hr after seeding on a plastic substratum. Aggregates of basal keratinocytes are forming growth centers (arrows). (Phase-contrast micrograph, × 140.)

The clarified suspension of basal keratinocytes was pelleted at $53 \times g$ for 10 min to obtain a packed volume and a 0.2% (vol/vol) suspension prepared in complete growth medium. These suspensions contained 5×10^5 cells/ml as determined by counting dissociated cells electronically. When these suspensions were seeded onto appropriate substrata and incubated at 35°C in an atmosphere of 5% CO_2 and 95% air with a humidity of 95%, the aggregates of basal cells and the few remaining spinous cells attached within 12 to 18 hr (22). Each aggregate of basal cells formed a growth center from which cells migrated and began proliferation as shown in Figure 1. Although the spinous cells attached initially, they did not proliferate and were detached and removed at the first change of growth medium (24 hr after seeding). Continued incubation for 7 days produced a confluent monolayer of closely connected epithelial cells (Fig. 2).

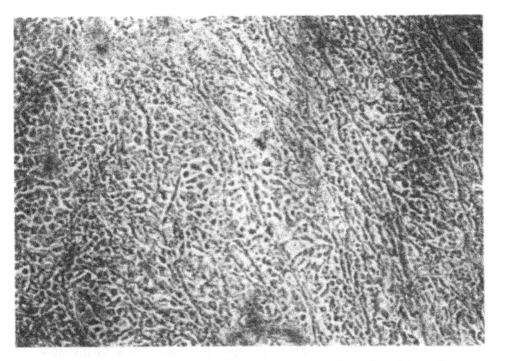

Figure 14.2 Proliferation of primary cultures of rat basal keratinocytes 7 days after seeding on a plastic substratum. A confluent
monolayer has developed. (Phase-contrast micrograph, × 140.)

B. Selection of an Appropriate Substratum for Attachment and Growth of Keratinocytes

1. *Plastic Substrata*

When sufficient amounts of cells were seeded initially on commercially
prepared plastic culture vessels, monolayers formed in 5 to 7 days
(example: 15×10^5 cells in a 60-mm plastic petri dish). The keratinocyte cultures just described (shown in Figs. 1 and 2) were
seeded on plastic vessels. Some stratification and differentiation
could also be observed 1.5 to 2 weeks after culture initiation.

2. *Collagen Substrata*

Various kinds of collagen gels were used as substrata and the resulting attachment and proliferation of keratinocytes after appropriate seeding observed. The gels were usually formed in the bottom

of plastic petri dishes and the cells seeded on the surface of the gels in sufficient medium to completely cover the cells.

When rat keratinocytes were seeded onto Vitrogen 100 bovine dermal collagen, which is composed of 95% type I and 5% type II collagens, firm attachment to this substratum resulted (24). However, 3 days after seeding, growth resulted in clusters of cells rather than in a uniform proliferation over the substratum surface as seen with plastic substrata. At this time, holes were also produced in the collagen gel, and the surface began to dissolve. If the culture was incubated longer, the entire collagen layer disappeared, and the cells were seen growing directly on the plastic surface of the container.

When rat tail collagen was used as a substratum for the growth of rat keratinocytes, attachment was observed 6 hr after plating, compared with 18 to 24 hr on commercial plastic vessels (24). By 24 hr, 75% of the surface of this substratum was covered with keratinocytes that had spread and proliferated. A confluent monolayer was formed in 4 days. However, gels formed from this collagen are very soft and difficult to handle.

Gels formed from mixtures of Vitrogen 100 and rat tail collagen were also investigated as appropriate substrata (24). A mixture containing equal amounts of the two types (1:1, vol/vol) supported attachment and growth of rat keratinocytes as well as rat tail collagen alone, and also resulted in a firm substratum that would allow relocation of the entire culture to the air—medium interface (a procedure that will be described in a later section). This collagen mixture also supported stratification and differentiation of submerged keratinocytes to a greater extent than plastic substrata.

3. Synthetic Membranes as Substrata

In a concentrated study, 11 commercially available and experimental synthetic membrane filters were selected and evaluated as appropriate substrata for rat keratinocytes (25). In Table 1, the source of each membrane and the material from which it was constructed are listed. Membranes TCM200, TCM440, HA-TF, and RA-TF were selected for study because they are specifically treated to remove detergent and are more suitable for tissue culture use. The Acroshield D (ACRO) and silicone—polycarbonate copolymer (MEM) were included because they are transparent and would allow cell growth to be continuously monitored by phase-contrast microscopy. All membranes used in this particular study were 13 mm in diameter and were placed in the wells of 24-well plastic culture vessels before seeding a suspension of keratinocytes on their surfaces. Wells without membranes were also seeded and served as a basis for comparing the resulting attachment and growth of cells on each membrane (plastic controls).

Table 14.1 Synthetic Membranes Selected as Substrata in Studies of Attachment and Proliferation of Rat Keratinocytes In Vitro

Membrane	Pore size (μm)	Material	Source
P200	0.20	Nylon (Puropor)	Gelman Sciences, Ann Arbor, Mich.
P450	0.45	Nylon (Puropor)	Gelman Sciences, Ann Arbor, Mich.
TCM200	0.20	Cellulose triacetate	Gelman Sciences, Ann Arbor, Mich.
TCM450	0.45	Cellulose triacetate	Gelman Sciences, Ann Arbor, Mich.
HA-TF	0.45	Cellulose nitrate / Cellulose acetate	Millipore, Bedford, Mass.
RA-TF	1.20	Cellulose nitrate / Cellulose acetate	Millipore, Bedford, Mass.
HT	0.20	Polysulfone (200W)	Gelman Sciences, Ann Arbor, Mich.
ACRO	0.00	Polyethylene	Gelman Sciences, Ann Arbor, Mich.
MEM	1.00	Silicon polycarbonate	General Electric, Schenectady, N.Y.

Initial attachment and proliferation of rat keratinocytes on selected membranes and plastic culture vessels have been compared
and reported (25). The extent of attachment and growth of cells on
opaque membranes were determined after fixing the cells and substratum in 10% phosphate-buffered formalin or Carnoy's solution (90%
butanol, 10% acetic acid). The fixed culture was stained with hematoxylin, dehydrated, cleared with xylene, and mounted on microscopic slides with Permount and coverslips. Quantification of initial
attachment and subsequent proliferation was accomplished by counting stained nuclei per unit area of the substratum. This was performed electronically with the Bioquant Image Analysis system on the
Apple IIE microcomputer and accessories. Cross sections for microscopic examination were obtained by first fixing selected cultures in
10% buffered formalin, embedding in paraffin, sectioning with a ro-
tary microtome, and mounting sections on microscopic slides and,
finally, staining with hematoxylin and eosin (H&E).

Epidermal cultures grown on synthetic membranes were also examined at the ultrastructural level. Cultures were fixed in Karnov-
sky's fixative, washed, and prestained with 2% osmium tetroxide,
stained enblock with uranyl acetate, dehydrated, and embedded in
resin. The embedded specimens were polymerized, sectioned, post-
stained with uranyl acetate and lead citrate, and observed with an
AEI Corinth 275 transmission electron microscope (TEM).

Figure 3 illustrates the initial attachment (see Fig. 3A) and en-
suing proliferation (see Fig. 3B) of rat keratinocytes on synthetic
membranes, compared with growth on a plastic substratum. The
data in this figure indicate that cultivation on some of the selected
membranes was superior to that on plastic culture vessels used as
controls. Observation of cultures after a 1-day incubation showed
that aggregates of keratinocytes had attached and spread out,
forming growth centers. The large variation in cell number per
unit area seen in Figure 3A reflects the different size and unequal
distribution of growth centers in early cultures. After 10 days of
incubation, monolayers had formed with varying degrees of compact-
ness, depending on the substratum (see Fig. 3B). Membranes spe-
cially prepared for tissue culture procedures (i.e., TCM200, TCM-
450, HA-TF, and RA-TF) supported attachment and proliferation
equal to or better than the plastic substratum. The increased attachment and proliferation of keratinocytes on Puropor nylon membranes (P200 and P450) over controls was statistically significant and
proliferation on P200 was superior to the others examined. Therefore, this membrane was selected for further study in producing
stratified, differentiated keratinocyte cultures in vitro (25).

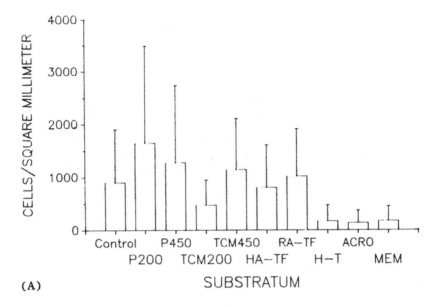

(A)

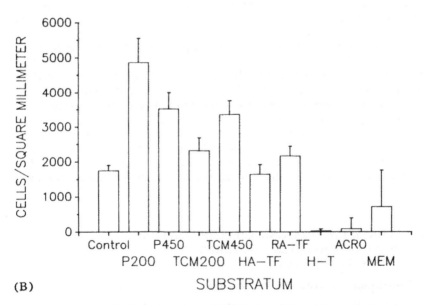

(B)

Figure 14.3 Attachment and proliferation of rat keratinocytes on sterilized synthetic membranes. (A) Attachment and growth for 1 day; (B) proliferation for 10 days. Data are presented as mean + SD, n = 4–5.

4. Attachment Factors Investigated in the Development of Epidermal
    Cultures

Two attachment factors were studied for their ability to enhance the
attachment and proliferation of rat keratinocytes seeded on synthetic
membranes (25). Human fibronectin (HFN) is a glycoprotein isolated
from human serum and was found to be essential for promoting the
attachment of epithelial cells in vivo (26). Laminin (LMN) is a protein that is part of the structure of the basement membrane of the
epidermis (27) and is commercially available, as is HFN.

The membranes precoated with HFN did not support the attachment and proliferation of keratinocytes better than untreated membranes ($P > 0.05$) as shown in Figures 4A and 4B. On membranes
P200, TCM200, and TCM450, there appeared to have been less proliferation on HFN-coated than on uncoated membranes as observed
after 10 days of incubation (see Fig. 4B). The coating of membranes with LMN enhanced the initial attachment, compared with untreated control membranes after 1 day of incubation (see Fig. 4A).
Enhanced attachment and spreading of seeded keratinocytes across
the substratum was particularly evident when LMN pretreated
Puropor nylon membranes were compared with untreated membranes.
However, only slight, although significant, differences in the total
number of cells on untreated or LMN pretreated nylon membranes
were observed after incubation for 10 days ($P < 0.01$; see Fig. 4B).
Therefore, precoating nylon membranes with LMN appeared to enhance initial attachment but did not support proliferation more than
did the untreated membrane. The LMN-coated HA-TF and RA-TF
membranes also supported keratinocyte growth substantially better
than uncoated membranes ($P < 0.001$). However, detachment of rat
keratinocytes as sheets from the membranes and culture vessels
precoated with LMN was observed in most cultures after incubation
for approximately 14 days. Because of these observations, neither
HFN nor LMN were used in further studies of the cultivation of rat
keratinocytes on synthetic membranes (25). Studies of the effect
of attachment factors on the cultivation of human keratinocytes are
incomplete.

III. STRATIFICATION AND DIFFERENTIATION
      OF KERATINOCYTES CULTURED ON
      NYLON MEMBRANES

A. Two-Stage Cultivation Procedures

Isolated keratinocytes were seeded on untreated nylon membranes
and incubated as submerged cultures until monolayers resulted, or
up to 7 days. The cultures were then raised to the air—medium in-

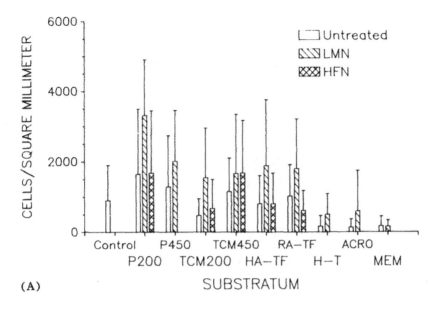

(A)

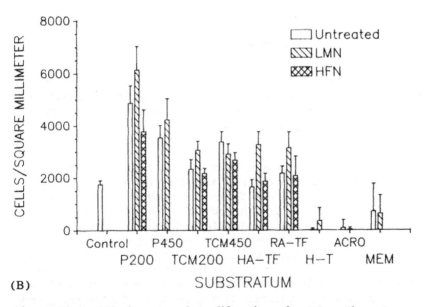

(B)

Figure 14.4 Attachment and proliferation of rat keratinocytes on synthetic membranes that were untreated (UNTR), precoated with LMN, or precoated with HFN. (A) Growth for 1 day; (B) growth for 10 days. Data are presented as mean + SD, n = 4—5.

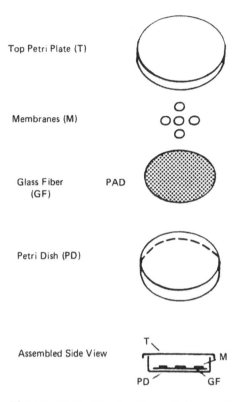

Top Petri Plate (T)

Membranes (M)

Glass Fiber   PAD
(GF)

Petri Dish (PD)

Assembled Side View

Figure 14.5 Illustration of the method used in lifting keratinocyte monolayers to the air—medium interface. Puropor nylon membranes containing the monolayer are placed on the surface of glass fiber filters saturated with growth medium.

terface by using the procedure illustrated in Figure 5. The membranes containing the monolayer of keratinocytes were transferred to the surface of a glass fiber filter positioned at the bottom of a petri dish and soaked in growth medium. This permits the culture to be fed from the underside while the cells are in contact with the atmosphere above (in the same manner as an organ culture).

This two-stage cultivation procedure (25) resulted in differentiated multilayers of keratinocytes on the synthetic nylon membrane. Various stages of differentiation were also observed that paralleled those observed in the epidermis formed in vivo. Multilayering and differentiation increased with the age of the lifted culture, up to approximately 12 to 14 days, after which time the number of layers remaining and the extent of differentiation did not change. Con-

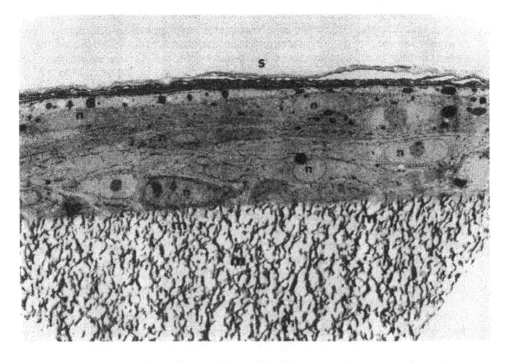

Figure 14.6 A TEM of an epidermal culture grown on a nylon membrane submerged for 7 days and lifted to the air—medium interface for 14 days. Several layers of nucleated cells (n) formed above the membrane (m) and numerous layers of enucleated cells formed a stratum corneum at the culture surface (s) (× 2200).

siderable stratification and differentiation took place between 4 and 8 days of incubation at the air—medium interface and after 8 days, four to five nucleated layers and 15 to 20 enucleated layers were produced.

B. Ultrastructural Examination of Resulting Cultures

Figure 6 is a transmission electron micrograph (TEM) of a cross section of a culture incubated submerged for 7 days and then lifted to the air—medium interface for an additional 14 days. Five to six nucleated and 15 to 20 enucleated layers of differentiating keratinocytes were produced.

Various markers of epidermal differentiation were clearly demonstrated in TEMs at higher magnifications. A culture lifted for 14 days showed abundant desmosomes, tonofilaments, and mitochondria in the basal layer (Fig. 7), and electron-dense granules, morphologically similar to in situ keratohyalin granules, and cells resembling those in the in situ stratum corneum were evident in the upper layers of lifted cultures at this age (Fig. 8). Many of these epidermal markers do not develop in submerged cultures.

IV. EXPERIMENTATION WITH SULFUR MUS-
TARD IN CUTANEOUS TOXICITY STUD-
IES USING THE EPIDERMAL CULTURE

A. Topical Application of a Xenobiotic to the Surface of the Lifted Cultures

This section describes the experimentation to measure the toxic response of epidermal cultures after topical application of the sulfur mustard, bis(β-chloroethyl)sulfide (BCES). Details of the methodology have been reported (28). Aliquots of BCES dissolved in methylene chloride (MC) (10 mg/ml) were further diluted to selected concentrations in one of the following solvents: 2,4-pentanedione (acetyl acetone), ethanol (ETOH), hexane, or dimethyl sulfoxide (DMSO). A total of 0.04 ml solvent or of solvent containing BCES was applied to the surface of an epidermal culture grown on a 13-mm disk of P200 and lifted for 14 days. This amount covered most of the culture surface without flowing over the side onto the supporting glass-fiber filter. The amount of material applied was expressed as nanomoles per square centimeter of the culture surface ($nmol/cm^2$). Solvent or solvent—BCES solutions were removed from the culture surface by gently washing in Earle's balanced salt solution (EBSS). The washed cultures were placed on the surface of supporting glass-fiber filters saturated with fresh growth medium and incubation continued for preselected periods. The cultures were next transferred to the appropriate medium containing radiolabeled precursors when experiments were conducted to observe effects on macromolecular metabolism subsequent to various treatment (28).

Data were obtained from preliminary experiments to select an appropriate solvent for topical application of xenobiotics to the surface of epidermal cultures. Acetyl acetone, hexane, and MC were all highly toxic as indicated by significant reductions in the incorporation of [3H] thymidine into DNA, when compared with untreated controls. In further studies, it was observed that dilutions of ETOH and DMSO in distilled water were much less toxic and were also appropriate solvents for BCES. Table 2 shows the results of both [3H] thymidine and [14C] leucine incorporation into DNA and

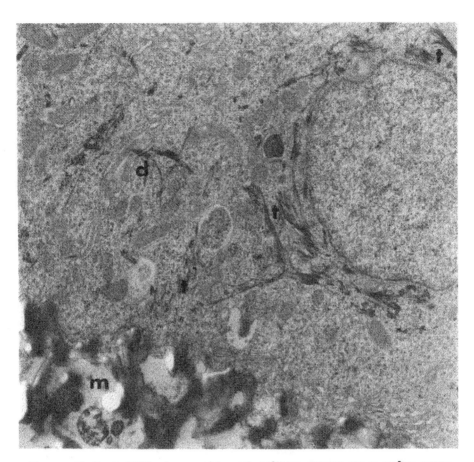

Figure 14.7 A TEM of an epidermal culture grown on a nylon membrane submerged for 7 days and lifted to the air—medium interface for 14 days. This is a higher magnification of the lower layers next to the membrane (m) showing tonofilaments (t), and desmosomes (d) (× 12,000).

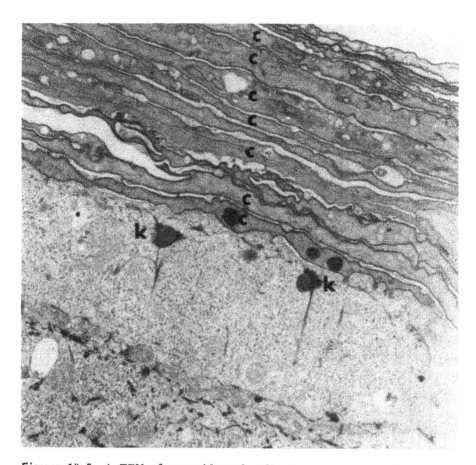

Figure 14.8 A TEM of an epidermal culture grown on a nylon membrane submerged for 7 days and lifted to the air—medium interface for 14 days. This is a higher magnification of the upper layers of the culture showing keratohyalin granules (k) and cornified cells (c) (× 12,000).

Table 14.2 Incorporation of [^{3}H] Thymidine and [^{14}C]Leucine by Epidermal Cultures After Topical Exposure to DMSO and ETOH

	Thymidine (CPM)		Leucine (CPM)	
	Exposure time (hr)		Exposure time (hr)	
Treatment[a]	2.5	24	2.5	24
Untreated control[b]	1190 ± 279	798 ± 125	2.69 ± 0.6	1.87 ± 0.19
70% DMSO	658 ± 138	564 ± 110	1.27 ± 0.03	1.42 ± 0.29
40% ETOH	836 ± 82	542 ± 162	2.05 ± 0.78	1.12 ± 0.24
Heat killed		83 ± 21		0.20 ± 0.06

[a]Cultures treated by topical application of 0.04 ml solvent.
[b]Culture controls treated by topical application of 0.04 ml saline. The values are the mean ± SD (n = 3).

protein, respectively, after topical application of these diluted solvents to culture surfaces (28). These data demonstrated that exposure of the cultures to 40% ETOH or 70% DMSO affected metabolic activity to a similar extent. There was no significant difference in the incorporation of either thymidine or leucine by cultures exposed to the two solvents for 2.5 hr or 24 hr (*P* > 0.01 in all comparisons). The table also compares the effects of the two solvents with those of untreated and heat-killed controls. It was decided to use 70% DMSO, which was not toxic at a higher percentage than ETOH, as the solvent in subsequent studies with BCES.

B. The Effect of Sulfur Mustard on Macromolecular Metabolism in Exposed Epidermal Cultures

In all experiments to determine the effect of BCES on macromolecular metabolism of exposed epidermal cultures, two controls were employed: untreated cultures handled identically with treated cultures but not exposed to the xenobiotic or to the solvent used to dissolve the agent; and solvent controls that were similarly handled including exposure to the solvent without BCES. Three cultures were used for each experimental observation. Counts per minute per milligram of protein or DNA were determined for each observation and results expressed as a percentage of untreated controls (28).

1. The Optimal Period for Exposing Epidermal Cultures to Sulfur
    Mustard

Cultures were exposed for 0.5, 1, and 2 hr to 50 to 500 nmol/cm^2
of BCES dissolved in 70% DMSO. The effect of this agent on DNA,
RNA, and protein metabolism, as detected by the use of the radiolabeled precursors [^{3}H]thymidine, [^{3}H]uridine, and [^{14}C]leucine,
respectively, is summarized in Figures 9A (thymidine), 9B (uridine),
and 9C (leucine). These data indicate that the major effect of
BCES takes place within 30 min after topical application. At that
time, the incorporation of all three precursors was decreased after
exposure to all concentrations of BCES when compared with solvent
controls (P < 0.001). The most drastic decrease in macromolecular
metabolism was in the incorporation of [^{3}H]thymidine (see Fig. 9A).
Decreases in the incorporation of the other two macromolecular precursors after exposure was also evident. Both uridine and leucine
incorporation was inhibited by exposure to 200 and 500 nmol/cm^2 of
BCES (P < 0.001) (see Figs. 9B and 9C). However, at lower concentrations, much less inhibition resulted, when compared with inhibition of DNA metabolism at those concentrations.

     These data demonstrate that a 30-min exposure to BCES was
sufficient time to initiate early lesions in cellular function of the
epidermal culture as determined by inhibition of macromolecular synthesis.

2. Dose-Dependent Inhibition of Macromolecular Metabolism After
    Exposure of Epidermal Cultures to Sulfer Mustard

Experiments were performed to identify the minimal concentration of
BCES that would cause measurable alteration in macromolecular metabolism. Concentrations of BCES were selected ranging from 0.01
to 500 nmol/cm^2, and the data resulting from several experiments
were combined. The cultures were exposed for 30 min to the selected concentrations of BCES dissolved in 70% DMSO, and the agent
washed away with EBSS. Treated cultures were transferred to the
labeling medium immediately (0 hr) or incubated in fresh growth
medium for 4, 8, or 24 hr after exposure (postexposure period) before transfer to labeling medium. The purpose of this approach
was to determine the minimum concentration of BCES that would result in a significant inhibition of metabolism in the treated culture
and to observe the extent of this inhibition, or possible recovery,
over a period. Figures 10A and 10B show the results of [^{3}H]thy-
midine incorporation. Figures 11A and 11B show results of incorporation of [^{3}H]uridine and [^{14}C]leucine, respectively.

     The data in Figure 10A show the incorporation of radiolabeled
thymidine by the cultures after exposure to BCES, 0.0, 0.01, 0.1,
and 0.5 nmol/cm^2. There was an initial stimulation of thymidine in-

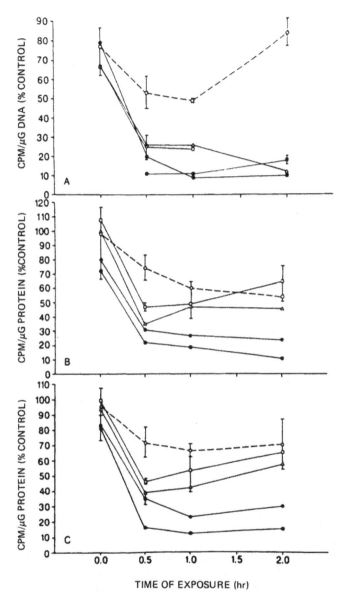

Figure 14.9 Incorporation of (A) [^{3}H] thymidine, (B) [^{3}H] uridine, and (C) [^{14}C] leucine by epidermal cultures after exposure to BCES in 70% DMSO for 30 min, 1 hr, and 2 hr. BCES concentrations (μmol/cm^2) were: 0.0 (solvent control), open circles; 50, open squares; 100, open triangles; 200, closed circle; 500, closed square. Results are expressed as percentage of untreated control ± SD (n = 9).

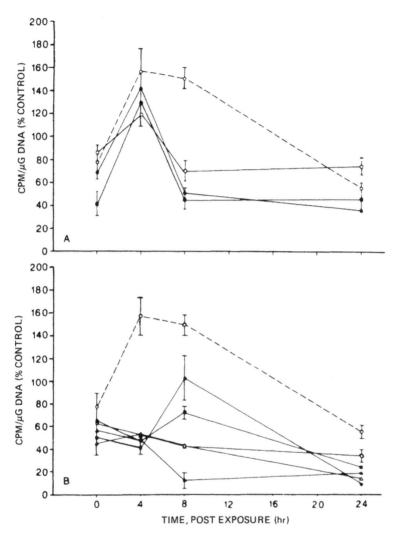

Figure 14.10 Incorporation of [³H] thymidine by epidermal cultures
after a 30-min exposure to BCES in 70% DMSO. The cultures were
pulse-labeled after removal of BCES followed by selected periods of
postexposure incubation. BCES concentrations (μmol/cm^2) were
(A) 0.0 (solvent control), open circles; 0.01, open square; 0.1,
closed circles; 0.5, closed squares: (B) 0.0, open circles; 1.0,
open squares; 10, open triangles; 50, closed circles; 200, closed
squares; 500, closed triangles. Results are expressed as percentage
of untreated control ± SD (n = 9).

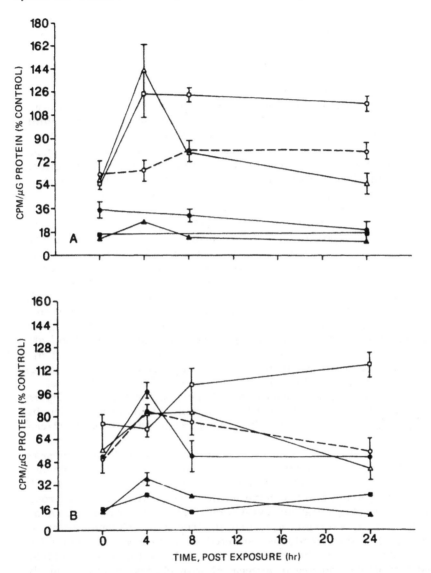

Figure 14.11   Incorporation of (A) [^{3}H] uridine and (B) [^{14}C] leucine by epidermal cultures after a 30-min exposure to BCES in 70% DMSO.   The cultures were pulse-labeled after removal of BCES followed by selected periods of postexposure incubation.   BCES concentrations (μmol/cm^2) were: 0.0 (solvent control), open circles; 0.01, open squares; 1.0, open triangles; 10, closed circles; 50, closed squares; 200, closed triangles.   Results are expressed as percentage of untreated control ± SD (n = 9).

corporation by both treated cultures and solvent controls 4-hr post-exposure, probably caused by stimulation by the solvent. By 8-hr postexposure, all concentrations of BCES included in this figure caused inhibition of incorporation ($P < 0.001$). A concentration of 0.01 nmol/cm^2 caused inhibition for at least 8-hr postexposure, but cultures exposed to this dose showed a recovery of metabolic activity by 24 hr, when compared with the solvent control. However, concentrations of 0.1 and 0.5 nmol/cm^2 of BCES resulted in significant inhibition of incorporation of radiolabeled thymidine as observed at 4- and 24-hr postexposure. Figure 10B includes concentrations of 1 to 500 nmol/cm^2 of BCES and summarizes their effects on thymidine incorporation by epidermal cultures. These concentrations caused significant inhibition at 4 hr and at 24 hr ($P < 0.001$ for both observations), compared with solvent controls.

Figure 11A shows the results of [^{3}H] uridine incorporation by cultures after exposure to BCES. Incorporation of this precursor after exposure to 0.01 nmol/cm^2 of BCES appeared to be stimulated, when compared with solvent controls ($P < 0.001$). This was evident 4 hr after exposure and remained for the full observation period (24 hr). A concentration of 1.0 nmol/cm^2 caused mild, but significant, inhibition of incorporation 24-hr postexposure, after an initial stimulation of incorporation 4-hr postexposure. The higher concentrations (10–200 nmol/cm^2) caused substantial inhibition of incorporation of this precursor. Similar results were obtained in the incorporation of radiolabeled leucine by lifted cultures after exposure to BCES (see Fig. 11B). Exposure to 0.01 nmol/cm^2 of BCES stimulated the incorporation of radiolabeled leucine, compared with solvent controls ($P < 0.001$). Significant inhibition of leucine incorporation occurred only after exposure to 50 and 200 nmol/cm^2 of BCES.

The data in these experiments indicated that only at comparatively high concentrations of BCES was there significant interference with RNA and protein metabolism. Inhibition in the uptake of radiolabeled uridine and leucine at these concentrations was probably a result of generalized cytotoxicity caused by massive DNA alkylation and irreversible molecular destruction. Thymidine metabolism, on the other hand, was affected at much lower concentrations. These data support the conclusion that DNA is the most important target of BCES and are in agreement with observations of a number of investigators. Crathorn and Roberts (29) found that exposure to a concentration causing DNA damage and interference with replicative ability in HeLa cells did not interfere with RNA or protein synthesis. Lawley and Brookes (30) observed that the cytotoxic action of sulfur mustard was not associated with the inhibition of growth of *Escherichia coli* as measured by RNA or protein synthesis, but with inhibition of DNA synthesis.

This study demonstrated that the epidermal culture formed with rat keratinocytes behaved similarly to other in vitro systems in its initial reaction to topically applied BCES. More directed studies may elucidate what happens to the alkylated DNA molecule upon exposure to the agent and how this initial molecular event is related to subsequent cellular destruction leading to vesication. The morphological markers (25) and biochemical markers (24) of this culture system provide the opportunity to simulate what happens in the intact epidermis. It may as a result be possible to identify the initial molecular event leading to vesication and, also, because of the ordered differentiation of the culture, there can be a determination of which cells of the epidermis are affected initially.

V. SUMMARY AND CONCLUSIONS

The studies described in this chapter demonstrated that keratinocytes attach, proliferate, and differentiate on selected synthetic membranes that are untreated except for sterilization. Growth on Puropor nylon membranes was superior to that on other membranes but there were others that supported growth equal to, or better than, commercially obtained plastic culture vessels. Membranes precoated with HFN gave inconsistent results, indicating that using this attachment factor offered little additional value in supporting rat keratinocyte attachment and proliferation in vitro. Precoating membranes with LMN initially supported attachment and growth of rat keratinocytes better than untreated membranes, but extended incubation of cells growing on that substratum proved unsuccessful.

Attachment and proliferation of keratinocytes on the proper substratum was essential in developing an epidermal culture for toxicological studies. However, differentiation of cultured keratinocytes was equally if not more important. The homeostatic relationship between keratinocytes in various stages of differentiation may have significant effects on the ability of an agent to cause tissue damage. Thus, establishing a microenvironment that fosters differentiation as well as controlled proliferation is mandatory.

Another characteristic a successful in vitro system must have if it is to be used for studying the toxic effects of xenobiotics on the epidermis is simple, reproducible construction. This is necessary if data obtained from different laboratories are to be used collectively in evaluating the effects of toxic agents. The successful cultivation and differentiation of rat keratinocytes on nylon membranes helps to fulfill these criteria. Puropor nylon is a synthetic material that can be manufactured with minimal variation in its chemical nature, whereas the preparation of collagen, for example, can vary significantly from lot to lot and, thus, is difficult to standardize. Comparatively

large amounts of the nylon membrane can be sterilized by various
methods that will not change its chemical structure, and no further
treatment is necessary for optimal attachment and growth of primary
keratinocytes.

The studies described in this chapter also demonstrated that
the epidermal culture formed from isolated rat keratinocytes behaved similarly to other in vitro systems in its initial reaction to
topically applied BCES. DNA was the main target of this agent in
this system, the morphological and biochemical structures of which
mimic the in situ epidermis. More directed studies may elucidate
what happens to the alkylated DNA molecule upon exposure to this
agent and how this initial molecular event is related to subsequent
cellular destruction leading to vesication. This approach can also
be used to verify earlier observations in whole skin that the first
cells destroyed after BCES exposure are basal and lower spinous
cells, and that this leads to blister formation.

Studies of the untoward actions of drugs on the skin after topical application are necessary before these chemicals can be released
for human use. Investigation of percutaneous absorption are frequently performed in vitro using membranes of animal or human cadaver skin that bear little similarity to the living human epidermis
and suffer enormous variability. As a result, they may show little
resemblance to the clinical performance of a given applied drug
(31). Therefore, what is needed is an in vitro model that possesses
barrier characteristics similar to those of the intact living epidermis
and that offers highly reproducible characteristics. The epidermal
culture developed in this laboratory resembles the in situ human
epidermis both morphologically and biochemically. With further re-
finement this system may become a valuable tool in investigations of
percutaneous characteristics pertaining to human health. Studies
have been initiated to evaluate the effectiveness of this in vitro
system in measuring transepidermal water loss (TEWL).

REFERENCES

1. J. H. Draize, G. Woodard, and H. O. Calvery, *Pharmacol.
    Exp. Ther.*, *82*:377, 1944.
2. C. S. Weil and R. A. Scala, *Toxicol. Appl. Pharmacol.*, *19*:
    276, 1971.
3. G. A. Nixon, C. A. Tyson, and W. C. Wertz, *Toxicol. Appl.
    Pharmacol.*, *31*:481, 1975.
4. K. J. Falahee, C. S. Rose, H. E. Seifried, and D. Sawhney,
    in *Product Safety Testing*   (A. M. Goldberg, ed.). Mary
    Ann Liebert, New York, p. 137, 1983.

5. J. L. Schmidt, F. C. McIntire, D. L. Martin, M. A. Hawthorne, and R. K. Richards, *Toxicol. Appl. Pharmacol.*, *1*: 454, 1959.
6. W. L. Guest, S. A. Rosenbluth, B. Schmidt, and J. Autian, *J. Pharm. Sci.*, *54*:1545, 1965.
7. F. Hu, C. S. Livinggood, and J. F. Hildebrand, *J. Invest. Dermatol.*, *26*:23, 1955.
8. F. Kao, J. Hall, and F. M. Holland, *Toxicol. Appl. Pharmacol.*, *68*:206, 1983.
9. W. R. Earl, *Fed. Proc.*, *17*:967, 1958.
10. M. Karasek and T. Moore, *J. Invest. Dermatol.*, *56*:205, 1971.
11. F. L. Vaughan, and I. A. Bernstein, *J. Invest. Dermatol.*, *56*:454, 1971.
12. F. L. Vaughan, R. S. Mitra, and I. A. Bernstein, *J. Invest. Dermatol.*, *66*:355, 1976.
13. M. Karasek and M. F. Charlton, *J. Invest. Dermatol.*, *56*: 205, 1971.
14. A. E. Freeman, H. J. Ingel, B. J. Herrman, and K. N. Kleinfeld, *In Vitro*, *12*:352, 1976.
15. J. G. Rheinwald and H. Green, *Cell*, *6*:331, 1975.
16. D. M. Peel and R. G. Ham, *In Vitro*, *16*:516, 1980.
17. E. Eisinger, J. S. Lee, J. M. Hefton, Z. Darzynkiewicz, J. W. Chiao, and E. DeHarven, *Proc. Natl. Acad. Sci. USA*, *76*: 5340, 1979.
18. D. M. Peehl and R. G. Ham, *In Vitro*, *16*:526, 1980.
19. S. Cohen, and C. R. Savage, *Recent Prog. Horm. Res.*, *31*: 551, 1974.
20. J. G. Rheinwald and H. Green, *Nature*, *265*:421, 1977.
21. I. Hayashi, J. Larher, and G. Sato, *In Vitro*, *14*:23, 1978.
22. F. L. Vaughan, L. L. Kass, and J. A. Uzman, *In Vitro*, *17*: 941, 1981.
23. J. H. Phillips and J. E. Terryberry, *Exp. Cell Res.*, *13*:341, 1957.
24. L. I. Bernstam, F. L. Vaughan, and I. A. Bernstein, *In Vitro*, *22*:695, 1986.
25. F. L. Vaughan, R. H. Gray, and I. A. Bernstein, *In Vitro*, *22*:141, 1986.
26. R. J. Klebe, *Nature*, *250*:248, 1974.
27. R. Timpl, H. Rhode, P. G. Robey, S. I. Rennard, J. M. Foidart, and G. R. Martin, *J. Biol. Chem.*, *254*:9933, 1979.
28. F. L. Vaughan, S. Zaman, R. Scavarelli, and I. A. Bernstein, *J. Toxicol. Environ. Health*, *23*:507, 1988.
29. A. R. Crathorn and J. J. Roberts, *Nature*, *211*:150, 1966.
30. P. D. Lawley and P. Brookes, *Nature*, *206*:480, 1951.
31. G. L. Flynn, E. M. Topp, and G. L. Amidon, in *Percutaneous Absorption* (R. L. Bronaugh, and H. I. Maibach, eds.), Marcel Dekker, New York, p. 17, 1985.

III

NONTRADITIONAL TOPICAL DRUG DELIVERY FORMULATIONS

15

The Microsponge: A Novel Topical Programmable Delivery System

SERGIO NACHT and MARTIN KATZ   *Advanced Polymer Systems, Inc.,
Redwood City, California*

I. THE NEED FOR BETTER SKIN DELIVERY SYSTEMS

Over the past 40 years, the ability to control the delivery rate of
active agents to a predetermined site in the human body has been
one of the biggest challenges—met by continued innovative solutions—that has faced the medical profession and drug industry.

During this time, some areas of pharmaceutical research have
been focused on the controlled delivery of systemic drugs. Several
predictable and reliable systems were developed for systemic drugs
under the heading of *transdermal delivery* using the skin as a por-
tal of entry (1). Transdermal patches, developed in the 1970s, improved the delivery of drugs such as nitroglycerin and scopolamine,
resulting in better control of therapeutic doses, simpler dosage regimens, and fewer side effects than the more traditional oral or parenteral administration of the same drugs. In general, these delivery systems have improved the efficacy and safety of many drugs
that now may be better administered through the skin.

Although transdermal delivery systems can be efficient in sup-
plying drugs for systemic effects, they are not practical for controlling the delivery of materials whose final target is the skin itself.

Controlled release of drugs onto the epidermis, with assurance
that the drug remains primarily localized and does not enter the
system in significant amounts, is an area of research that has only
recently been addressed with success. No efficient vehicles have
been developed for the controlled and localized delivery of drugs

into the stratum corneum and underlying skin layers. Yet, there
are many instances when epidermal localization of a drug is desirable, but absorption beyond the epidermis is not.

Corticosteroids are a typical example. Although effective for
skin disorders, their topical application can result in significant
systemic absorption; this may lead to unwanted side effects such as
adrenal suppression or interference with immune functions (2).

Similarly, with sunscreens it is necessary to maximize the amount
of time that the active ingredient is present on the skin surface or
within the outer layers of the epidermis while minimizing its transepidermal penetration into the body.

Another problem with the application of topical drugs is that
many vehicles, such as ointments, often prove aesthetically unap-
pealing. Greasiness, stickiness, or even discolorations in clothing
can make daily wear unpleasant. This frequently results in the patient's lack of compliance with treatment.

Many of these conventional vehicles require high concentrations
of active agents for effective therapy because of their low efficiency
as delivery systems. As a consequence, irritation or allergic responses can be elicited in a significant percentage of users. Other
disadvantages of existing topical drug formulations can be uncon-
trolled evaporation of the active ingredient, unpleasant odor, and
the potential incompatibility of one or more drugs with each other or
with the vehicle.

Thus, the need exists for systems to maximize the amount of
time that an active ingredient, such as a sunscreen, is present,
either on the skin surface or within the epidermis, while minimizing
its transdermal penetration into the body.

Such a new system would possibly increase the efficacy of topically active agents while enhancing product safety.

The Microsponge® polymeric microsphere system uniquely fulfills
these requirements. These tiny spongelike spherical particles (Fig.
1) can entrap active ingredients and then release them onto the
skin over time and in response to a trigger. They are biologically
inert, nonirritating, nonmutagenic, nonallergenic, nontoxic, and
nonbiodegradable. They can extend product stability because of
their unique configuration and improve its aesthetic properties.

II. MICROSPONGE TECHNOLOGY

Microsponges are patented (3) polymeric delivery systems consisting
of porous microspheres that can entrap a wide range of active ingredients, such as emollients, fragrances, essential oils, sunscreens,
and anti-infective, antifungal, and anti-inflammatory agents. These

entrapped active agents can then be incorporated into many product forms, such as creams, lotions, powders, and soaps. After the product is applied, the entrapped materials are then delivered to the skin in a controlled time-release pattern or a preprogrammed manner through the use of several different "triggers": rubbing or pressing the microsponge after it has been applied to the skin; elevating skin surface temperature; introducing solvents for the entrapped material, such as water, alcohol, or even perspiration; and controlling the rate of evaporation.

Like a true sponge, each microsphere consists of a myriad of interconnecting voids within a noncollapsible structure with a large porous surface. The size of the Microsponge can be varied, usually from 5 to 300 μm in diameter, depending upon the degree of smoothness or after-feel required for the end formula. Although the Microsponge size may vary, a typical 25-μm sphere can have up to 250,000 pores and an internal pore structure equivalent to 10 ft in length, providing a total pore volume of about 1 ml/g. This results in a large reservoir within each Microsponge, which can be loaded with up to its own weight in active agent.

III. MICROSPONGE SYNTHESIS

Microsponge particles are conveniently formed by suspension polymerization in a liquid—liquid system (3).

In general, a solution is made comprising the monomers and the functional or active ingredient which is immiscible with water. This phase is then suspended with agitation in an aqueous phase, usually containing additives, such as surfactants and dispersants, to promote suspension.

Once the suspension is established with discrete droplets of the desired size, polymerization is effected by activating the monomers either by catalysis, increased temperature, or irradiation.

The result is a series of polymer "ladders" wrapping around one another into solid microspheres (Fig. 2). As the polymerization process continues, a spherical structure is produced containing thousands of microspheres bunched together like grapes, forming interconnecting reservoirs in which the porogen is entrapped. These reservoirs open onto the surface of the sphere as pores through which the active ingredient can be released when triggered.

Once polymerization is complete, the solid particles that result from the process are recovered from the suspension. The particles are then washed and processed until they are substantially ready for use.

Particle formation and incorporation of the functional substance is thus performed as a single step. This may be termed a one-step procedure.

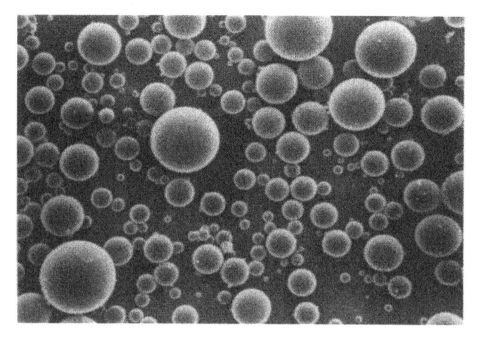

(A)

Figure 15.1 Scanning electron micrographs of Microsponges. Magnification: (A) × 240; (B) × 1200; (C) × 6000; (D) × 24,000; (E)
× Freeze-fracture section × 6000; (F) × Freeze-fracture section ×
24,000.

Functional substances that can be entrapped in this manner
must meet the following criteria:

1. They are either fully miscible with the monomer mixture or ca-
    pable of being made fully miscible by the addition of a minor
    amount of a non—water-miscible solvent.
2. They are immiscible with water or, at most, only slightly solu-
    ble in it.
3. They are inert to the monomers and stable when in contact
    with any polymerization catalyst used and when subjected to
    conditions needed to induce polymerization (such as temperature
    and radiation).

Because of these requirements, often the functional substance
itself may not serve as the porogen or pore-forming agent. In

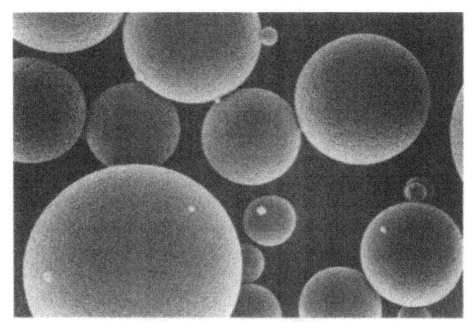

(B)

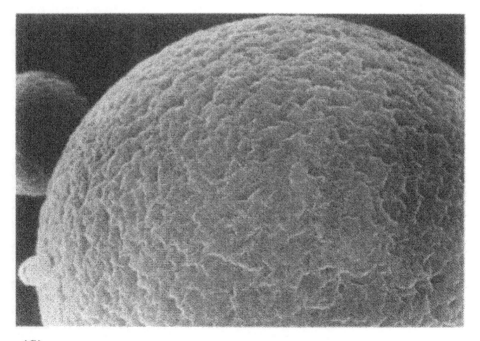

(C)

Figure 15.1 (Continued)

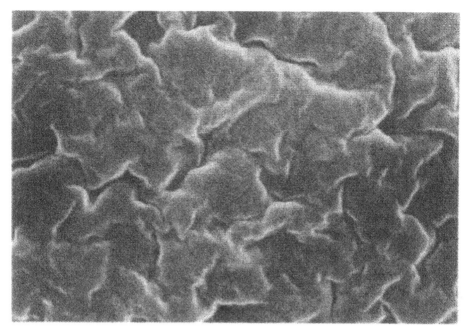

(D)

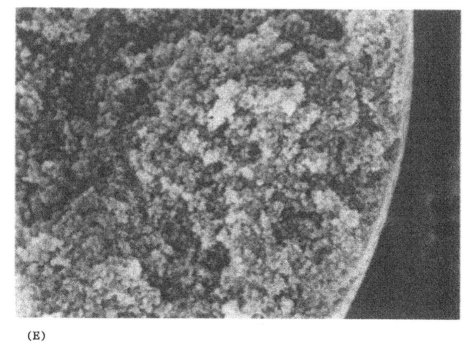

(E)

Figure 15.1 (Continued)

(F)

Figure 15.1 (Continued)

these cases, an inert liquid, fully miscible with the monomers but immiscible with water, is used during polymerization to form the pore network. The liquid that has served as a porogen and occupies the pores of the formed particles can then be removed leaving preformed dry microspheres.

Subsequently, placement of the functional substance inside the reservoirs may be achieved by impregnation of the preformed dry porous polymer particles according to techniques, such as contact absorption, assisted, when necessary, by solvents to enhance the absorption rate. The final product is thus prepared in two sequential steps. First, polymerization is performed using the substitute porogen. This substitute is then removed and replaced by the functional substance.

Materials suitable as substitute porogens are liquid substances meeting the foregoing criteria and having the further characteristic of being readily extracted from the particles' pore network once polymerization is complete. This covers a wide range of substances, in particular inert, nonpolar organic solvents.

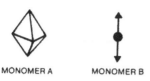

MONOMER A   MONOMER B

Step One: Selecting the Monomers

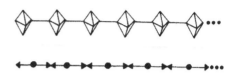

Step Two: Making Monomer Chains

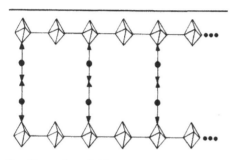

Step Three: Crosslinking
        the Chains (Polymers)

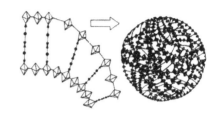

Step Four: Wrapping-up the Polymer Ladders
           into Microspheres

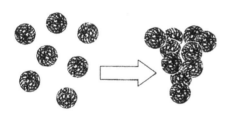

Step Five: Gathering Bunches of Polymeric
           Microspheres

Step Six: Binding the Bunches
          to Form a Polymeric
          Microsponge™, the
          Basis of Command
          Release™ Technology

Figure 15.2 The making of a polymeric Microsponge system.

IV. PHYSICAL CHARACTERIZATION OF MICROSPONGES

The Microsponge is physically characterized through the use of quality assurance techniques to measure particle size and pore structure parameters, as well as diameter, volume, and release characteristics.

The most valuable of these techniques is mercury intrusion porosimetry (4). During testing, a random sampling of polymer beadlets is placed in a vacuum chamber and submerged under a pool of mercury contained within a volume-calibrated cell. As pressure is gradually increased on the cell, mercury is forced into progressively smaller pores of the microspheres. Thus, the apparent volume of mercury within the calibrated cell is reduced as it penetrates into the Microsponge pores. Values comparing the applied pressure to the apparent mercury volume are reduced by computer into data on pore distribution by volume and diameter (Fig. 3).

Changes in surface area and pore diameter can be easily detected with the use of mercury intrusion porosimetry, providing a reproducible method to monitor these parameters consistently and to correlate them with processing variables or release rates of the active ingredient from the Microsponge.

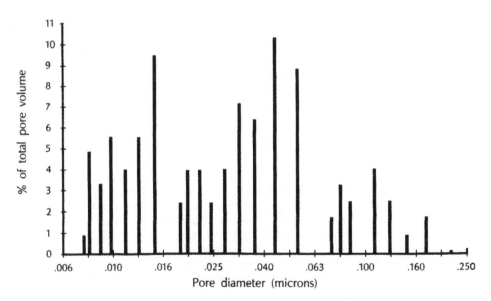

Figure 15.3 Pore volume distribution versus pore diameter as determined by mercury intrusion porosimetry.

Table 15.1   Typical Microsponge Entrapment
System Physical Profile

Characteristic	Value
Sample: CH-274	
Total intrusion volume	1.22 cc/g
Total surface area	74.5 m^2/g
Average pore diameter	0.023 μm
Bulk particle density	0.708 g/cc

Table 1 illustrates a typical set of values for a sample of porous microspheres.

V.   PROGRAMMABLE PARAMETERS

Several primary characteristics, or parameters, of the Microsponge
can be altered during the production phase to obtain spheres tai-
lored to specific product applications and vehicle compatibility.
These parameters include particle size, pore diameter and volume,
resiliency, and monomer composition.   Each can be predesigned at
the time of manufacture by altering the components and polymerization conditions required to produce a Microsponge.
Preselection of the programmable parameters of the Microsponge
results in a truly "custom-made" particle designed to meet the specific needs and requirements of the product formulator.

A.   Particle Size

Free-flowing powders with fine aesthetic attributes can be created
by tightly controlling the diameter of the microspheres at the time
of polymerization.   Typically, diameters between 5 and 300 μm are
feasible.   Small particle-sized systems provide the feel of a super-
fine talcum powder, even when they contain substances such as
mineral oil at a 50% payload.   Larger-sized particles can be prepared according to the ultimate needs for the products.   Particle
size also has some influence on the release rate of the active ingredient from the Microsponge.   The smaller the particle size, the
slower the rate of release (Fig. 4).

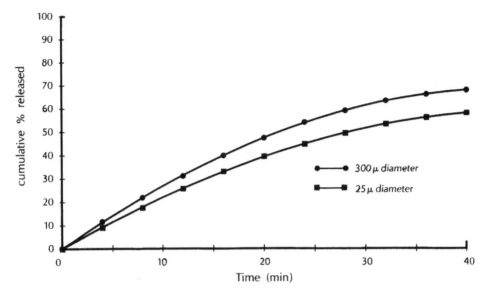

Figure 15.4 Effect of particle size on release rate. Release rates of decane from two Microsponge systems of different particle size: 1.47%/min (25 μm. squares) and 1.81%/min (300 μm, circles).

B. Pore Diameter and Volume

Pore volume (void volume) determines the amount of active ingredient that can be entrapped within the microsphere. Conversely, pore diameter can have a significant effect on the rate of release of the ingredient. Figure 5 demonstrates the effect of pore diameter on the volatilization rate of menthol. Both pore volume and diameter are vital in controlling the intensity and duration of effectiveness of the active ingredient. Pore diameter also affects the migration of the active ingredient from the Microsponge into the vehicle in which the material is dispersed. As a result, the diameter of the pores can have direct impact on the stability of the final formulation.

C. Resiliency

By altering the degree of monomer cross-linking at the time of polymerization, resiliency (viscoelastic properties) of the microsphere can be modified to produce a beadlet that is softer or firmer according to the needs of the final formulation. Cross-linking in excess of 10% is efficient in most systems. This allows sufficient strength for the Microsponge to retain its shape after some or all of the ac-

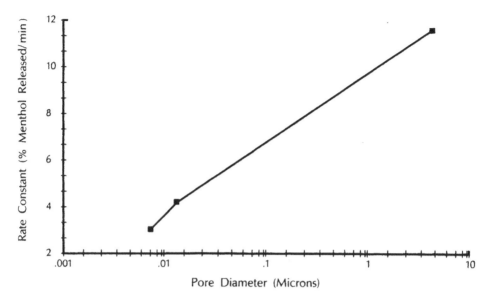

Figure 15.5 Effect of pore diameter on rate of release. Rates of
release of menthol from Microsponge systems of same polymeric composition and particle size with varying pore diameters.

tive ingredient has been removed from the pore network. Increased
cross-linking tends to slow down the rate of release (Fig. 6).

D. Monomer Composition

Selection of the monomer is dictated both by the characteristics of
the active ingredient ultimately to be entrapped and by the vehicle
into which it will be dispersed (Fig. 7). Polymers with varying
electrical charges or degrees of hydrophilicity or lipophilicity may
be prepared to provide flexibility in the release of such materials as
lipids, humectants, moisturizers, sunscreens, vitamins, insect repel-
lants, fragrances, and a variety of pharmacologically active ingredients. Once entrapped, these ingredients can be formulated into
virtually any product form: powders, gels, ointments, lotions,
creams, liquids.

VI. RELEASE MECHANISMS

A. Sustained or Time Release

In the development of a sustained-release Microsponge, many variables must be considered.

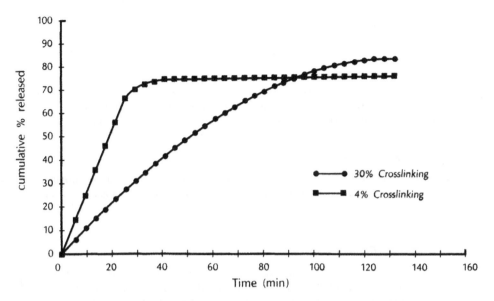

Figure 15.6 Release as a function of cross-linking. Release rates of decane from two polymeric systems with different cross-linking: 2.12%/min (4% cross-linking, squares) and 0.91%/min (30% cross-linking, circles).

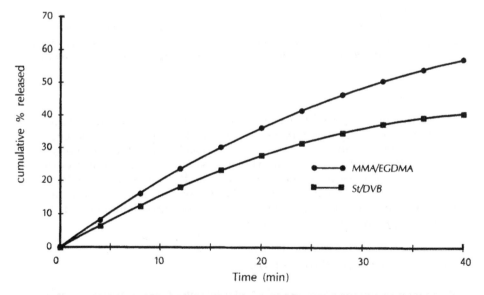

Figure 15.7 Release of decane from two Microsponge systems of different polymeric composition (MMA/EGDMA : Methyl-methacrylate/ Ethyleneglycoldimethacrylate; St/DVB : Styrene/Divinylbenzene).

Physical/chemical properties of the entrapped active ingredient, in-
    cluding volatility, viscosity, and solubility of the entrapped ma-
    terial

Physical/chemical properties of the polymeric microsponge, such as
    pore diameter and volume and resiliency of the polymeric sphere

Vehicle and product form in which the Microponges will be used
    (oils, lotions, creams, powders and other physical forms)

B. Release on Command

By proper manipulation of the aforementioned programmable parameters, Microsponges can be designed to release given amounts of active ingredients over time in response to one or more external triggers.

1. Pressure Release

Like an ordinary sponge, Microsponge systems release fluid when
pressed or "squeezed," thereby replenishing the level of entrapped
active ingredient onto the skin. This action results in renewed
product efficacy (Fig. 8). The amount of ingredient released may
also depend upon the pressure applied to the sponge and the resiliency of the microsphere.

2. Temperature Change

Some entrapped materials, such as sunscreens and emollients, can
be too viscous at room temperature to flow spontaneously from the
Microsponge onto the skin. When warmed by the skin temperature,
the sun or other heat source, their viscosity may decrease, resulting in an increased flow rate (Fig. 9).

3. Solubility

By taking into account the solubility of the entrapped ingredient,
the Microsponge system can be programmed to respond to water,
perspiration, or other solvents. For example, dry Microsponges
loaded with a water-soluble ingredient, such as antiperspirants or
antiseptics, will release that ingredient in the presence of water
(Fig. 10). Release can also be activated by diffusion, taking into
consideration the partition coefficient of the ingredient between the
Microsponge and the outside system.

VII. SAFETY SUBSTANTIATION

Microsponge systems are made of biologically inert polymers. More
than 30 safety studies, including skin irritation (in rabbits and hu-

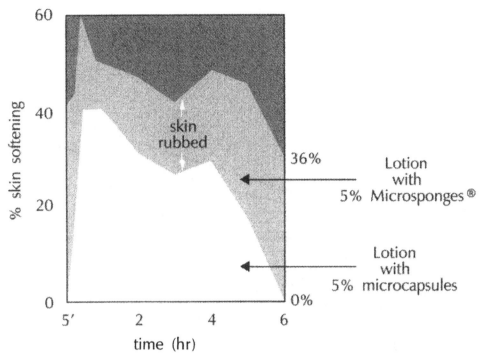

Figure 15.8 Comparison of human skin emolliency induced by mineral oil in a Microsponge system vs. microcapsules. Measured with the gas bearing electrodynamometer (Ref. 5).

mans), eye irritation (in rabbits), oral toxicity (in rats), mutagenicity (in bacteria), and allergenicity (in guinea pigs) demonstrate that the polymers are nonirritating, nonmutagenic, nonallergenic, nontoxic, and nonbiodegradable—the body cannot convert them into other substances. Furthermore, preliminary data indicate that upon injection into the skin, some Microsponge systems are not recognized as foreign bodies and cause no reactions by the body in response to their presence.

Because of their size, these microspheres are too large to pass through the stratum corneum. They remain on the skin surface, gradually releasing their contents on command or over time. This release pattern prevents excessive accumulation of active agents in the epidermis and, as a result, may enhance the safety of topical drugs.

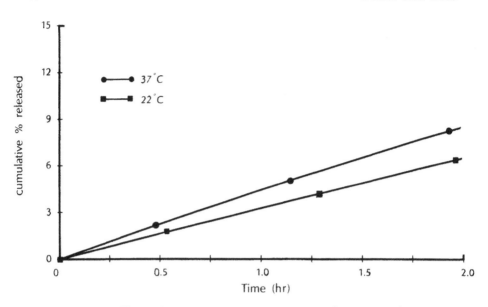

Figure 15.9 Effect of temperature on the rate of release of octyl
dimethyl PABA.

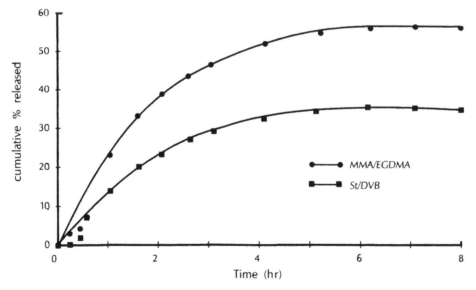

Figure 15.10 Solvent-activated release: release of an antiperspirant in the presence of water from two different Microsponge systems.

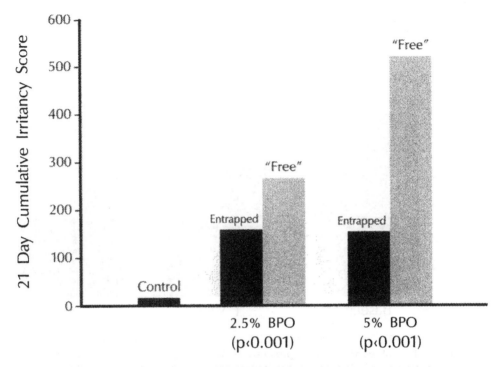

Figure 15.11 Reduction in skin irritancy. Results from a 21-day cumulative irritancy test in rabbits with formulations containing two different levels of benzoyl peroxide, either freely dispersed or entrapped in a Microsponge system.

A reduction in the irritancy potential of the antiacne drug benzoyl peroxide is a demonstration of the increased safety achievable by the Microsponge entrapment system. In a 21-day cumulative irritancy study, lotions containing 2.5% and 5% concentrations of free or entrapped benzoyl peroxide were prepared and applied to rabbits. Free benzoyl peroxide at both concentrations produced much higher irritancy than the entrapped drug (Fig. 11). Effectiveness of the benzoyl peroxide was not impaired as demonstrated by an in vivo human antimicrobial test (Fig. 12). Both *Propionibacterium acnes* and aerobic bacteria showed reductions in their skin number well within the range required in the over-the-counter (OTC) Food and Drug Administration (FDA) monograph.

Allergenicity can also be reduced when the sensitizing ingredient is entrapped in a Microsponge system. Cinnamic aldehyde, a component of cinnamon oil commonly used in pharmaceutical and cos-

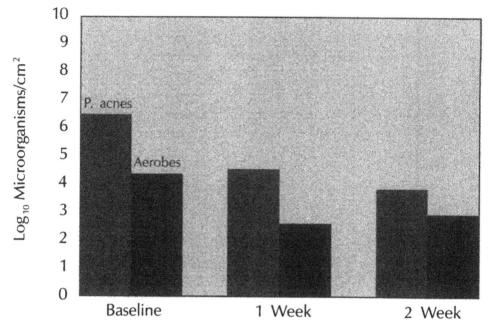

Figure 15.12 *In vivo* antimicrobial efficacy of benzoyl peroxide lotion. (Lotion containing 5 percent BPO entrapped in a Microsponge
system tested as described by Williamson and Kligman (from Ref. 8).

metic applications, is a known, albeit mild, sensitizer. Free and entrapped cinnamic aldehyde was tested at 3%, 4.5%, and 6% levels in
the same base. At each concentration, the number of animals sensitized was significantly lower with the entrapped system (Fig. 13).

Another safety concern is the potential bacterial contamination
of the materials entrapped in the Microsponge. Because the size of
the pore diameter is smaller than bacteria, ranging from about 0.007
μm to 0.2 μm, bacteria cannot penetrate into the tunnel structure
of the Microsponge. Microbiology tests conducted over a 12-month
period show that Microsponge systems are nonsupportive of microbial
growth and require no preservative system.

Therefore, formula stability may also be extended by entrapping
unstable or bacteria-sensitive ingredients in the Microsponge, entrapment will keep the ingredient protected until release from the
Microsponge is effected.

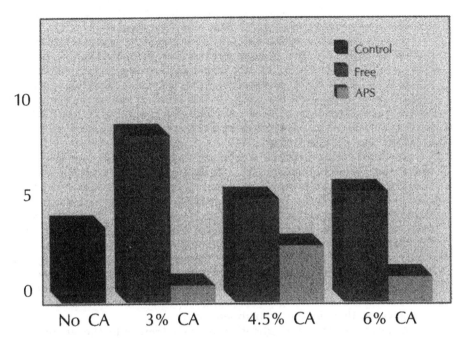

Figure 15.13   Reduction in skin sensitization. Results from a Draize skin sensitization test in guinea pigs with preparations containing three different levels of cinnamic aldehyde, either freely dispersed or entrapped in a Microsponge system.

VIII. INCREASED EFFICACY AND AESTHETIC APPEAL OF DRUG FORMULATIONS

Skin care formulations can have increased efficacy and higher aesthetic appeal to the consumer with the use of active ingredients entrapped in Microsponge systems.

In an experiment, the gas-bearing electrodynamometer (GBE) technique was used to measure skin softening (5). Microsponges containing 50% mineral oil were dispersed at a 5% concentration in a typical oil-in-water skin lotion. A similar preparation containing a urea—formaldehyde microcapsule system with mineral oil was used as a control. Both products were applied to specimens of excised human skin and readings of skin softening were made with the GBE. At the end of 3 hr, both sites of application were gently rubbed. The Microsponge system produced a greater initial increase and a repeated enhancement of skin softening. This softening effect was

significant even 6 hr after application, the time at which the microcapsule system had returned to baseline (see Fig. 8).

A dose-response relationship was obtained in another test conducted in vivo. A Microsponge system was prepared with a propri-
etary emollient entrapped at 50% concentration and incorporated into
a standard oil-in-water emulsion at 0.5%, 1%, and 3% levels. These
creams were applied to the back of the hand of human subjects and
periodic measurements of skin softening were made with the GBE.
An increase in skin suppleness proportional to the concentration of
emollient was obtained (Fig. 14).

Corticosteroids produce a localized blanching effect on the skin
caused by vasoconstriction (6). To determine the release of corticosteroids entrapped in Microsponges, an in vivo vasoconstriction
assay was conducted. A system containing 0.05% entrapped fluocinolone acetonide was applied to intact skin on both forearms of all
test subjects. Blanching effects were measured at 8, 24, and 32
hr after application. One-half hour before reading, one arm was
gently rubbed to effect a renewed release of the entrapped ingredient. Observations on the nonrubbed arm indicate that sustained
release was effective even at 32 hr (Fig. 15). However, at each
time-point, the blanching observed in the site rubbed 30 min earlier
was greater than in the nonrubbed site.

Although most topical drugs must penetrate into the stratum
corneum to be effective, sunscreens have greater efficiency when
they remain on the skin surface to absorb ultraviolet light from the
sun. After penetration, sunscreens can absorb radiation below the
skin surface, but damage is already occurring that reduces sunscreen efficacy. Sunscreens can also be irritants and even skin
sensitizers. If these ingredients can be kept on the skin surface
for a longer length of time, and in controlled quantities, an increased efficacy and a reduced sensitivity to these materials would
result. This could overcome the need for repeated applications and
provide enhanced aesthetic appeal for the consumer.

To determine whether or not sunscreens entrapped in a Microsponge system would still be available for sun protection, octyl dimethyl PABA (Padimate 0) was entrapped at 50% payload; the Microsponges were incorporated into an oil-in-water emulsion at a 2.8%
level to provide a 1.4% sunscreen concentration. The sun protection
factor (SPF) of this formulation was compared with that of a similar
formulation containing the same amount of sunscreen that was freely
dispersed in the base. To assess the possible additive effects of
free and entrapped sunscreen, a product containing 1.4% free and
1.4% entrapped Padimate 0 was also prepared. The lotion base alone
and one containing 8% free homosalate were used as negative and
positive controls. The SPF testing was performed on the backs of

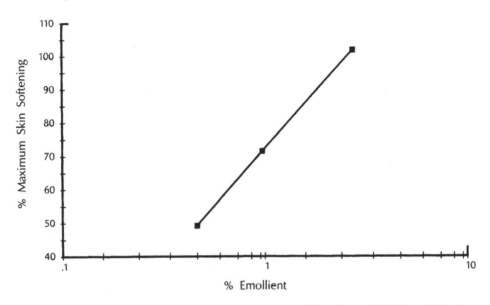

Figure 15.14 Maximum skin softening as a function of total emollient concentration. Measured on human skin with gas-bearing electro-dynamometer.

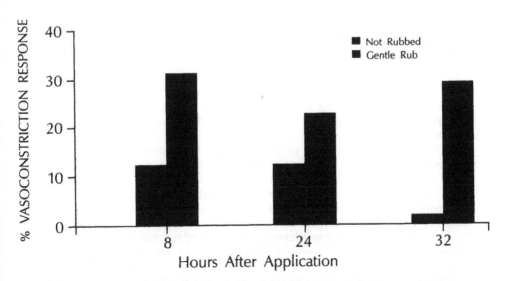

Figure 15.15 Demand and sustained release of a corticosteroid from a Microsponge system as measured by the vasoconstrictor effect.

SPF

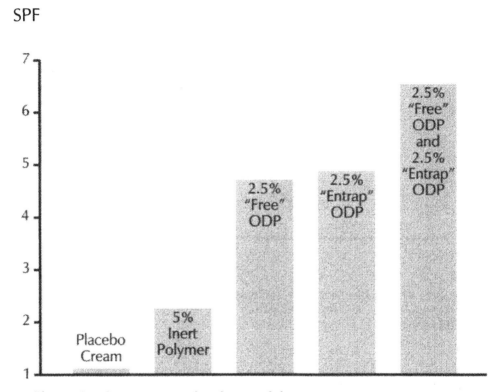

Figure 15.16 Sun protection factor of formulations containing octyl-dimethyl PABA (ODP) either freely dispersed in the vehicle or entrapped in a Microsponge system.

human volunteers using a xenon-arc solar simulator as recommended in the corresponding FDA OTC tentative final monograph for sunscreens (7).

Results obtained from this test (Fig. 16) indicate (1) entrapped sunscreen is indeed available for protection and (2) the addition of free and entrapped drug results in additive SPF.

Further tests have confirmed that higher SPF values (up to 15 or more) can be obtained by preparing suitable formulations containing mixtures of UVB and UVA absorbers.

An in vitro study was conducted on excised human skin to determine the amount of sunscreen remaining on the skin surface when formulations containing [14C]octyldimethyl PABA, either free or entrapped in a Microsponge system, were applied to it. The test was

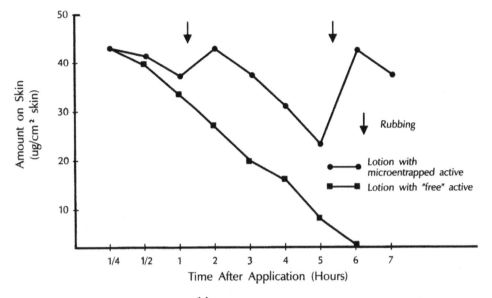

Figure 15.17 Amount of [^{14}C]octyldimethyl PABA recovered from the superficial skin layers of disks of excised human skin after application of lotions containing sunscreen either freely dispersed (squares) or entrapped in a Microsponge system (circles). At each time point, the surface of individual skin disks was quickly rinsed with a stream of 95% ethanol and the stratum corneum was then stripped three times and counted.

conducted as described elsewhere (9). Disks of excised human skin were utilized and one disk was used for each data point. The amount of sunscreen remaining on the skin surface was determined by quickly rinsing the surface of each disk with a stream of ethanol to remove nonabsorbed material. For mass balance, the rinses were collected and counted for radioactivity. The stratum corneum was then stripped three times with cellophane tape and the strips were counted for radioactivity. The results presented in Fig. 17 show that the amount of sunscreen present on the epidermal layers of these skin disks treated with the "free" sunscreen product gradually decreased as a consequence of percutaneous absorption. Conversely, those disks treated with the sunscreen in the Microsponge system consistently had a higher surface concentration of active ingredient and, when the skin surfaces were rubbed just before assay, that amount went back almost to baseline.

Table 15.2   Consumer Panel Evaluation of Two Final Products

	% Preference	
Attributes	Microscopic sponge entrapment system	Encapsulation system
Less oily	72	28
Less greasy during application	78	22
Disappears faster	72	28
Less greasy afterfeel	78	22
Less residue on skin	67	33

Aesthetic appeal plays an important role in the use of topical drugs. Users tend to reject those products they judge to be too oily, too greasy, slow in absorption, or leave a noticeable residue on the skin.

A double-blind crossover panel test with 200 subjects was performed with the same two products used in the experiment described in Figure 8. Subjects tested the two formulations on a blind basis for aesthetic appeal (Table 2). The product with the Microsponge system rated significantly higher in each of the attributes evaluated by the panel.

IX.   POSSIBLE APPLICATIONS FOR MICRO-
SPONGE SYSTEMS

Designed expressly for applications in which the skin itself is the target organ, Microsponge topical delivery systems offer the formulator a range of alternatives limited only by one's own imagination to develop drug or cosmetic products with enhanced safety or efficacy.

A.   Sunscreens

High levels of sunscreens can be utilized for longer-lasting product efficacy to increase protection against sunburn and sun-related injuries (aging and skin cancer). Even at elevated sunscreen concentrations, oiliness, irritancy, and sensitization could be reduced.

B. Antiacne

Efficacy of antiacne ingredients, such as benzoyl peroxide, can be maintained, whereas skin irritation and sensitization are decreased.

C. Skin Depigmentation

Skin bleaches, such as hydroquinone, can be stabilized and protected from oxidation as evidenced by the lack of discoloration of the product. This results in improved efficacy and aesthetic appeal.

D. Anti-Inflammatory

Long-lasting anti-inflammatory activity can be attained for drugs, such as hydrocortisone, to provide extended benefits in the reduction of skin allergy responses and dermatoses.

E. Antipruritics

Extended activity is achieved for anti-itching compounds and local anesthetics used to treat pruritus caused by dry skin, eczema, hemorrhoids, or poison ivy.

F. Antidandruff

The unpleasant odors of zinc pyrithione and selenium sulfide, as well as other antidandruff ingredients, are reduced. Irritation can also be lowered, whereas safety and efficacy are extended.

G. Antifungals

All-day treatment and symptomatic relief of fungal problems, such as athletes foot, can be provided by the sustained release of the active ingredients.

H. Rubefacients

Prolonged activity with less irritancy are obtained with these active agents, accompanied by a reduction in odor and greasiness of the product.

In each of the foregoing cases, the final product can be designed to give distinct and perceivable benefits to the user that were not previously feasible because of formulation restrictions. Such statements as "greaseless," "all-day relief," and "new pleasing fragrance" can now be achievable consumer benefits for these types of products.

X. ADVANTAGES FOR INDUSTRY

For the pharmaceutical industry, Microsponge topical delivery systems provide a wide range of formulating advantages.

Entrapment offers a way to control undesirable ingredient characteristics that have restricted the use of some ingredients in the past. Liquids with such undesirable properties as greasiness or stickiness, can be transformed into nonoily powders. Formulations can be developed using otherwise incompatible ingredients through separate entrapment of one or more of those ingredients. New product forms can be developed as a result of entrapment.

Shelf life and product stability can be prolonged without the use of chemical preservatives. And, because the pores of the Microsponge are smaller than bacteria, the entrapped ingredients are automatically protected from bacterial contamination.

Finally, safety can be increased when irritating and sensitizing ingredients are entrapped. Programmed release can both control the amount of delivery to the targeted site and minimize entry into the lower skin layers, where undesirable side effect are most likely to occur.

Microsponge topical delivery systems clearly answer the needs of industry to maximize the amount of time an active ingredient is present on the skin's surface or within the epidermis, while minimizing its penetration through the dermis and, therefore, into the body.

In summary, these systems can provide increased efficacy for topically active agents, enhanced safety, extended product stability and improved aesthetic properties in an efficient and novel manner.

REFERENCES

1. A. F. Kydonieus, and B. Berner, *Transdermal Delivery of Drugs*. CRC Press, Boca Raton, 1987.
2. J. P. Nater, and A. C. De Groot, *Unwanted Effects of Cosmetics and Drugs Used In Dermatology*. Elsevier, New York, 1985.
3. R. Won, U.S. Patent 4,690,825, Sept. 1, 1987.
4. Poresizer Model No. 9310, Micromeritics Instrument Corp., Norcross, Georgia.
5. M. S. Christensen, C. W. Hargens III, S. Nacht, and E. H. Gans, *J. Invest. Dermatol.*, *69*:282, 1977.
6. M. Katz, and B. J. Poulsen, *J. Cosmet. Chem.*, *23*:565, 1972.
7. *Sunscreen Drug Products for Over-the-Counter Human Drugs*, Fed. Reg., *43*(166):38259, (Aug. 25, 1978), Dept. of Health, Education and Welfare, Food and Drug Administration.

8. P. Williamson, and A. Kligman, *J. Invest. Dermatol.*, 45:498, 1965.

9. D. Yeung, S. Nacht, D. Bucks, and H. I. Maibach, *J. Am. Acad. Dermatol.*, 9:920, 1983.

16
Liposome-Based Vehicles for Topical Delivery

PAUL S. USTER   *Liposome Technology, Inc., Menlo Park, California*

I. INTRODUCTION

Liposomes are microscopic vesicles consisting of amphipathic lipids arranged in one or more concentric bilayers.   These thermodynamically stable, lamellar structures form spontaneously when lipid is brought into contact with an aqueous phase.   Unlike micelles, emulsions, and microemulsions, liposomes have an entrapped, discontinuous aqueous phase separated by 4-nm thick, bilayered lamellae from the continuous aqueous phase.   Liposomes were developed originally to model biological membrane functions such as transport phenomena (1).   However, it was soon recognized that the phenomenon of spontaneous compartmentation might be used to improve the therapeutic index of drugs by targeting active ingredient to the appropriate site of action and modulating drug release from the vehicle.

Because liposomes are a new vehicle to the pharmaceutical industry, the scope of this review will be to introduce the concept of liposomes, discuss some aspects of formulating a topical product with liposomes, and review the literature on liposome-based topical delivery systems.

II. "LIPOSOMOLOGY"

Liposome preparations can be distinguished according to size and number of lamellae, lipid composition, the encapsulation volume, and the percentage entrapment (Table 1).

Table 16.1 Liposome Nomenclature

Acronym	Vesicle diameter (μm)	Number of lamellae	Captured volume (ml/g)	Percent entrapped[a]
MLV	> 0.1	$\geq$ 5	$\geq$ 0.6	6 $\leq$ < 70
SUV	0.03 < 0.06	1	0.6–1.9	6 < < 19
LUV	> 0.06	1	$\geq$ 1.9	19 $\leq$ < 70
OLV	> 0.06	1 < $\leq$ 5	$\geq$ 0.6	6 $\leq$ < 70

[a]Calculated for 10% lipid (w/v).

A. Liposome Specific Parameters

The *encapsulation volume* or *captured volume* is defined as the volume of entrapped discontinuous aqueous phase per unit mass of lipid phase, and is typically expressed in units of microliters per
micromole lipid (μl/μmol) or milliliters per gram lipid (ml/g). It is
often desirable to maximize drug entrapment with minimum lipid,
and this can be achieved by formulating for the greatest possible
captured volume. For a given quantity of lipid, the captured volume is increased by increasing the diameter of the liposomes and reducing the number of lamellae. The theoretical relationship between
liposome diameter and captured volume of ideal, uniformly sized
unilamellar liposomes is illustrated in Figure 1. These values represent the upper limit of captured volume. For liposomes with more
than one lamella per vesicle, the captured volume will be reduced as
a function of number of bilayers and the interbilayer repeat distance. In practice, the captured volume is determined by using a
water-soluble marker (such as radiolabeled sucrose, inulin, or the
drug of interest) of known concentration in the hydrating buffer.
This buffer is added to a known amount of lipid. After the lipid is
completely hydrated, the liposome-entrapped marker is separated
from unentrapped material by centrifugation, column chromatography,
dialysis, or some other means. Final marker concentration and lipid
concentration in the cleaned up preparation a.'e determined by suitable methods. The captured volume is calculated from

$$M/(C \times L) = \text{captured volume} \qquad [1]$$

where M is the total mass quantity of marker in the cleaned up
liposome preparation, C is the original concentration of marker in

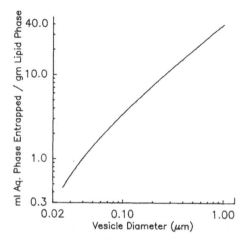

Figure 16.1 Theoretical relationship of captured volume to diameter of unilamellar vesicles.

the hydrating buffer, and L is the total mass quantity of lipid in the cleaned-up preparation.

Another liposomal property, *percent entrapment* or *percent encapsulation*, is an important parameter for characterizing the efficiency of the hydrating process. It is defined as the fraction of the total aqueous compartment sequestered within liposomal membranes and is calculated from

Captured volume × lipid concentration × 100% = % entrapment

[2]

There is a practical upper limit (approximately 70%, see Table 1) to the percent entrapment of a highly water-soluble marker in purely aqueous phase. This is due to geometric limits of packing spheroids into a given volume during the hydrating process.

This geometric constraint does not apply if the active ingredient adsorbs to the bilayer or partitions into the lipid phase. Under these conditions, it is more accurate to describe the apparent efficiency of drug association with liposomes as the *percent incorporation* of active ingredient.

In addition, estimates of parameters such as the average number of lamellae per vesicle and interlamellar spacing (2), and the number of vesicles and total surface area of a liposome suspension (3) can be derived from the captured volume, average vesicle size, and lipid concentration.

Liposome size is usually determined by Coulter counting for vesicles larger than 1 µm diameter, or quasi-elastic laser light scattering if smaller than 1 µm diameter. Relative particle size of small, unilamellar vesicles and the degree of contamination with heterogeneous large particles can also be measured turbidimetrically (4). Electron microscopy remains the only accurate, albeit time-consuming method of determining absolute liposome size and mean number of lamellae per lipid vesicle. However, a great deal can be learned with a good phase-contrast microscope equipped with polarizing filters. It is an indispensable tool of liposomology. Both visual and microscopic inspection provide the investigator with a good idea of the approximate size and "lamellarity" of the preparation.

B. Liposome Classes

Size and number of lamellae have guided the standardization of liposome nomenclature. Each liposome category, general methods for producing them, and visual properties will be discussed later, and the reader is referred to appropriate references for details of each method. Also, recent reviews are available that discuss methods of liposome preparation and characterization in greater detail than will be discussed here (5–8; and also Vol. 1–3 of *Liposome Technology* (G. Gregoriadis, ed.).

Multilamellar vesicles, known as MLVs, are distinguished by being larger than 0.1 µm in diameter and generally having more than five lamellae enclosing the aqueous core. They are prepared simply by drying a thin film of lipid on a surface and hydrating with buffer (1). Thin films are used preferably because the large surface area of exposed lipid facilitates liposome formation and efficient entrapment of aqueous phase. Alternatively, MLV dispersions can be formed by homogenizing or sonicating dried lipid granules suspended in aqueous solution. Care must be taken to ensure that such a preparation is not contaminated with unhydrated lipid. To the naked eye, an MLV dispersion appears milky and opaque at 0.1% lipid (w/v) or greater. The MLVs appear as cross-sectioned, onionlike structures of variable size and shape under high magnification in the phase-contrast light microscope. These preparations can display a Maltese cross birefringent pattern when observed under crossed polarizing filters, if there are sufficient lamellae per vesicle and the interbilayer distance is relatively small and evenly spaced.

The limiting size of a liposome is about 0.03 µm diameter because of the radius of curvature imposed by lipid geometry (Fig. 2). Of necessity, these smallest of vesicles have only one lipid bilayer and, thus, are known as small unilamellar vesicles (SUVs). The

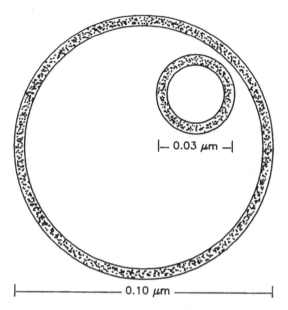

Figure 16.2 Size relationship between a SUV of limiting radius and a LUV of 0.1-μm diameter. Bilayer thickness is 0.004 μm.

SUVs are produced by sonicating an MLV preparation to terminal clarity (9,10). Because they are so small and do not scatter visible light, an SUV preparation at about 0.1% to 1% lipid (w/v) will appear perfectly transparent to the eye. It is not possible to observe them by light microscopy.

During the 1970s considerable effort went into developing a new class of liposomes that were much more efficient (per unit mass of lipid) at entrapping aqueous solutes. This is because the lipid phase displaces 60% of the total volume of a 0.03-μm diameter SUV, but only 2.4% of a 0.1-μm diameter liposome. The goal was to make the largest possible diameter liposome with the fewest number of lamellae, preferably a large, unilamellar vesicle (LUV). For LUVs 0.1 μm or larger in diameter, doubling the vesicle diameter approximately doubles the captured volume (see Fig. 2). Note that for vesicles smaller than 0.1 μm in diameter, this relationship is not linear because of the volume displaced by the lipid phase. From the literature, SUVs are generally considered to be no more than 0.06 μm in diameter, and 0.1 μm is often accepted as an arbitrary lower size limit for LUVs. I propose that the line of demarcation should be at the limiting diameter of a vesicle of two or more lamellae (about 0.06 μm) because it is necessary to distinguish between LUVs and oligolamellar vesicles (OLVs).

Numerous methods have been developed that are capable of making LUV preparations on the research bench scale, and can be grouped into three major approaches. *Fusion-extrusion* techniques often start with SUVs that are then fused into larger structures by addition of divalent cations (11), freezing (12,13), dehydration (7), or rapid pH shift (14). These larger, usually oligolamellar vesicles are usually sized through polycarbonate filters to reduce vesicle size heterogeneity. Another general strategy can be characterized as *organic phase addition—removal.* Dissolved lipid in organic solvent is introduced to an aqueous buffer. The organic solvent is removed during or after liposome formation by evaporation (15,16) or dialysis (17). Still a third approach, *detergent—dialysis*, is to make mixed micelles of lipid and detergent, followed by detergent removal by column chromatography (18) or dialysis (19–21). A 0.1% to 1% lipid (w/v) LUV preparation is clear and opalescent to the eye. True LUVs, having only one lipid bilayer with which to scatter light, are essentially impossible to observe by phase-contrast microscopy.

It is noteworthy that unless experimental parameters are carefully characterized and optimized, the casual application of these LUV methods will result in a preparation of predominantly oligolamellar, large vesicles known as OLVs. Other acronyms have been developed that derive from the specific method used (REVs, LUVETS, SPLVs, etc.) but these methods all result in preparations that are predominantly LUV to OLV in character. Under phase-contrast light microscopy, OLVs larger than 0.4 μm in diameter can be resolved as phase gray "ghosts" that are not birefringent under crossed polarizing filters.

III.   FORMULATING TOPICAL PRODUCTS
WITH LIPOSOMES

The commercialization of liposome technology began in the early 1980s, and it is a rapidly evolving field. There are now four publicly held companies in the United States exclusively devoted to product development using liposome-based vehicles. Many large pharmaceutical houses and cosmetic firms are also engaged in active research and development.

Cosmetic skin creams containing liposomes have been available for several years; however, the first therapeutically acceptable product remains to be introduced. The milestones for a successful liposome-based pharmaceutical product are no less than those of conventional dosage forms. Efficacy, safety, scaleup, stability, raw materials supply and purity, process control, and validated analytical methods must all be achieved. Selected parameters that are germane to formulating with liposomes will be noted in the following section.

A. Vehicle Selection for the Active Ingredient

An active ingredient can be situated in the aqueous compartments, hydrophobic phase, or oriented at the water—bilayer interface. The selection of the appropriate liposome vehicle (MLV, SUV, or other) should be based on a careful analysis of the physical properties of the drug and its tendency to associate with the lipid components. Additional factors, such as buffer composition, pH, ionic strength, and lipid composition, all will influence the distribution of active ingredient. Cost of goods is also germane when selecting a liposome vehicle. A very costly, water-soluble ingredient may require maximum captured volume of the formulation. A process that results in an LUV to OLV preparation should be employed. However, MLVs are more cost-effective at incorporating lipophilic active ingredients at considerable savings, because a simpler process can be used and a clean up step to remove unincorporated drug may be omitted.

B. Process Selection

Liposomes have their own unique problems of scaleup and process control. These are concerned with adequate supplies of phospholipid raw materials, lipid hydration to achieve the desired capture volume and particle size, use of pharmaceutically acceptable organic solvents, removal of solvents (organic phase addition—removal processes; 22) or detergent (detergent dialysis; 20), and clean up of unentrapped drug.

Selection of a process should be based on the solubility properties of the drug in aqueous solution and appropriate solvents, the liposome / buffer partition coefficient, and the required percentage entrapment or incorporation.

Topical liposome formulations have the advantage of not requiring the stringent particle size specifications of parenteral products. If liposome size or heterogeneity does not affect in vivo performance, the specifications can be kept quite broad and the process simplified considerably.

C. Characterization

The characterization of bulk properties of liposome formulations, such as color, pH, rheology, are very similar to those commonly employed for other types of dispersions and suspensions.

When formulating emulsions and microemulsions, considerable effort is placed on making appropriate blends of oil and surfactant—
cosurfactants to achieve a stable dispersion. With phospholipid-
based liposomes that spontaneously form a stable structure, more
emphasis is placed on characterizing liposome-specific properties.

Determination of vesicle size, captured volume, percent entrapment and percent incorporation have already been discussed. One
other property unique to liposomes is the active ingredient *release
rate* or *efflux rate*, which has important implications for the performance and stability of a formulation. It is most simply measured
by monitoring the change in percent entrapment or percent incorporation of a formulation over time under a given set of conditions
(temperature, other excipients, ionic strength, presence of biological fluids). Average liposome size of a formulation is an important
factor in determining the release rate of small nonelectrolytes (23).
However, mathematical model fitting from experimental determinations
of glucose release indicates that the degree of liposome size heterogeneity (polydispersity) does not affect the release rate significantly (24).

D. Antimicrobial Preservatives

Topical products are usually preserved by the addition of one or
more antimicrobial preservatives. Preservatives are intrinsically no
different from an active ingredient, which is to say they can be entrapped in liposomes or incorporated into the bilayer or interfacial
region. The end result is that microbial deterrence at a given concentration will be different from that of a similar aqueous solution
without liposomes.

For instance, the commonly used parabens are inactivated by
liposomes as a consequence of partitioning into the bilayer (25).
The extent of inactivation will be a function of partition coefficient
and the lipid concentration. Consequently, the total quantity of
lipophilic preservative must be increased to achieve a satisfactory
germicidal effect. If the "free" concentration (C_{free}) required for
the germicidal effect of preservative P, the formulation's liposome—
water partition coefficient (K_p) and the total lipid concentration are
known, the total concentration of preservative (C_{total}) required in
the formulation can be calculated from a modification of Bean's
equation (26):

$$C_{total} = C_{free} \times (\phi \times K_p + 1)/(\phi + 1) \qquad [3]$$

where ϕ $\approx$ ratio of lipid to aqueous phase (w/w), and

K_p $\approx$ (g P/g lipid)/(g P/g aqueous phase)

Note that this equation may also be used to predict loading of lipophilic active ingredients in a liposome vehicle.

For water-soluble compounds, inactivation can occur if sufficient preservative of interest is entrapped and has such a low bilayer permeability that, after the initial microbial challenge, it is unavailable for subsequent challenges. If such is determined to be true, addition of the preservative should be delayed until liposome formation is completed.

E. Stability

Formulation stability can be subdivided into chemical, physical, and liposome-specific characteristics.

1. Chemical

For chemical stability of phospholipid-based liposomes, lipid peroxidation and fatty acyl—ester hydrolysis are the primary concerns. Phospholipid peroxidation is accelerated by oxygen, transition metal ions, elevated temperature, and light. It is usually controlled by the addition of chelating agents to bind iron salts (27), antioxidants such as α-tocopherol (28), by using phospholipid that has saturated or mono-unsaturated fatty acids, or combinations. Cholesterol, another primary component of liposome formulations, is minimally affected by oxidation in liposomes because the phosphatidylcholines are very effective inhibitors (29).

Fatty acid ester hydrolysis is a time-, temperature-, and pH-dependent process. Hydrolysis results in the formation of lysophosphatidylcholine and free fatty acid. Depending on the active ingredient, changes in liposome size, efflux rate, percutaneous absorption, and in vivo performance may be affected. Frokjer et al. (30) have published a very excellent, systematic study of the chemical, physical, and release rate stability of phosphatidylcholine—cholesterol liposomes. The pseudo—first-order rate constants for phosphatidylcholine hydrolysis at 70°C showed a pH minimum at 6.5. Only 6% hydrolysis of distearylphosphatidylcholine to lysophosphatidylcholine was observed after 300 days at pH 6.5 and 25°C. Thus, ester hydrolysis can be minimized by formulating in the vicinity of pH 6.5.

2. Physical

Changes in the bulk physical properties of a liposome preparation usually reflect changes in vesicle aggregation or increasing liposome size.

Gross aggregation of liposomes can be observed macroscopically as "creaming" or flocculation of the originally homogeneous disper-

sion. Occasionally, controlled flocculation of liposomes can be put
to good purpose in controlling drug suspensions (31), but, for the
most part, aggregation is a nuisance from the viewpoint of cosmetic
acceptability, and it may have undesirable effects on formulation
properties.

Vesicle aggregation can be controlled by imparting an electrostatic charge to the liposomes. Typically, negatively charged lipids,
such as phosphatidylglycerol and phosphatidylserine, are added in
small quantities (about 1%–10% by weight of total phospholipid).
The addition of amphipaths, such as stearylamine or cetyltrimethyl-
ammonium bromide, impart a positive surface charge to liposomes,
but cationic lipid toxicity usually precludes use of positively
charged lipids in pharmaceutical products. A chelating agent, such
as ethylenediaminetetraacetic acid (EDTA) (0.01%–0.1% w/v) should
also be included to prevent divalent metal cation induced aggregation.

Growth of mean liposome size can also be inhibited by imparting
an electrostatic charge, the inclusion of trace amounts of chelating
agent and by avoiding formulations based on SUVs. Because the
high radius of curvature in SUVs cause them to be thermodynamically strained, such preparations show net vesicle growth, which is
particularly rapid when the formulation is incubated at, or thermally
cycled through, the crystalline–liquid crystalline phase transition
of the lipid (32).

3. Liposome-Specific Parameters

Changes in the chemical or physical stability of a liposome formulation will oftentimes be reflected by changes in the liposome-specific
parameters. Appreciable changes in mean particle size may affect
the release rate of active ingredient (23). The captured volume
and the percentage entrapment will be changed if vesicle diameter
increases or decreases.

Thermodynamic equilibrium will drive the flux of entrapped material into the continuous aqueous phase of a liposome vehicle.
Therefore, the desired percent entrapment shelf life of a liposome
formulation may be difficult to achieve if unentrapped material must
be removed and the release rate in the storage buffer is not markedly different from that of the in vivo release rate. By example,
consider a liposome formulation of a water-soluble drug that is designed to release half its contents within 2 days of application to an
ophthalmic surface. A cleaned up formulation must have a storage
buffer release rate at least three orders of magnitude slower than
the in vivo rate to achieve 2 years of *percent entrapment* stability.

IV. RESEARCH STUDIES

Therapeutic applications of liposomes have been envisioned since the early 1970s, but research on the topical route of administration was not published until the 1980s. The current body of literature concerned specifically with topical applications of liposomes can be grouped into ophthalmic and dermal studies.

A. Ophthalmic Studies

In 1981, Smolin et al. (33) used free and liposome-entrapped idoxuridine, an antiviral agent, to study the treatment of herpes simplex keratitis in rabbits. An aqueous solution of 0.1% (w/v) idoxuridine, 0.1% idoxuridine in phosphatidylcholine—phosphatidylglycerol—cholesterol LUV—OLV liposomes (35% entrapped drug), or the liposome placebo, were applied three times daily to rabbits that had been infected previously by corneal abrasion and application of a titer of virus. Liposome-entrapped idoxuridine reduced the size of viral-induced corneal defects significantly, compared with free drug or liposome placebo.

Lee and colleagues (34,35) investigated the delivery of free and liposome-entrapped inulin and epinephrine HCl, both water-soluble model compounds of substantially different permeability. In vitro release experiments revealed that epinephrine HCl crossed liposome bilayers readily; 50% of originally entrapped epinephrine HCl was released from MLVs within 1 hr, whereas more than 90% of the entrapped inulin remained entrapped after 4 hr. In vivo studies in a rabbit model indicated that liposomes did not slow down epinephrine HCl removal from tear fluid. Furthermore, 45 min after instillation of the liposome vehicle, epinephrine concentrations in eye tissues (conjunctiva, cornea, aqueous humor, and iris—ciliary body) were significantly lower than aqueous solutions of epinephrine HCl.

However, when using liposome-entrapped inulin in vivo, marker levels were increased many-fold over aqueous solutions in the conjunctiva, cornea, and iris—ciliary body. Inulin could not be detected at all in the aqueous humor when entrapped in liposomes. Thus, depending on the drug, liposome entrapment changed marker distribution significantly. Liposomes were able to increase localized concentrations of a water-soluble marker with low membrane permeability, perhaps by adhering to ocular tissues.

These results were verified by Ahmed and Patton (36) who also found that transcorneal uptake of MLV liposome-entrapped inulin liposomes is suppressed and noncorneal uptake was enhanced when compared with an aqueous inulin solution. Aliquots of free or liposomal-entrapped tritiated inulin were instilled in rabbit eyes and assayed for radiolabel 20 min later. Increased inulin levels were found in the cornea and iris—ciliary body when liposome-entrapped inulin was instilled. Whereas free inulin was able to penetrate into the aqueous humor to some small degree, liposome-entrapped inulin was not detectable in the aqueous humor. The authors suggested that liposome dosage forms can be used to promote drug absorption selectively to noncorneal ocular tissues.

In subsequent studies, Lee and co-workers found that liposome-entrapped inulin uptake into intraocular tissues was indeed dependent on adhesion of liposomes to corneal epithelium (37). The retention of MLV (phospatidylcholine—cholesterol) liposomes in tears, and their interaction with corneal and conjunctival surfaces was studied using tritiated-inulin (5%, w/v) as a marker for entrapped contents. The rate constant for clearing these neutral MLVs was only 30% slower than that compared with free inulin, and this rate constant was insensitive to volume of instilled dose. Inulin-loaded liposomes competed with empty liposomes for binding sites in the corneal and conjunctival surfaces. Pretreatment with empty MLVs reduced inulin-loaded liposome binding by greater than 75% for both surfaces. Nevertheless, liposomes did not bind avidly to these surfaces; rinsing removed them effectively. The authors suggest that desorption into tear fluid is the major pathway of clearance.

Another factor influencing the reduced concentrations of inulin in various eye tissues was the slow liposomal release rate of inulin (about 1% per hour in this study). Inulin is a large-molecular-weight (about 5000 MW), water-soluble compound and is not expected to have a rapid efflux rate.

Lee et al. (37) concluded that topically applied, neutral MLV liposomes are removed from the corneal surface by tear drainage. The extent of entrapped inulin absorption seemed to be controlled by the number of *adsorbed* liposomes as opposed to the number of *instilled* liposomes. Adsorption was not sufficiently strong, and efflux from liposomes was too slow, to sustain inulin concentration in intraocular tissues. Thus, improved ocular delivery of water-soluble entrapped materials requires enhanced liposome-binding affinity and a rapid release rate.

Krohn and co-workers attacked the problem of improving ocular retention of liposomes by pretreating isolated rabbit corneas with wheat germ agglutinin (WGA) and instilling phosphatidylcholine—mixed brain ganglioside MLV or SUV liposomes (38). Here, lectin-mediated adhesion of liposomes enhanced binding to rabbit corneal epithelium 2.5-fold. Transcorneal flux of radiolabeled carbachol entrapped in such ganglioside-containing liposomes was studied in an in vitro model using excised corneas with simulated "tear flow." Unentrapped carbachol was not removed from the liposome formulation. Thirty minutes after instillation there were no significant dif-

ferences in carbachol activity in cornea or transcorneal receiver
compartment for liposome-entrapped or free carbachol. However,
90 min after instillation the liposome vehicle was able to maintain
significantly higher carbachol concentrations in the cornea and in
the transcorneal receiver compartment. The authors propose that
the liposome vehicle's ability to sustain drug levels at 90 min in the
receiver compartment was due to the continuing presence of drug at
the corneal surface.

Schaeffer and Krohn (39) studied in vitro liposome—corneal
binding mediated by electrostatic adsorption. Stearylamine or di-
cetyl phosphate was used to modify the net charge of phosphatidylcholine—cholesterol liposomes. Rank ordering of liposome binding
to excised rabbit corneas showed significantly greater affinity
(about twofold) of positively charged liposomes compared with negatively charged liposomes. In turn, negatively charged liposomes
bound twofold greater than neutral liposomes.

Recently, Guo and co-workers (40) have been able to demonstrate that the in vivo retention of radioiodinated liposomes in rabbit eyes is increased many-fold by using MLVs containing positively
charged cholesteryl esters. They observed that liposome binding in
the eye was saturable, and the half-life of liposome retention could
be manipulated by varying the charged cholesteryl ester/phospholi-
pid ratio, the length of the spacer arm separating amino function
from the lipid head group, and lipid phase transition (fluidity).

Schaeffer and Krohn (39) investigated the in vitro transcorneal
flux of the water-soluble drug penicillin G in MLV or LUV liposome
vehicles of qualitatively different electrostatic charge. All liposome
vehicles tested showed significantly improved drug delivery over an
aqueous penicillin G solution. Penicillin G flux across rabbit corneas 1 hr after instillation exhibited the same charge-dependent
rank ordering as lipid binding. Free drug added to preformed,
empty liposomes did not show increased drug flux, suggesting that
drug encapsulation and liposome binding was a prerequisite for improved delivery.

The transcorneal flux of the lipophilic drug indoxole incorporated in neutral phosatidylcholine liposomes or in a polysorbate 80 dispersion was evaluated in vivo with rats (39). At 1 hr after instillation, a 1.0 mg indoxole per milliliter liposome vehicle preparation
provided the same drug concentrations in rat aqueous humor (about
20 µg/ml) as a 10.0 mg indoxole per milliliter polysorbate 80 vehicle.
Furthermore, Schaeffer and Krohn discovered that the liposome vehicle caused none of the histological changes induced by polysorbate
80 ocular toxicity (unpublished results). Thus, in addition to improving the solubility of the anti-inflammatory agent indoxole, vehicle safety was improved.

Shiota et al. (41) reported a several-fold improvement of lipophilic vitamin A delivery in vivo. Increased specific activity in various rabbit eye tissues persisted for up to 8 hr after instillation of vitamin A in positively charged liposomes, when compared with vitamin A in a simple vegetable oil vehicle.

The consistent conclusion from ophthalmic studies is that liposomes provide improved local concentrations of active ingredients provided the compound is lipophilic, or it is water-soluble but not very membrane-permeable. The indoxole study also suggests that a liposome vehicle may improve safety by replacing other more toxic excipients used for drug solubilization. The foregoing studies also indicate that liposomes improve transcorneal flux of lipophilic molecules and for, at least, some water-soluble molecules such as penicillin G.

B. Dermal Studies

The first results of topically applied liposomes were reported by Mezei and Gulasekharam (42) in which triamcinolone acetonide entrapped within dipalmitoylphosphatidylcholine—cholesterol MLVs resulted in significantly increased drug levels in rabbit dermis and epidermis, reduced blood and reduced urine concentrations. In a series of reports (43—45) Mezei and co-workers used MLVs to demonstrate the in vivo transdermal delivery of corticosteroid drugs, such as triamcinolone acetonide, triamcinolone acetonide 21-palmitate, and progesterone, in rabbits and guinea pigs. Controls were conventional dosage forms such as ointments, although it is not clear that control and liposome preparations had similar thermodynamic activity to drive steroid partitioning out of the vehicle. Preparations were compared tissue by tissue. For triamcinolone acetonide and its more lipophilic palmitate ester derivative, liposome preparations were reported to consistently give significantly higher epidermal and dermal drug concentrations, no significant differences in systemic tissues (except for blood, which was significantly lower), and reduced urinary excretion. Mezei argues that this is indicative of selective triamcinolone acetonide delivery to skin tissues.

This generalization was also reported for econazole base and econazole nitrate (44). For this active ingredient, all skin tissues showed increased concentrations of radiolabeled drug. The pattern of in vivo disposition of seven different econazole—liposome preparations is similar to triamcinolone acetonide; but different lipid compositions delivered different quantities. Multiple dosing with control and econazole-loaded liposome preparations increased econazole accumulation significantly when liposome vehicles were applied.

However, progesterone-loaded liposome preparations showed increased concentrations in all skin tissues, and with the exception of blood, *increased* drug concentrations in biopsy specimens from all systemic organs (44).

Mezei concludes that liposomes can provide a vehicle that is most useful for selective delivery of drug to local sites, with reduced systemic activity. This would be a highly desirable feature for certain drugs (corticosteroids by Mezei's example) in which systemic effects must be avoided. At least for triamcinolone acetonide and econazole, it seems possible to achieve significantly higher skin levels of drug without increased systemic levels.

Liposome preparations of progesterone and hydrocortisone have been studied in some detail. Flynn and colleagues (46,47) made in vitro flux measurements to determine the mechanism of percutaneous absorption of progesterone, hydrocortisone, and glucose across hairless mouse skin. A mathematical treatment was derived, which was later validated by Ertel and Carstensen (48). Incorporation into a liposome vehicle did not change the skin permeability coefficient of progesterone and hydrocortisone significantly, but the skin permeability coefficient of glucose was reduced several orders of magnitude when entrapped in liposomes. Active ingredient flux across lipid bilayers was the rate-limiting step of percutaneous absorption for highly water-soluble molecules like glucose. Membrane-permeant lipophilic molecules such as progesterone and hydrocortisone were hypothesized to move across by a "direct transfer" mechanism from the bilayer hydrophobic phase to skin hydrophobic phase.

Knepp et al. (49) investigated the transfer mechanism of progesterone in more detail by comparing the transdermal flux of free progesterone and steroid in liposome vehicles. The liposome vehicles exhibited zero-order kinetics of transdermal flux for over 24 hr. Furthermore, there were no appreciable differences in transdermal flux of progesterone from liposomes in suspension or progesterone-containing liposomes immobilized in a gel. Because of these similar kinetics and the physical impossibility of liposomes in a gel coming into intimate contact with the skin, the authors concluded that interfacial release of progesterone into the surrounding aqueous medium was actually the rate-limiting step of the transfer mechanism.

A similar mechanism may be operating for hydrocortisone as well (46,47). Lasch and Wohlrab (50,51) studied the concentration profile of hydrocortisone in human cadaver skin after treatment with hydrocortisone ointment or hydrocortisone-loaded liposomes. At 30 or 300 min after application of hydrocortisone in ointment or liposome vehicle, the skin was tape-stripped to remove layers of the stratum corneum, or it was dermatomed to section the epidermis and dermis. As a whole, there were no differences in the quantity of steroid found in the stratum corneum with either vehicle. However, at 30 min, the liposome vehicle provided about an eightfold and 14-

fold improved hydrocortisone concentration in the epidermis and dermis, respectively. At 300 min, the differences between liposome and ointment were about four-fold for the epidermis and nine-fold for dermis. Lasch and Wohlrab concluded that liposomes provide increased concentrations of corticosteroid in skin tissues.

It is worth noting one study that is in disagreement with the general results on steroid-containing liposomes. Vermorken et al. (52) investigated the biological effect on hamster flank organs of topically applied 5α-dihydrotestosterone. An acetone or liposome vehicle (uncharacterized) was compared at equal androgen concentrations for its ability to increase the flank organ size. The acetone vehicle, it was discovered, elicited a greater effect than the dihydrotestosterone-loaded liposomes. The authors suggest several reasons for the disparity between their results and that of Mezei's work with triamcinolone acetonide. Different steroids, schemes of application, and animal species were used; Vermorken and colleagues also measured a biological response, not tissue concentration of drug.

Also, it is not known whether or not the liposome vehicle was saturated with drug. The actual drug/lipid ratio may not have favored significant partitioning of dihydrotestosterone from the vehicle into skin layers. The importance of considering the effect of this ratio on quantity of active ingredient absorbed was illustrated with butylparaben by Komatsu et al. (25,53). The quantity of butylparaben absorbed per unit time across excised guinea pig skin was inversely proportional to the liposome lipid/paraben ratio.

C. Miscellaneous Studies Using Phospholipid-Based Vehicles

Nishihata et al. (54) studied the in vitro and in vivo percutaneous absorption of the nonsteroidal anti-inflammatory compound sodium diclofenac with aqueous drug solutions and "phospholipid gels." Both in vitro and in vivo transport across rat skin was increased by the addition of hydrogenated soy phospholipid. The in vitro penetration rate of diclofenac increased with increasing soy phospholipid content up to 0.5% (w/w) lipid. The phospholipid gel systems also promoted significant accumulation of diclofenac in skin tissue; the cumulative amount was proportional to the penetration rate. The authors proposed that the improvement was due to the surface-active nature of the phospholipids. Intravenous administration of diclofenac resulted in no significant skin tissue concentrations, and topically applied diclofenac was not as bioavailable systemically as intravenous injection. The authors concluded that aqueous phospholipid gels of diclofenac are more suitable for localized than for systemic treatments.

In this study, the phospholipid used was a hydrogenated soy phospholipid, identified as 30% phosphatidylcholine and 70% phosphatidylethanolamine. The preparation was not characterized and identified as liposomes. Other investigators have used freeze-fracture, x-ray diffraction, and phosphorus nuclear magnetic resonance (NMR) techniques to characterize similar lipid mixtures as in a transition phase between the bilayer (lamellar) phase of pure phosphatidylcholine and nonbilayer (reversed hexagonal phase) of pure phosphatidylethanolamine preparations (55). Nishihata's study is unique in that lecithin (phosphatidylcholine) was not the major lipid species, which raises interesting questions about the role of lipid composition and nonbilayer phases in promoting percutaneous absorption.

Phospholipid effects on percutaneous absorption have also been studied in organic solvent vehicles in which liposomes are not present. Kato et al. (56) studied the effects of 1% lecithin (w/v) in propylene glycol on the in vitro percutaneous absorption of bunazosin, theophylline, or isosorbide dinitrate across excised hairless mouse skin. The addition of lecithin increased the amount of drug penetrating in vitro at 24 hr, for all drugs, with the greatest effect for bunazosin (55-fold). Lecithin, at concentrations used in this study, did not change bunazosin solubility in propylene glycol.

The in vivo penetration of bunazosin (about 23 mg/g solvent) dissolved in propylene glycol with 3% lecithin was compared with bunazosin—propylene glycol alone applied to rabbits. In vivo penetration was undetectable in the drug—solvent vehicle, but the addition of 3% lecithin was able to effect significant plasma levels during the 24 hr of observation. Because the formulations used were all saturated solutions of drug (similar thermodynamic activity) and because lecithin did not change bunazosin saturation, the authors propose that the mechanism of lecithin action is by changing the permeability barrier of the skin. This is plausible, but, since the chemical purity of the lecithin was not characterized, it is impossible to separate the action of phosphatidylcholine from that of its degradation products, free fatty acids, and lysophosphatides. These known penetration enhancers would indeed be expected to change the permeability barrier of the skin.

D. Conclusions

The results of many studies may be summarized by the general observations that

1. Liposome binding to some topical surfaces can be mediated by electrostatic interactions.

2. Topical application of drug-loaded liposomes promotes delivery of the active ingredient to local tissues and, for at least some cases, may reduce systemic drug levels.

3. Transdermal and transocular delivery of lipophilic molecules is definitely enhanced by liposome incorporation.

4. Transdermal or transocular topical delivery of hydrophilic materials is considerably more problematic. Most small, membrane-impermeant markers show no improvement, yet the occasional success with some molecules like penicillin G suggests that more studies to discover structure—activity correlations are warranted.

It is perhaps not surprising that transdermal delivery of water-soluble membrane-impermeant compounds is further impeded by adding yet another permeability barrier, namely, the liposome. Certainly, there is no evidence for penetration of intact liposomes with their payload through the skin and, indeed, evidence to the contrary. Specific monitoring of radiolabeled phospholipid markers shows no evidence for phospholipid penetration through the skin under in vitro (47; Uster, unpublished data) or in vivo conditions (25). The penetration of intact liposomes through the stratum corneum can be likened to the probability of passing a basketball through a chain-link fence without rupturing it.

V. FUTURE DIRECTIONS

A. Basic Research

Topically applied, drug-loaded liposomes can substantially improve drug loading, drug delivery, and sustained release, thereby offering clearcut advantages over traditional dosage forms. These advantages are particularly evident for the more lipophilic active ingredients, in which elevated local concentrations are consistently demonstrated and that for, at least, some drugs reduces the systemic concentrations. Because there is no evidence for liposome penetration through intact topical surfaces, such as the stratum corneum, it is perfectly logical that water-soluble, membrane-impermeant active ingredients face an additional barrier to increased bioavailability when entrapped within liposomes. For topical surfaces, the structural integrity of which is not compromised, a second generation of liposome vehicles will have to be developed.

However, water-soluble drugs entrapped in conventional liposomes may play an extremely useful role in treating conditions in which the integrity of the stratum corneum or other topical surface has been damaged. The study by Smolin and co-workers on viral infection of abraded rabbit corneas is especially encouraging (33).

The entire area of drug—liposome treatment of "broken" skin or other compromised topical surfaces has yet to be explored in depth. Almost all of the research investigations that have been cited on topical applications have focused on phosphatidylcholine or phosphatidylcholine—cholesterol mixtures. Usually, measurements of important liposome variables such as mean size, captured volume, and percent encapsulation are not reported. For the most part, lipid raw materials and the liposome formulations themselves have not been carefully characterized for actual lipid composition or for degradation products, such as free fatty acid and lysophosphatides. Careful biochemical and physicochemical characterization is warranted because differences in these properties may lead to substantial discrepancies in results from preparations that nominally appear to be the same. Indeed, the precise role of lipid composition in facilitating drug delivery is another research area that requires systematic charting.

Liposome interactions with the immediate skin microenvironment also have not been explored in any detail. Fundamental questions, such as the structural integrity of liposomes in intimate contact with skin components, remain unanswered.

The potential discrepancy between bioassay and marker monitoring, as highlighted by Vermorken et al. (52), raises a general issue about liposome formulations that is especially pertinent to water-soluble, membrane-impermeant active ingredients. Namely, do tissue concentrations of drug, as measured by marker or chemical assay, represent biologically available drug, or is the drug permanently masked within its liposome carrier? Future research should carefully evaluate, either by use of by an appropriate bioassay or by demonstrating the physical separation of active ingredient from liposome carrier at the site of action, whether or not the increased tissue concentrations of the active ingredient actually is bioavailable.

B. Liposome Formulations in Product Development

From the literature, it seems clear that lipophilic drugs are the most likely candidates for conventional liposome vehicles in potential efficacy. Lipophilic drug candidates also have advantages during formulation and process development, most notably by improving drug loading and obviating the need for a cleanup step.

As with any new technology, there have been unique problems that must be overcome to introduce a pharmaceutically acceptable product to market. Many of these challenges have already been met successfully because there are now more than a half-dozen parenteral liposome formulations in clinical trials worldwide. The rapid progress in liposome technology that has been made during the past

few years suggests that topical liposome products are practical as
well as useful. It appears to be simply a matter of time before a
topical formulation will also be in clinical trials.

ACKNOWLEDGMENTS

I would like to thank Drs. R. Abra, L. Guo, and M. Woodle for a
critical reading of this manuscript and for their many useful sug-
gestions.

REFERENCES

1. A. D. Bangham, M. M. Standish, and J. C. Watkins, *J. Mol.
    Biol.*, *13*:238, 1965.
2. C. Pidgeon, A. H. Hunt, and K. Dittrich, *Pharm. Res.*, *3*:23,
    1986.
3. C. Pidgeon, and C. A. Hunt, *J. Pharm. Sci.*, *70*:173, 1981.
4. D. A. Barrow, and B. R. Lentz, *Biochim. Biophys. Acta*, *597*:
    92, 1980.
5. F. Szoka, Jr., and D. Papahadjopoulos, *Annu. Rev. Biophys.
    Bioeng.*, *9*:467, 1980.
6. F. Szoka, and D. Papahadjopoulos, in *Liposomes: from Physi-
    cal Structure to Therapeutic Applications* (G. Knight, ed.).
    Elsevier/North-Holland Biomedical Press, Amsterdam, p. 51,
    1981.
7. D. W. Deamer, and P. S. Uster, in *Liposomes* (M. J. Ostro,
    ed.). Marcel Dekker, New York, p. 39, 1983.
8. P. R. Cullis, M. J. Hope, M. B. Bally, T. D. Madden, L. D.
    Mayer, and A. S. Janoff, in *Liposomes: from Biophysics to
    Therapeutics* (M. J. Ostro, ed.). Marcel Dekker, New York,
    p. 39, 1987.
9. L. Saunders, J. Perrin, and D. B. Gammack, *J. Pharm.
    Pharmacol.*, *14*:567, 1962.
10. M. B. Abramson, R. Katzmann, and H. B. Gregor, *J. Biol.
    Chem.*, *239*:70, 1964.
11. D. Papahadjopoulos, W. J. Vail, K. Jacobsen, and G. Poste,
    Biochim. Biophys. Acta, *4*:483, 1975.
12. U. Pick, *Arch. Biochem. Biophys.*, *212*:186, 1981.
13. M. J. Hope, M. B. Bally, G. Webb, and P. R. Cullis, *Bio-
    chim. Biophys. Acta*, *812*:55, 1985.
14. W. Li, and T. H. Haines, *Biochemistry*, *25*:7477, 1986.
15. D. W. Deamer, and A. D. Bangham, *Biochim. Biophys. Acta*,
    443:629, 1976.

16. F. Szoka, Jr., and D. Papahadjopoulos, *Proc. Natl. Acad. Sci., USA*, 75:4194, 1978.
17. J. M. H. Kremer, M. W. J. v. d. Esker, C. Pathmamanohoran, and P. H. Wiersema, *Biochemistry*, 17:3932, 1977.
18. H. G. Enoch, and P. Strittmatter, *Proc. Natl. Acad. Sci., USA*, 76:145, 1979.
19. J. Philippot, S. Mutaftschiev, and J. P. Liautard, *Biochim. Biophys. Acta*, 734:137, 1983.
20. R. A. Schwendener, *Cancer Drug Delivery*, 3:123, 1986.
21. O. Zumbuehl, and H. G. Weder, *Biochim. Biophys. Acta*, 640:252, 1981.
22. R. J. Klimchak, and R. P. Leuk, *Biopharm. Manufact.*, 1:18, 1988.
23. R. H. Guy, J. Hadgraft, M. J. Taylor, and I. W. Kellaway, *J. Pharm. Pharmacol.*, 35:12, 1983.
24. Z. T. Chowhan, T. Yotsuyanagi, and W. L. Higuchi, *Biochim. Biophys. Acta*, 266:320, 1972.
25. H. Komatsu, K. Higaki, H. Okamoto, K. Miyagawa, M. Hashida, and H. Sezaki, *Chem. Pharm. Bull.* 34:3415, 1986.
26. H. S. Bean, and S. M. Heman-Ackahl, *J. Pharm. Pharmacol.*, 16:58t, 1964.
27. J. M. C. Gutteridge, *FEBS Lett.*, 172:245, 1984.
28. C. A. Hunt, and S. Tsang, *Int. J. Pharm.*, 8:101, 1981.
29. N. D. Weiner, W. C. Bruning, and A. Felmeister, *J. Pharm. Sci.*, 62:1202, 1973.
30. S. Frokjer, E. L. Hjorth, and O. Worts, in *Optimization of Drug Delivery*, Alfred Benzon Symposium 17 (H. Bundgaard, A. B. Hanson, and H. Kofod, eds.). Munkagaard, Copenhagen, p. 384, 1982.
31. S.-L. Law, W.-Y. Lo, and G.-W. Teh, *J. Pharm. Sci.*, 76:545, 1987.
32. F. J. Martin, and R. C. MacDonald, *Biochemistry*, 15:321, 1976.
33. G. Smolin, M. Okumoto, S. Feiler, and D. Condon, *Am. J. Ophthalmol*, 91:220, 1981.
34. R. E. Stratford, Jr., D. C. Yang, M. A. Redell, and V. H. L. Lee, *Int. J. Pharm.*, 13:263, 1983.
35. R. E. Stratford, Jr., D. C. Yang, M. A. Redell, and V. H. L. Lee, *Curr. Eye Res.*, 2:377, 1983.
36. I. Ahmed, and T. F. Patton, *Int. J. Pharm.*, 34:163, 1986.
37. V. H. L. Lee, K. A. Takemoto, and D. S. Iimoto, *Curr. Eye Res.*, 3:585, 1984.
38. H. E. Schaeffer, J. M. Breitfeller, and D. L. Krohn, *Invest. Ophthalmol. Vis. Sci.*, 23:530, 1982.
39. H. E. Schaeffer, and D. L. Krohn, *Invest. Ophthalmol. Vis. Sci.*, 22:220, 1982.

40. L. S. S. Guo, C. T. Redemann, and R. Radhakrishnan, *Invest. Ophthalmol. Vis. Sci.*, *28*:72a, 1987.
41. R. Shiota, A. Sarris, R. Radhakrishnan, and L. S. S. Guo, *Invest. Ophthalmol. Vis. Sci.*, *28*:159a, 1987.
42. M. Mezei, and V. Gulasekharam, *Life Sci.*, *26*:1473, 1980.
43. M. Mezei, and V. Gulasekharam, *J. Pharm. Pharmacol.*, *34*:473, 1982.
44. M. Mezei, in *Topics in Pharmaceutical Sciences*   (D. D. Breimer and P. Speiser, eds.). Elsevier Science Publishers, Amsterdam, p. 345, 1985.
45. B. B. Harsanyi, J. C. Hilchie, and M. Mezei, *J. Dent. Res.*, *65*:1133, 1986.
46. M. G. Ganesan, N. D. Weiner, G. L. Flynn, and N. F. H. Ho, *Int. J. Pharm.*, *20*:139, 1984.
47. N. F. H. Ho, M. G. Ganesan, N. D. Weiner, and G. L. Flynn, *J. Controlled Release*, *2*:61, 1985.
48. K. D. Ertel, and J. T. Carstensen, *Int. J. Pharm.*, *34*:179, 1986.
49. V. M. Knepp, R. S. Hinz, F. C. Szoka, Jr., and R. H. Guy, *J. Controlled Release*, *5*:211, 1987.
50. J. Lasch, and W. Wohlrab, *Biomed. Biochim. Acta*, *45*:1295, 1986.
51. W. Wohlrab, and J. Lasch, *Dermatologica*, *174*:18, 1987.
52. A. J. M. Vermorken, M. W. A. C. Hukkelhoven, A. M. G. Vermeesch-Markslag, C. M. A. A. Goos, P. Wirtz, and J. Ziegenmeyer, *J. Pharm. Pharmacol.*, *35*:334, 1984.
53. H. Komatsu, H. Okamoto, K. Miyagawa, M. Hashida, and H. Sezaki, *Chem. Pharm. Bull.*, *34*:3423, 1986.
54. T. Nishihata, K. Kotera, Y. Nakano, and M. Yamazaki, *Chem. Pharm. Bull.*, *35*:3807, 1987.
55. S. W. Hui, T. P. Stewart, P. L. Yeagle, and A. D. Albert, *Arch. Biochem. Biophys.*, *207*:227, 1981.
56. A. Kato, Y. Ishibashi, and Y. Miyaki, *J. Pharm. Pharmacol.*, *39*:399, 1986.

17

Surfactant Association Colloids as Topical Drug Delivery Vehicles

DAVID W. OSBORNE *The Upjohn Company, Kalamazoo, Michigan*

ANTHONY J. I. WARD* and KILIAN J. O'NEILL *University College Dublin, Dublin, Ireland*

I. INTRODUCTION

The addition of a surface active agent to a drug delivery vehicle can result in improved drug stability and clinical potency, increased drug absorption, and decreased drug toxicity (1). If the drug vehicle containing a surfactant is further optimized to form a microemulsion or lyotropic liquid crystalline system, the additional advantages of potentially increased solubilization of a poorly soluble drug and thermodynamic stability are realized (2). For microemulsions, these advantages stem from an apparent particle size in the range of 100 to 1000 A compared with the 5000 to 10,000 A minimum particle size, typical of macroemulsions. The use of micellar systems, microemulsions, and lyotropic liquid crystals as topical drug delivery vehicles will be the focus of this chapter. Collectively, these molecularly structured vehicles are called *surfactant association colloids.* A description of these vehicles and current progress concerning their formulation will be followed by a discussion of in vitro percutaneous absorption experiments that have been used to evaluate surfactant association colloid vehicles.

Before proceeding with this chapter, it is important to define the term *microemulsion.* Unfortunately, this term has different meanings depending upon whether the investigator has a colloidal chemistry background or a pharmaceutical background. In this chapter we will consider a microemulsion to be a system of water, oil, and amphiphile(s) that forms a single phase, transparent or translucent (opalescent), fluid, isotropic dispersion that is presum-

---

Current affiliation: Clarkson University, Potsdam, New York

ably thermodynamically stable (3). Lyotropic liquid crystals are
likewise formed from a system of water and amphiphile(s) that may
include an oil. These solvent-dependent liquid crystals may be either viscous anisotropic or stiff isotropic dispersions. Frequently,
a water, oil, and amphiphile system will contain concentration ranges
that can be defined as *micellar/microemulsion* and also contain concentration ranges that form liquid crystals (i.e., higher surfactant
or lower oil concentrations).

II. DESCRIPTION OF SURFACTANT
ASSOCIATION COLLOIDS

Surfactant molecules may be described in terms of their hydrophilic
and lipophilic natures. It is the balance between these two conflict-
ing properties that determines the range of phenomena from surface
effects to colloidal behavior found in these systems. Overall, the
basic thermodynamic requirements of minimum free energy and maximum entropy in the system apply. Briefly, the unfavorable entropy
consideration associated with the aggregation of amphiphile molecules
is balanced by the lower system free energy obtained by maximizing
the hydrophilic—water contacts.

Advances in understanding the phase behavior of surfactants
and lipids have been made (4—7) by using thermodynamic approaches
incorporating the geometric requirements of the molecular packing.
Observations of aggregated systems that use novel optical (8) and
electron microscopic (9) techniques both support the nature of the
aggregates expected and confirm their highly dynamic nature.

In Chapter 10 of this text (Sect. II.A.2) the ternary representation of phase behavior was introduced, and the usefulness of this
representation described. In light of the thermodynamic approaches
just given as references, it is useful to systematically consider the
molecular events that occur to cause phase changes. For the surfactant systems considered in this text, the solvent is water and
will be positioned on the lower left corner of the triangular plot,
whereas the surface-active component (surfactant) will be positioned
at the lower right corner. The remaining component, usually an
oil, will be positioned at the top of the triangle. First consider the
binary system in which a surfactant is added to water.

Although sodium dodecyl sulfate (SDS) is not acceptable for use
in topical products (10), the well-characterized, definite series of
phase changes that occur for this surfactant, as the surfactant/
water ratio increases, serves as an instructive example. Note that
increasing the surfactant/water ratio means the composition changes
from the lower-left corner of the ternary diagram to the lower-right
corner of the ternary diagram. Initially, the addition of SDS to

water at 25°C results in a clear, colorless isotropic solution (11). Macroscopically this mixture does not change appearance until a 36% SDS solution is obtained. Molecularly, however, dramatic changes are occurring. The first change occurs at the critical micellization concentration (CMC). Before the CMC, the SDS molecules in the bulk water phase can be considered as individually dispersed and as acting as simple electrolytes. Above the CMC, many of the solution's physical properties change. These correspond to the formation of normal micelles (Fig. 1) in which approximately 60 amphiphiles form roughly spherical micelles with a hydrated radius of 25 A. Between the CMC and 25% SDS the micelles are approximately constant in size and maintain a relatively constant number of bound counterions (80%). Above 25% SDS, however, a transition from spherical micelles to rodlike (prolate ellipsoid) shapes rapidly occurs. This effect, sometimes known as the third CMC, is temperature-sensitive (12), occurring at 15% SDS when the temperature is raised to 70°C.

The addition of electrolyte also causes a transition from spheres to rods to occur at lower SDS concentrations (13). At 25°C, a 6.9×10^{-10} M SDS in 0.15 M NaCl solution contains rods with a semiminor axis equal to the radius of spherical micelles (25 A) and a semimajor axis slightly larger, with a length of 33 A. This is dramatically contrasted with a 6.9×10^{-10} M SDS solution in 0.6 M NaCl in which the semimajor axis increases to a length of 208 A, whereas the semiminor axis remains unchanged. When NaCl is present in the SDS—water system, temperature has the opposite effect. Here, decreasing temperature causes micelles to elongate with the greatest distortion occurring at the highest NaCl concentrations (13).

The next phase encountered with further addition of SDS follows directly from the formation of the rodlike micelles. Before the addition of 36% SDS, the rod-shaped micelles are disordered and, therefore, isotropic. After the addition of 36% SDS, the disordered rodlike micelles come into equilibrium with the similar, ordered hexagonal liquid crystal phase (see Fig. 1). From 39% to 50% SDS, all of the SDS amphiphiles become ordered to produce the one-phase hexagonal liquid crystal region. At soap concentrations above 50% SDS, there is no longer enough water to act as a solvent, and the crystalline soap is in equilibrium with the hexagonal liquid crystal. Thus, the baseline of ternary-phase diagrams for crystalline ionic surfactants frequently show this progression of micelles, to hexagonal liquid crystals, to crystalline surfactant. For nonionic surfactants of the polyoxyethyleneglycol type the baseline phase progression is from normal micelle, to lamellar liquid crystal, to inverse micelle or surfactant-rich solution (3).

Upon addition of a third component to an anionic surfactant system, a different characteristic phase progression results. The

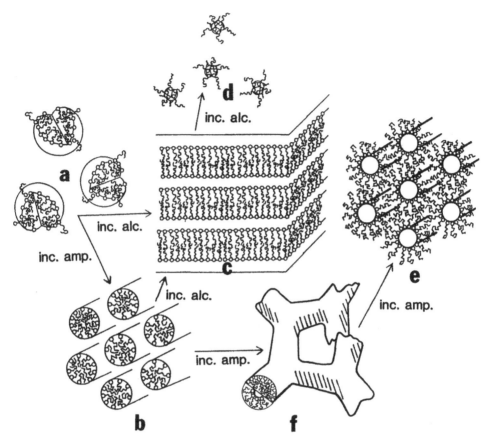

Figure 17.1   Idealized static structures for surfactant association colloids: (a) normal micellar, (b) hexagonal liquid crystal, (c) lamellar liquid crystal, (d) inverse micellar, (e) reversed hexagonal liquid crystal, and (f) a cubic liquid crystal in which the cross section has been split to show that either a normal or inversed orientation can be visualized.   Other structures have been proposed for cubic liquid crystalline phases.   Additions to a water-rich single-tailed anionic surfactant system that will result in moving from one phase to another are noted by the arrows.

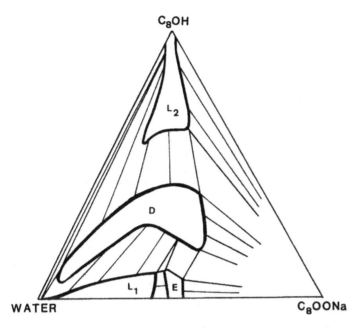

Figure 17.2 Phase diagram according to Ekwall (11) for the system water—octanol—sodium octanoate at 25°C. The single-phase regions are denoted by L_1 for the normal micellar, E for the hexagonal liquid crystal, D for the lamellar liquid crystal, and L_2 for the inverse micellar. This phase behavior is typical for a single-tail anionic surfactant in combination with a medium chain alcohol. (From Ref. 11, used with permission of Academic Press).

thoroughly characterized three-component system water—octanol—sodium caprylate shows the typical phase behavior in Figure 2 (11). At high water content, normal micelles form. As octanol is added, the lamellar liquid crystalline phase is encountered and, ultimately, at highest octanol content, the inverse micelle region occurs (see Fig. 1). This trend, which is characteristic of ionic surfactant systems with medium-chain length alcohols as cosurfactants, can be reasonably explained by packing considerations (6).

The packing ratio (PR) for these systems is defined as

$$v/(l_c a_o) = PR$$

where v and l_c are approximately the volume and the length of the hydrocarbon portion of the amphiphile, and a_o is the optimum area

per headgroup. The method used to calculate the packing ratio
uses Tanford's relations to calculate v and l_c as shown in the following (7):

$$v = (27.4 + 26.9n) A^3$$

$$l_c = (1.5 + 1.265n) A$$

where n is slightly less than the number of carbon atoms per amphiphile chain. Determination of a_0 is accomplished by x-ray diffraction measurements on the lamellar liquid crystal.

Defined this way, the values for packing ratios related to the
different structures are:

Structure	Ratio
Spherical micelles	$PR < 1/3$
Cylindrical micelles	$1/3 < PR < 1/2$
Bilayers (or vesicles)	$1/2 < PR < 1$
Inverted structures	$PR > 1$

The use of these principles can explain the transition of normal
micelles, to bilayers, to inverted structures upon cosurfactant (octanol) addition. Ninham argues that a cosurfactant acts principally
to increase the volume per surfactant molecule, without significantly
affecting a_0 or l_c (6). This is because the uncharged cosurfactant
is able to pack between the ionic head groups without causing repulsion. Thus, as octanol is added to the normal micellar region,
the packing ratio increases until it falls in the range of the lamellar
liquid crystal, thereby implying that packing considerations find
this association structure more favorable. Analogously, further addition of octanol increased the packing ratio toward and above unity, indicating that inverted structures are more favored.

To further investigate this idea, Friberg and Flaim (14) evaluated the packing ratio for a number of lamellar systems at their
highest cosurfactant concentrations. For all systems studied, the
packing ratio approached unity at that point at which the system
entered into equilibrium with the inverted phase. Despite the simple geometric arguments used in the packing ratio concept, the
method provides a straightforward, conceptual method of relating
association structure formation to the physical characteristics of the
component compounds. From these arguments, the phase progres-

sions of normal micelle, to cylindrical micelle, to hexagonal phase, to crystalline soap and normal micelle, to lamellar (bilayer) liquid crystals, to inverse micelles, can be related to a series of molecular events that are based on the molecular structures of the surface-active materials involved. Therefore, it is very reasonable that water—alcohol—ionic surfactant systems exhibit similar, but not identical, phase behavior, whereas nonionic surfactant systems will show different behavior, but similar phase progression. These trends become very useful for predicting how surfactants will interact with solvents, even when previous combinations of these components have not been mixed. An empirically derived schematic description of the transition from normal to reversed structures (15) is shown in Figure 1.

III. FORMULATION OF TOPICAL SURFAC-TANT ASSOCIATION COLLOIDS

Although brief, the foregoing description of surfactant association colloids does provide the background necessary for a more detailed review of the literature that discusses topical microemulsions and liquid crystals. This section will describe formulations that are suitable for topical use, but that have not been evaluated by per-cutaneous transport methods to measure release of a particular drug or model compound. Some of these formulations may have been evaluated for parental administration and, thus, could also be considered for use topically. Formulations that have been evaluated using in vitro percutaneous transport techniques will be described in Section IV.

A. Microemulsions

As stated in the introduction of this chapter, microemulsion vehicles have the potential advantages of both improved stability and solu-bilization characteristics compared with macroemulsion topicals. These advantages stemmed from the smaller particle size character-istic of microemulsions. It is noteworthy that particles of 1000 A and smaller are less than one-fourth the wavelength of visible light, resulting in transparent or translucent (opalescent) systems. Al-though transparency may provide a conceptual appeal to the con-sumer, it provides other advantages to the formulator. A clear system guarantees mixing has been adequate to solubilize all of the drug or other solid materials in the formulation. Milky white creams typical of macroemulsion-based topicals obscure the presence of un-dissolved solids. This is particularly important when the drug is a

solid, because bioavailability of a crystalline drug suspension may
vary dramatically from that of a solubilized drug. For milky white
macroemulsion creams or lotions it is difficult to determine if the
final product requires additional mixing time to properly disperse or
solubilize the drug. However, for a microemulsion formulation, mixing until all of the solid is dissolved or solubilized, as evidenced
by a clear product, provides a readily discernible endpoint for
product manufacture.

The ability of microemulsion systems to solubilize poorly soluble
materials is fundamental to the extensive study of these systems
over the years. Because drugs frequently cannot be delivered to a
therapeutic dosage because of insufficient solubility in solvents that
can be ingested, injected, or applied topically; solubilization of the
drug within a microemulsion may be particularly important. The alternative prodrug approach of improving solubility by molecularly
changing the drug to a more soluble analogue requires toxicological
studies to be completed on each analogue, which may result in unacceptable time delays or increased expense. Thus, solving problems of low solubility by using surfactants can complement prodrug
approaches by satisfying solubility or solubilization demands in the
short-term, while the ideal prodrug is being developed, characterized, and screened for safety.

Thermodynamic stability is important because this characteristic
results in (1) a formulation the properties of which are not dependent upon process (i.e., shear rate, forced cooling, and such), and
(2) a formulation that will not phase-separate, provided temperature
and pressure conditions remain reasonably constant. It is useful to
further consider what it means for a system to be thermodynamically
stable. For a microemulsion, *thermodynamically stable* means the
system is conceptually analogous to an equilibrated salt solution below saturation. As long as the system remains closed (i.e., no
evaporation or chemical degradation) and the temperature and pressure remains the same, the properties of the system will remain unchanged indefinitely. However, if the temperature of the solution
is decreased sufficiently to cause salt crystals to precipitate, then
the single-phase solution changes to the two-phase system of solid
crystals in equilibrium with saturated salt solution. Returning this
system to the original temperature will reverse the phase change,
and at equilibrium will result in a solution with properties and composition identical with the original salt solution. Analogously, a
microemulsion once mixed (equilibrated) will not separate into its
component phases provided that the temperature and pressure remain constant. If changes in storage conditions cause the microemulsion to split, it will re-form once it returns to the original conditions. Importantly the original formation or reformation of the

single-phase microemulsion is independent of process, being solely
dependent upon adequate mixing to bring the system to equilibrium.
Comparison of thermodynamically stable microemulsions with kinetically stabilized macroemulsions emphasizes the processing advantages
gained by formulating microemulsions. Macroemulsions are generally
very process-dependent, as such require the use of homogenizers
or scrapped surface coolers to produce stable products. If conditions change, resulting in separation, simple mixing will not be sufficient for reformation of a macroemulsion system. Likewise, even
the slightest change in processing may produce a difference in the
properties of the macroemulsion system. Based upon these manufacturing considerations, thermodynamically stable microemulsion formulations require less capital investment because only the most basic
mixing equipment is required.

In addition, microemulsions that have an inverse micellar structure may be less comedogenic than either creams or solutions. This
speculation is based on studies of the lipids characteristic of hair
follicles (16). Such follicular lipids are rich in fatty acids. When
viewed in the light of recent information on the effect that partially
neutralized fatty acids have on bilayer-structured epidermal lipids
(17), it seems possible that bilayers may also form from follicular
lipids. The increase in viscosity that accompanies bilayer formation
could be the beginning of the comedo "plug." If bilayer formation
were to occur, water-based emulsions would tend to swell the bilayers, whereas an inverse micellar system may tend to destabilize
the bilayers.

The current disadvantages of microemulsion systems for topical
drug delivery can be related to the limited number of topical microemulsion systems studied. Because most published microemulsion
studies are concerned with detergency (18) or tertiary oil recovery
(19), the oil phases tend to be either short- to medium-chain alcohols (butanol to dodecanol) or hydrocarbons (e.g., decane, hexadecane, benzene). Only a few surfactant association colloid systems
have been characterized that are pharmaceutically applicable. Thus,
rather than current disadvantages, it is more accurate to list these
as current challenges in formulating topical microemulsions, namely,
(1) determining systems that are nontoxic, nonirritating, noncome-
dogenic, and nonsensitizing, and (2) formulating cosmetically elegant microemulsion topicals.

Although only a few of the aforementioned pharmaceutically applicable studies are limited to topical systems, more general texts
provide useful discussions concerning the implications of drug delivery systems that contain surfactants. In particular, the text by
Attwood and Florence (20) and a very readable review by Bhar-
grave et al. (21) concerning use of microemulsions as oral, injecta-

Table 17.1 Safety Considerations of Short- to Moderate-Chain
Length Alcohols

Alcohol	Oral LD_{50} in rat (mg/kg)	Eye irritation	Skin irritation
2-Butanol (least irritating of four isomers)	6480	Moderate	Threshold conc. 7.8%
Hexanol	720	Severe	Mild
Octanol	1790[a]	Moderate	Threshold conc. 12%
Decanol	472	Severe	Severe

[a]Oral LD_{50} for mouse, oral LD_{50} for humans 0.5 to 5 g/kg.

ble, and topical dosage forms. Studies that have been completed to
identify a microemulsion formulation that will be pharmaceutically acceptable, that is, nontoxic, nonirritating, and so on, will be the
focus of this section.

The formation of microemulsions has traditionally been dependent
upon the addition of a medium-chain length alcohol to function as a
cosurfactant. Because alcohols of this chain length range tend to
be skin or eye irritants (Table 1), their use in topical formulations
is very limited. An early study that attempted to find a more suitable topical microemulsion compared the solubilization of hydrocortisone in a traditional microemulsion with the solubilization in a microemulsion containing pharmaceutically acceptable surfactants (22).
The traditional microemulsion consisted of sodium stearate—sodium
myristate surfactants and moderate-chain length alcohols (i.e., n-
butanol through n-heptanol) whereas the pharmaceutical microemulsion was a mixture of Brij 35 and Arlacel 186 combined with isopropanol. For both the traditional and pharmaceutical microemulsions, solubility of hydrocortisone was found to be independent of
the water/oil ratio studied (0.1—0.5) and independent of the oil
chain length (n-alkanes C_8, C_{14}, C_{15}, C_{16}). The pharmaceutical
formulation specified in this study consisted of 10 ml n-alkane (oil),
4 ml isopropanol, 4 g Arlacel 186, 2 g Brij 35, and 1 to 5 ml water.

A more recent study by this research group (23) indicated that
the combination of the anionic surfactant docusate sodium (dioctyl
sodium sulfosuccinate, which is the USP grade of the substance sodium 1,4-bis(2-ethylhexyl)-sulfosuccinate and is marketed as Aero-
sol-OT, most commonly abbreviated as AOT in the chemical litera-

ARLACEL 20

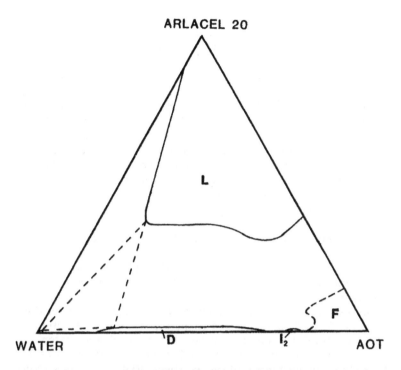

WATER   D   I₂   AOT

Figure 17.3 Ternary phase behavior for the water—Arlacel 20—AOT system at 25°C. The single-phase regions are denoted by L for a clear, colorless, isotropic liquid, D for the lamellar liquid crystalline phase, I_2 for the cubic liquid crystalline phase, and F for the inverse hexagonal liquid crystalline phase.

ture) and the nonionic surfactant sorbitan monolaurate results in a microemulsion capable of obtaining a molar water/surfactant ratio of approximately 50. Both of these surfactants are suitable for topical use, and the phase behavior for AOT has been extensively studied (24,25).

To further evaluate this system, a comprehensive study of the phase behavior resulting from this surfactant mix when combined with water and oil was completed (26,27). Figure 3 provides the base diagram for the water—Arlacel 20—AOT system. The most striking feature of this diagram is that less than 5% (w/w) of Arlacel 20 can be solubilized in the water—AOT lamellar structure at equilibrium. For the water/AOT weight ratios from 4:1 to 3:7, addition of 4% to 8% Arlacel 20 resulted in separation of an isotropic phase that could be detected microscopically after approximately 1 week. The addition of larger amounts of Arlacel 20 had a more im-

mediate destabilizing effect. This base diagram provides insight
concerning the surfactant interactions that allow the formation of an
alcohol-free microemulsion. The addition of Arlacel 20 (sorbitan
laurate) destabilized the structure of the AOT—water lamellar phase.
This disordering effect decreases the lamellar region, whereas com-
plementarily increasing the size of the isotropic single-phase region.
Thus, for the water—Arlacel 20—AOT base diagram, it appears that
Arlacel 20 functions as a cosurfactant similar in mechanism to that
of medium-chain alcohols in traditional anionic soap systems. Addition of the cosurfactant destabilizes the bilayer-structured lamellar
liquid crystalline phase, resulting in an increase in the compositional range over which the microemulsion is present.

In contrast with the anionic soap systems (11), medium-chain
alcohols do not function as cosurfactants when added to the aqueous
AOT system. As seen in Figures 4 through 6 (28), the size of the
lamellar liquid crystal phase is not reduced to the degree seen upon
addition of Arlacel 20. Although the microemulsion region is large
when either hexanol or octanol is added as the third component,
the most dominant features of these diagrams are the large cubic
and inverse hexagonal liquid crystalline regions. The stability of
these stiff phases shape the boundaries of the microemulsion region.
The water—butanol—AOT diagram shown in Figure 4 is characterized
by a continuous solubility region between the butanol and water
corners. This phase behavior is typical for butanol systems, and
the resulting fluid bicontinuous phase is speculated to be less likely
to have the desired ability to solubilize both lipophilic and hydrophilic drugs.

The maximum water solubilization obtainable for three Arlacel
20/AOT ratios in hexadecane is shown in Figure 7. The 73:27 Arlacel 20/AOT surfactant ratio that gave maximum water solubilization
for the base diagram (see Fig. 3) did not give maximum water solubilization after addition of more than 20% hexadecane. Similarly,
the 85:15 Arlacel 20/AOT surfactant ratio had limited ability to solubilize water. However, increasing the amount of AOT to a 60:40
Arlacel 20/AOT surfactant ratio provided a large isotropic region
capable of solubilizing more than 60% water. More importantly, this
system was able to solubilize significantly more water at higher hexadecane concentrations than the systems with larger Arlacel 20/sur-
factant ratios. Thus, addition of hexadecane to the AOT—Arlacel
20 system provides for a systematic characterization of the alcohol-
free microemulsion. As seen in Figure 7, the surfactant ratio has
a significant influence on the water solubilization capacity of the
system. If high water solubilization over a range of mixed surfac-
tant/oil ratios is required, then the Arlacel 20/AOT ratio should be
less than 73:27 (i.e., the surfactant ratio that results in the maximum water solubility in the base diagram; see Fig. 3).

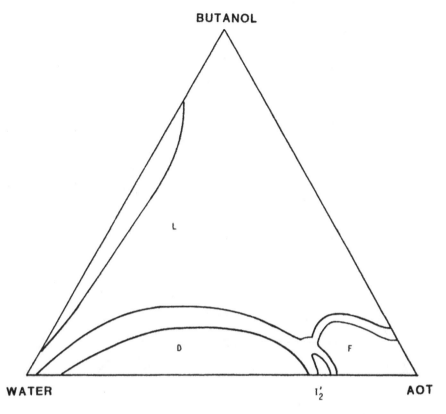

Figure 17.4 Phase diagram for the system water—butanol—AOT at
24 ± 2°C. L is continuous solution region, D is the lamellar liquid
crystal, I_2 is the cubic liquid crystal, and F is the reversed hexagonal liquid crystal.

The low solubility of the waxy semisolid AOT in water and the
subsequent formation of liquid crystalline phases at higher AOT
compositions has been well characterized in the literature (24,25).
Initial phase behavior studies of the AOT—sorbitan laurate microemulsions indicated a severe practical limitation of the system. Long
mixing times were required for the waxy semisolid surfactant AOT
to mix with the viscous liquid Arlacel 20. In attempts to avoid the
slow mixing of the surfactants, the commercially available liquid surfactant AOT-75 was used instead of AOT. AOT-75 is a solution
containing 75% AOT, 20% water, and 5% ethanol. It is a liquid because the addition of 5 wt% ethanol to the 75:20 AOT/water ratio

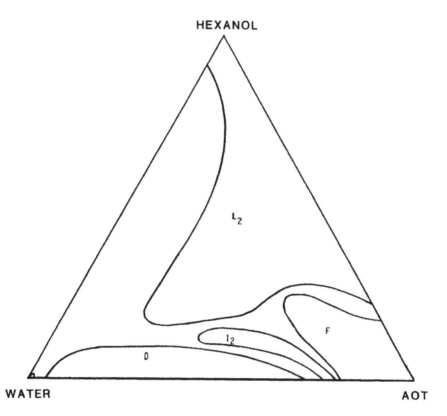

HEXANOL

L₂

F

I₂

D

WATER   AOT

Figure 17.5 Phase diagram for the system water—hexanol—AOT at 24 ± 2°C. L₂ is the inverse micellar (microemulsion) region. Other phases are labeled the same as in Fig. 4.

prevents the formation of the cubic liquid crystalline phase that exists from 75% to 80% AOT in water. Thus, AOT-75 remains a liquid (viscosity approximately 200 cp), provided ethanol is not lost through evaporation. Unlike the medium-chain alcohols, ethanol shows very low irritation and toxicity. Thus, in these small amounts, ethanol is completely acceptable as a pharmaceutical ingredient. By determining the rheological properties of the miscible AOT-75—sorbitan laurate mixture, and characterizing the phase behavior of this surfactant mixture when combined with hexadecane and water, optimal compositions for a pharmaceutical low-alcohol microemulsion were determined.

As seen in Figure 8, addition of Arlacel 20 (viscosity approximately 5700 cp) does not dramatically increase the viscosity of AOT-

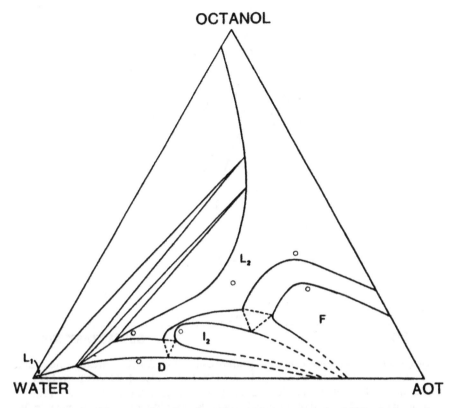

Figure 17.6 Phase behavior for the water—octanol—AOT system at 24 ± 2°C. Phases are labeled as in Fig. 4. Compositions circled are those described in Table 4.

75 until less than 40% AOT-75 is present (i.e., greater than 60% Arlacel 20). Thus, the mixing problems characteristic of the waxy 100% AOT are not encountered with AOT-75. Furthermore, the rheology studies indicate that the most readily combined binary surfactant mixtures will contain greater than 40% AOT-75. Because microemulsions are thermodynamically stable and their formation is reversible with temperature, the mixing of 100% AOT could be facilitated by heating. However, studies indicating hydrolysis of AOT upon long-term storage (29) discourages this approach.

The extent of the microemulsion region between 100% AOT-75 and 60:40 AOT-75/Arlacel 20 is shown in Figure 9. This one-phase region grows in area smoothly. All surfactant ratios except the

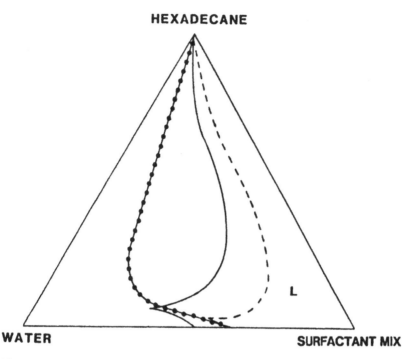

Figure 17.7 Pseudo—ternary-phase behavior for the water—hexadecane—Arlacel 20—AOT system, in which the Arlacel 20/AOT weight
ratio is 60:40 (—●—●—); 72.5:27.5 (———); and 85:15 (---). The
single-phase clear, isotropic microemulsion side of the diagram is labeled by L.

100% AOT-75 are completely miscible with hexadecane, and the
amount of water soluble in the surfactant—hexadecane solutions
grows as the amount of Arlacel 20 increases. Figures 10 and 11
show the transition region near the maximum water solubility microemulsion. The maximum water solubility appears to occur between
55:45 and 60:40 AOT-75/Arlacel 20 ratios and is approximately the
same for hexadecane/surfactant mixture ratios between 20:80 and
50:50. At 54% AOT-75, a two-phase region appears inside the one-
phase microemulsion region. The samples in this two-phase region,
and nearby in the one-phase region, were temperature-sensitive and
were examined in a temperature-controlled air chamber. As the
fraction of AOT-75 decreases further, the microemulsion region also
decreases in size (Fig. 12).

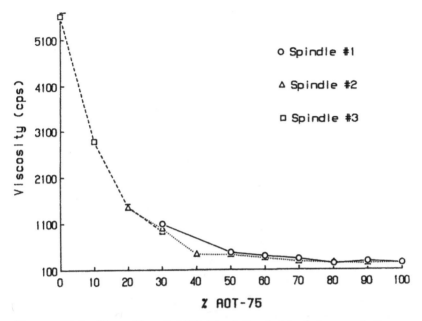

Figure 17.8 Viscosity of AOT-75—Arlacel 20 mixtures at the constant apparent shear rate of 64.6 sec^{-1}.

Comparison of Figures 9 through 12 with Figure 7 shows that replacing the waxy semisolid AOT with the liquid AOT-75 does not alter the trends in the phase behavior. The small amount of ethanol present in the microemulsion appears to have minimal effect on the microemulsion. Although the narrow extensions found in Figure 9 can incorporate approximately 55% water, the broad-rounding boundary characteristic of 57% to 70% AOT-75 is much more desirable because maximum water content can be maintained over a wider range of added hydrocarbon. For AOT-75 concentrations above the 70%: 30% AOT-75/Arlacel 20 mixture, the amount of water solubilized decreases, once again indicating that it is the blend of these two surfactants that produce the microemulsion region capable of the greatest water incorporation.

The unique phase behavior seen for both Figures 10 and 11 deserves further comment. It is not surprising that a mixed surfactant system would be characterized by an ideal surfactant ratio for incorporating water into the microemulsion. As shown in Figure 11, the two-phase region found within the single-phase microemulsion suggests that critical points may be present in this region of the

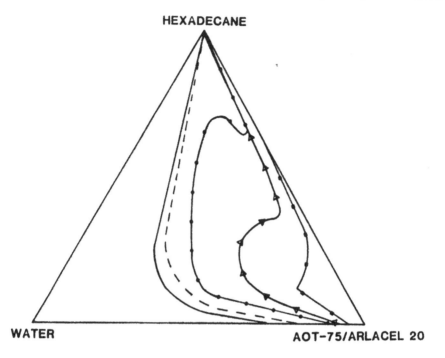

Figure 17.9 Pseudoternary phase behavior for the water—hexadecane—AOT-75—Arlacel 20 system at 24° ± 2°C. The ratios of AOT-75
(75% sodium-1,4-bis(2-ethylhexyl)sulfosuccinate, 20% water, 5% ethanol) to Arlacel 20 (sorbitan laurate) are as follows: ▬■▬■▬■▬ 100%
AOT-75; ▸▸▸▸▸▸ 90:10 AOT-75/Arlacel 20; ●●● 80:20 AOT-75/
Arlacel 20; − − − 70:30 AOT AOT-75/Arlacel 20; ——— 60:40 AOT-
65/Arlacel 20.

quaternary plot. It is also the 54:46 ratio of AOT-75/Arlacel 20
that provides the maximum incorporation of water. However, the
extreme temperature sensitivity of the phase behavior at this surfactant ratio makes the 50% to 60% AOT-75 in Arlacel 20 systems of
limited practical use. Because single-phase systems are desired for
stable pharmaceutical formulations, slightly larger amounts of AOT-
75 (60%—70%) should be mixed with Arlacel 20 to formulate a pharmaceutical microemulsion capable of incorporating large amounts of
water. By thoroughly characterizing the phase behavior of this
multiple-component system, further efforts toward the formulation
of a low-alcohol pharmaceutical microemulsion is possible. It should
be noted that such thorough characterization of a four-component

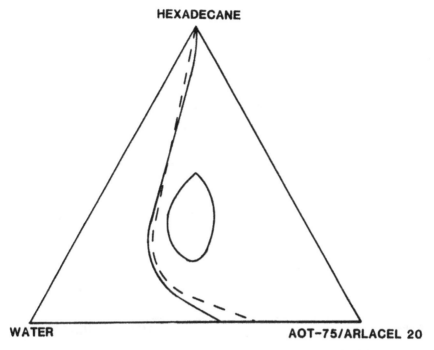

Figure 17.10 Pseudoternary phase diagram for the water—hexadec-ane—AOT-75—Arlacel 20 system at 25° ± 0.2°C. Dashed line is for 57:43 AOT-75/Aracel 20 ratio, whereas the solid line is for the 54:46 AOT-75/Arlacel 20 ratio. Note the two-phase region within the single-phase region for the 54:46 AOT-75/Arlacel 20 system.

system, especially one of practical concern, is relatively rare in the open literature (2,30,31). Also, note that AOT does not require the presence of a cosurfactant to form a microemulsion. The micro-emulsion region resulting from water—hydrocarbon—AOT systems has been extensively studied by scattering techniques (32,33). These systems do not tend to incorporate as much water as when Arlacel 20 is used as a cosurfactant. Furthermore, the highest water incorporation for these three-component microemulsions tends to be relatively narrow extensions toward the water corner. Thus, the three-component microemulsions, although probably well suited for some applications, are not expected to be as widely useful as the Arlacel 20—AOT microemulsions.

Another surfactant association colloidal system that has received active investigation as a pharmaceutical vehicle consists of lecithin—

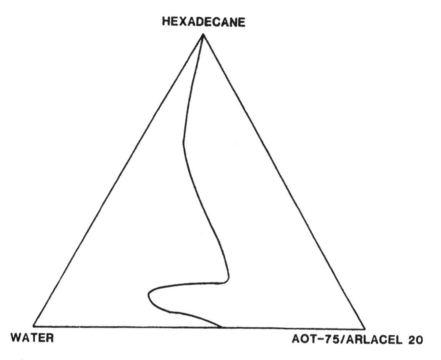

HEXADECANE

WATER   **AOT-75/ARLACEL 20**

Figure 17.11 Pseudoternary phase diagram for the water—hexadecane—AOT-75—Arlacel 20 system at 24 ± 2°C for the 50:50 AOT-75/
Arlacel 20 mixture.

bile salt mixtures. These systems have been studied for use as in-
jectables (34,35) and topicals (36). Additional studies on the phase
behavior of these systems have been completed in efforts to better
understand dissolution—solubilization of cholesterol gallstones (37,
38). Evaluation of this system for topical drug delivery included
comparison of totally hydrogenated lecithin with natural lecithin that
had been "purified" but not hydrogenated, and replacement of water in part by other polar solvents (39).

The results of this study are shown in Figure 13. The striking
feature is the difference between the totally hydrogenated lecithin
and the unsaturated purified lecithin. The distinct decrease in the
ability of the sodium cholate micelles to incorporate the totally hydrogenated lecithin can be attributed to the stiffness of the lecithin's saturated chains. This stiffness would be expected to raise
the transition temperature of the bilayer, thus preventing the for-

HEXADECANE

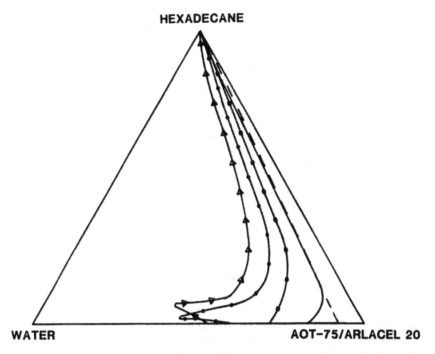

WATER   **AOT-75/ARLACEL 20**

Figure 17.12   Pseudoternary phase diagram for the water—hexadecane—AOT-75—Arlacel 20 system at 24 ± 2°C. The AOT-75/Arlacel
20 ratios are as follows: ▷–▷–▷ 40:60, AOT-75/Arlacel 20; ●–●–●
30:70 AOT-75/Arlacel 20; ■–■–■ 20:80 AOT-75/Arlacel 20; ——— 10:
90 AOT-75/Arlacel 20; and — — — 100% Arlacel 20.

mation of the L_α phase at 30°C. This is confirmed by the phase
behavior of the system, because the excess phase for the totally
hydrogenated lecithin is crystalline rather than liquid crystalline.

The micellar phase region for the purified lecithin was similar
to that found in the literature (37). The transition from spherical
to oblate, ellipsoidal structure (38) was noted in the purified lecithin system for lecithin/sodium cholate ratios greater than unity.
This was inferred by observing the extremely slow solubilization of
the lecithin in this transition region.

The use of a 1:1 mixture of water/propylene glycol as the solvent allowed slightly greater solubilization of the lecithin. This is
readily attributed to the bilayer destabilizing nature of propylene
glycol (40). This same effect accounts for the rapid solubilization
of lecithin into the sodium cholate micelles for the propylene glycol—

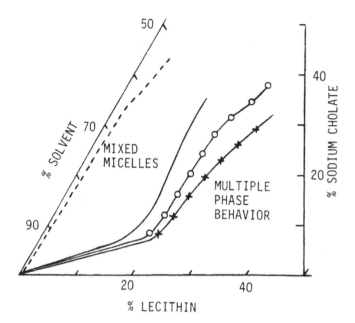

Figure 17.13 Partial ternary diagram for the system water—sodium cholate—totally hydrogenerated lecithin (– – – – – –), and for three systems using purified lecithin--sodium cholate and either water (————), water/propylene glycol 1:1 (–o–o–), or water/butanediol 1:1 (–*–*–). The micellar region was determined at 30°C.

water system, compared with having water as the solvent. For the water system, a lamellar liquid crystalline surface film immediately formed around the added lecithin, severely hindering solubilization. The bilayer destabilizing action of the propylene glycol prevented this film formation resulting in rapid dissolution of the lecithin in the sodium cholate—water micellar system. Replacement of half the water with butanediol had an effect similar to the use of propylene glycol.

B. Topical Liquid Crystals

Lyotropic liquid crystalline vehicles have generally the same advantages that were listed for microemulsions. These mesophases are thermodynamically stable and have unique, often superior, solubilization properties. Unlike microemulsions, some of the liquid crystalline systems may be slightly cloudy, rather than clear. In addition to the advantages listed for microemulsions, liquid crystals may

release drugs over a sustained period either because of slower diffusion from the liquid crystal or because of an increased solubilized
residence time for the drug after topical application.

Most work published, to date, describing delivery from liquid
crystals focuses on cubic-structured phases. This liquid crystal
has been described (41) as being either erodible or nonerodible depending upon its phase behavior in excess water. For the polyoxyethylene surfactant—water system (42) two liquid crystalline systems
were described. The cubic phase containing low amounts of surfactant was formed from closely packed micelles and, thus, readily
"eroded" into a micellar system when added to water. The cubic
phase that formed at high surfactant content passes through a viscous hexagonal phase upon addition of water, then forms the closely
packed micellar cubic liquid crystal upon further addition of water.
Thus, erosion will be much slower for the cubic phase containing a
greater amount of surfactant. Although this concept of erosion is
more pertinent to parental or oral liquid crystalline delivery, the
use of the complete phase diagram to understand the delivery of a
drug is equally important for topicals.

The cubic liquid crystalline systems studied above (41) included
60% monoolein—40% water, 48% monoolein—12% lecithin—40% water, and
60% monoolein—4% to 10% glycerol—36% to 30% water. (All percentages were based upon weight.) Methylene blue was used as the
marker to establish sustained release from each of these vehicles.
Other pharmaceutical applications include the use of a cubic liquid
crystals to act as a parenteral depot for peptides that protects
against biodegradation (43) and to depress the rate of drug hydrolysis even in the presence of a gastrointestinal-like bulk phase (44).

Lamellar liquid crystals have also been evaluated for possible
use as topical drug delivery systems. Work by Wahlgren et al.
(45) showed that lecithin—water systems could dissolve up to 5 wt%
hydrocortisone, and that the diffusion rate of the drug within this
liquid crystalline vehicle was four orders of magnitude higher than
the diffusion rate of hydrocortisone across the skin. Additional
lamellar liquid crystalline systems were characterized by El-Nokaly
et al. (46,47) in which the water was replaced by either alkanediols
or oligomers of polyethylene glycols. Although some of these solvents (e.g., ethylene glycol) are not suitable for topical use, many
of the compounds are, and these systems should be considered for
evaluations as topical drug delivery formulations. An additional
nonaqueous lamellar liquid crystalline system that was evaluated for
topical use combined triethanolamine with oleic acid and other polar
and nonpolar solvents (48). This study characterized the interaction of various components of human sebum with the liquid crystalline vehicle.

IV. DRUG DELIVERY FROM TOPICAL SUR-
FACTANT ASSOCIATION COLLOIDS

Although relatively few microemulsion and liquid crystalline systems
that are suitable for topical use have been described in the literature, even fewer systems have been evaluated by percutaneous
transport techniques. In this section, the systems that have been
studied will be described.

A. Microemulsions

One of the first, and certainly most interesting studies of topical
drug delivery from a microemulsion, compared with a gel and cream
of similar composition, was reported by Ziegenmeyer and Fuhrer in
1980 (49). In this study, the three formulations evaluated contained
the same amount of dodecane and water. The amount of decanol was
varied, and polyoxyethylene 7—lauryl ether was the surfactant used.
In vitro experiments using skin membranes and 1% tetracycline HCl
formulations showed that much less active agent penetrates through
the skin from the cream than from the gel and that both of these
formulations are definitely exceeded in effect by the microemulsion.
In vivo studies using fluorescent tetracycline confirmed that the microemulsion was superior to the other systems in ability to promote
penetration.

A more comprehensive study of topical drug delivery from a microemulsion was recently undertaken by our research groups (50—
52). This study focused on the microemulsion and liquid crystalline
regions that result from the water—octanol—AOT system. As seen
in Figure 6, the phase behavior of this system is truly unique, especially considering the large extension of the cubic liquid crystalline (i.e., ringing gel) phase. The close proximity of the liquid
crystalline regions to the microemulsion region allows evaluation of
the effect of vehicle structure upon percutaneous transport of a
drug. For this unique system, this evaluation can be completed for
vehicles that vary in component composition by only a few weight
percent. Also noteworthy is the ability of the 58:42 AOT/octanol
weight ratio to incorporate greater than 70% water into the microemulsion region. Thus, microemulsions with the same AOT/octanol
ratio, but with widely ranging (15—70 wt%) amounts of water were
evaluated for drug delivery.

The initial study (50) limited itself to examining the flux of
tritium-labeled water from four microemulsion vehicles across human
cadaver skin. Each of the vehicles has the set ratio of 58:42 AOT/
octanol and ranged in water content as shown in Table 2. In an
attempt to separate vehicle effects upon water transport from the

Table 17.2 Water Self-Diffusion and Normalized In Vitro Transdermal Flux Values for the Microemulsion Formed by Addition of Water to a 58:42 Weight Ratio of AOT/Octanol

% Water	X Water	D/Dw	Normalized flux (ave)
15	0.68	0.035	0.89, 0.78 (0.84)
35	0.87	0.072	2.9, 2.1 (2.5)
45	0.91	0.100	
52	0.93	0.200	
58	0.94	0.353	4.9, 5.4 (5.2)
67	0.96	0.363	6.2, 2.9 (4.5)

skin barrier effects associated with the stratum corneum, the diffusion of water within the vehicle was measured using a pulsed field gradient fourier transform NMR technique (53). The unitless value D/Dw is the ratio of the self-diffusion coefficient for water in the microemulsion vehicle divided by the self-diffusion coefficient for water in water. Therefore, a D/Dw value of 0.100 means than, on the average, water in the 45% water microemulsion environment diffuses at one-tenth the rate that a water molecule would diffuse in a totally aqueous environment. The initially low values for D/Dw for microemulsions of low water content, and subsequent increase with addition of water, can be attributed to binding of the water molecules to the surfactant headgroup. Thus, for the 15% water microemulsion, most of the water in the microemulsion is bound to the surfactant headgroup and is not available for transport across the skin. The result is that transport of water across the skin from this particular microemulsion is less than the transport of water from neat water, that is, normalized flux is less than unity. Pretreatment studies using octanol, AOT, and AOT/octanol (58:42 ratio; Table 3) further indicated that the enhancement in water penetration from the high water content microemulsions was a result of a synergistic effect between AOT and octanol and was not due to the vehicle's microemulsion structure.

Additional studies that evaluated the in vitro transdermal transport of labeled glucose from the microemulsions across cadaver skin (51) showed that the glucose essentially crossed with the water. Thus, microemulsion systems that showed high water transport, also demonstrated high glucose transport when an infinitely small amount of labeled glucose was spiked into the microemulsion.

Table 17.3 Normalized In Vitro Transdermal Flux Values for Neat
Water Following Pretreatment

Substance	Duration (hr)	Normalized flux
Octanol	0.5—6	1.7
AOT (from EtOH)	2	2.1
AOT/Octanol 58:42	1—6	3.8
	18—22	7.3

B. Liquid Crystals

The close proximity of the liquid crystalline regions to the microemulsion region allows comparison of drug transport from the structurally different, but compositionally similar, surfactant association
colloids. A summation of this comparison is given in Table 4.   Two
different study designs were used to evaluate in vitro transdermal
flux of water across human skin after application of a microemulsion
or liquid crystal. Study design A utilized surgical skin obtained
from radical mastectomy cases, with the skin being used within 24
to 48 hr after procurement. This skin was full thickness, stored on
wet ice, and used after removal of the subcutaneous fat (50). Resulting flux values were normalized against flux values for tritiated
water using skin from the same donor in an attempt to eliminate individual variation. For study design A, the vehicles were prepared
by dilution of tritium-labeled water with distilled water, and then
addition of the appropriate amount of AOT dissolved in octanol. A
single-label liquid scintillation-counting study was used to assay the
receiver phase of the flow through transdermal cell. In contrast,
study design B used dermatomed cadaver skin that was shipped on
dry ice, and that remained frozen until the time of use. Five rep-
licates of each formulation were tested using skin from a single individual. Although normalization of flux values was not required to
minimize variation, the necessary transport data was determined to
allow direct comparison with the results from study design A. Samples were prepared with tritiated water as described in design A
with the exception that a 4-μCi spike of glucose D-[^{14}C(U)] (3.7
mCi/mmol) was dissolved in the water before addition of the AOT—
octanol solution. For study design B, a dual-label liquid scintillation-counting study was used to assay the receiver phase of the
flow through transdermal cell.

Considering the differences in experimental design between
these two studies, the normalized flux of water across the skin after

Table 17.4 Water Transport from Structurally Different Surfactant
Association Colloids

Colloid type	Normalized flux values	
	Design A	Design B
15% water microemulsion	0.8	0.4
35% water microemulsion	2.5	2.0
67% water microemulsion	4.5	5.0
Lamellar liquid crystalline	2.4	1.1
Inverse hexagonal liquid crystalline	0.7	0.3
Cubic liquid crystalline	2.5	1.3

delivery from microemulsions or liquid crystals is remarkably consistent. Note that delivery from the microemulsions follows the same trends for both designs and can be described in terms of water mobility, as discussed earlier. The high water content liquid crystalline phases give approximately equivalent water transport that is approximately half the value characteristic of the highest water content microemulsions. This decrease in transport from the liquid crystalline phases is attributed to the involvement of octanol and, although probably to a smaller extent, AOT in the formation of the liquid crystalline structure. The cubic liquid crystalline phase illustrates this point. Without a doubt, the stiff amphiphilic films that are necessary to produce the cubic liquid crystal results in the dramatic decrease in self diffusion of octanol and AOT. The diffusion of octanol within the vehicle likely becomes slower than the diffusion of octanol across the skin. Thus, a component of the vehicle that was required to produce enhancement is no longer limited in its effect by slow diffusion across the stratum corneum—dermis barrier, rather the enhancer is limited by slow diffusion out of the cubic liquid crystalline vehicle. Note that the self-diffusion coefficient of water within the cubic phase is equivalent to the self-diffusion coefficient of water within the high water content microemulsions, whereas the transport of water across the skin is approximately half for the cubic phase compared with the microemulsion.

The low water transport value for the lamellar liquid crystal is not readily explained in terms of decreased diffusivity of octanol. The amphiphilic film for the water—octanol—AOT bilayer region is probably more fluid than for the cubic liquid crystal. For the la-

mellar phase the diminished water transport may be attributed to
the AOT/octanol ratio being less than the synergistic optimum ratio.
Thus, the 58:42 AOT/octanol ratio provides greater enhancement
than the 83:17 AOT/octanol ratio necessary to form the lamellar liquid crystal.

Explanation of the very low water transport from the inverse
hexagonal liquid crystal follows directly from the preceding discussion of the 15% water microemulsion. For these systems, the water
is bound to the headgroups of the surfactant and, thus, is less
available for transport. The glucose also will presumably reside in
this polar region of the vehicle, and will similarly be limited in its
transport.

Sokoloski et al. (54) have also examined the skin transport of
drugs of different polarities delivered from a water—propylene glycol—AOT lamellar liquid crystal. For this liquid crystalline vehicle,
hairless mouse skin permeabilities of minoxidil and p-nitrophenol
were determined and compared against delivery of the same two model drugs from a PG—water mixture. The use of the liquid crystalline vehicle increased the transport of minoxidil, but decreased the
transport of p-nitrophenol. These results emphasize the need to
understand, in detail, how each component of the formulation interacts with the skin, and how the drug interacts with each component
of the formulation.

V. CONCLUDING REMARKS

Drug delivery from surfactant association colloids appears to be
unique when compared with delivery from traditional formulations.
The mobility of the drug within the vehicle, and the mobility of vehicle components that enhance percutaneous transport must be considered. Highly structured vehicles, such as inversed hexagonal
and cubic liquid crystals, may exhibit sustained-release properties,
either by binding the water (i.e., internal phase) or by stiffening
the amphiphilic film (i.e., the penetration enhancer components)
within the formulation. Alternatively, microemulsions, if properly
formulated, can provide optimum drug delivery. For microemulsion
systems the phase behavior should be adequately characterized to
assure that sufficient solvent is available to solvate the amphiphiles.

In summary, the studies that have evaluated topical drug delivery by microemulsions or liquid crystalline formulations indicate
that surfactant association colloids are often superior vehicles for
topical drug delivery. Microemulsions can be formulated so that
their molecularly dynamic nature is maximized, thus allowing maximized delivery of both drug and penetration enhancer. Liquid

crystals can be formulated with either fluid or bound-water cores, and with surfactant films of varying stiffness. Because of their structure, liquid crystals provide a range of release properties for both the "solvent" portion of the structure (i.e., the drug) and the penetration enhancer. As traditional enhancers are themselves absorbed into the systemic circulation, substained release from the vehicle could provide percutaneous penetration enhancement for extended periods. These drug delivery results when combined with increased drug solubilization and thermodynamic stability strongly suggest that association colloids will become increasingly important as specialized topical drug delivery formulations.

REFERENCES

1. A. T. Florence, *Drugs Pharm. Sci.*, *12*:15, 1981.
2. S. E. Friberg and R. L. Venable, in *Encyclopedia of Emulsion Technology*, Vol. 1. (P. Becher, ed.) Marcel Dekker, New York, pp. 287–356, 1985.
3. K. Shinoda and S. Friberg, *Emulsions and Solubilization*, Chap. 1. John Wiley & Sons, New York, 1986.
4. J. N. Israelachvili, D. J. Mitchell, and B. W. Ninham, *J. Chem. Soc. Faraday Trans. II*, *72*:1525, 1976.
5. J. N. Israelachvili, *Proc. Int. Sch. Phys. "Enrico Fermi"*, (*Phys. Amphiphilas*), *90*:24, 1985.
6. B. W. Ninham, and D. J. Mitchell, *J. Chem. Soc. Faraday Trans. II*, *77*:601, 1981.
7. C. Tanford, *J. Phys. Chem.*, *76*:3020, 1972.
8. D. D. Miller, J. R. Bellare, D. F. Evans, Y. Talmon, and B. W. Ninham, *J. Phys. Chem.*, *91*:674, 1987.
9. J. R. Bellare, T. Kanedo, and D. F. Evans, *Langmuir*, *4*: 1066, 1988.
10. A. W. Fulmer, and G. J. Kramer, *J. Invest. Dermatol*, *86*:598, 1986.
11. P. Ekwall, in *Advances in Liquid Crystals*, Vol. 1. (G. H. Brown, ed.). Academic Press, New York, 1975.
12. F. Reiss-Husson, and V. Luzzati, *J. Phys. Chem.*, *68*:3504, 1964.
13. N. A. Mezer, G. B. Benedek, and M. C. Carey, *J. Phys. Chem.*, *80*:1075, 1976.
13. E. Toultou, and A. H. Goldberg, European Patent Application 84109963.1, June 22, 1984.
14. S. E. Friberg, and T. D. Flaim, in *Inorganic Reactions in Organized Media*. (S. L. Holt, ed.). ACS Symposium Series 177, Washington, D.C., 1982.

15. P. Ekwall, L. Mandell, and K. Fontell, *Mol. Cryst. Liquid Cryst.*, *8*:157, 1969.
16. K. M. Nordstrom, J. N. Labows, K. J. McGinley, and J. J. Leyden, *J. Invest. Dermatol.*, *86*:700, 1986.
17. S. E. Friberg, and D. W. Osborne, *J. Dispersion Sci. Tech.*, *6*:485, 1987.
18. F. Comelles, C. Solans, N. Azemae, J. Leal, and J. L. Parrra, *J. Dispersion Sci. Technol.*, *1*:369, 1986.
19. D. O. Shah, ed., *Surface Phenomena in Enhanced Oil Recovery*. Plenum Press, New York, 1981.
20. D. Attwood, and A. T. Florence, *Surfactant Systems: Their Chemistry, Pharmacy and Biology*. Chapman and Hall, New York, 1983.
21. H. N. Bhargava, A. Narurkar, and L. M. Lieb, *Pharm. Technol.*, *11*(3):46, 1987.
22. A. Jayakrishnan, K. Kalaiarasi, and D. O. Shah, *J. Soc. Cosmet. Chem.* *34*:335, 1983.
23. K. A. Johnson, and D. O. Shah, *J. Colloid Interface Sci.*, *107*:269, 1985.
24. J. Rodgers, and P. A. Winsor, *J. Colloid Interface Sci.*, *30*: 247, 1969.
25. E. I. Frances, and T. J. Hart, *J. Colloid Interface Sci.*, *94*: 1, 1983.
26. D. W. Osborne, C. A. Middleton, and R. L. Rogers, *J. Dispersion Sci. Tech.* *9*:415, 1988.
27. D. W. Osborne, R. J. Chipman, and C. V. Peschek, *Int. J. Pharm. Tech. Prod. Manuf.* (in press).
28. D. W. Osborne, A. J. I. Ward, and K. J. O'Neill, unpublished results. Docusate sodium USP was obtained from American Cyanamide; butanol, hexanol, and octanol were the highest purity available from Aldrich. These materials were used as received. Samples were mixed and equilibrated for at least 2 weeks before phase behavior characterization using polarized light microscopy.
29. P. Delord, and F. C. Larche, *J. Colloid Interface Sci.*, *98*: 277, 1984.
30. S. M. Ng, and S. G. Frank, *J. Dispersion Sci. Tech.*, *3*:271, 1982.
31. R. C. Baker, A. T. Florence, T. F. Tardos, and R. M. Wood, *J. Colloid Interface Sci.*, *100*:311, 1984.
32. B. H. Robinson, C. Toprakcioglu, J. C. Dore, and P. Cheiux, *J. Chem. Soc., Faraday Trans. I*, *80*:13, 1984.
33. M. Kotharchyk, S. H. Chen, J. S. Huang, and M. W. Kim, *Phys. Rev. A*, *29*:2054, 1984.
34. Hoffmann-LaRoche AG, European Patent 133-258-A, February 20, 1985.

35. American Lecithin, U.S. Patent 4,252,793, February 24, 1981.
36. I. Lyon, and H. Lyon, U.S. Patent 4,115,313, September 19, 1978.
37. G. Lindblom, P. O. Erikson, and G. Arvidson, *Hepatology 4* (suppl. 5):1295, 1984.
38. K. Muller, *Hepatology*, 4(suppl. 5):1345, 1984.
39. D. W. Osborne, Unpublished results. Lecithin was obtained from American Lecithin Company, Atlanta, Ga., sodium cholate was obtained from Sigma Chemical Co., St. Louis, Mo.
40. A. J. I. Ward, C. Marie, L. A. Sylvia, and M. A. Phillipi, *J. Dispersion Sci. Technol.*, 9:149, 1988.
41. S. Engström, K. Larsson, and B. Lindman, International Patent Application WO 84/02076, June 7, 1984.
42. D. J. Mitchell, G. J. T. Tiddy, L. Waring, T. Bostock, and M. P. McDonald, *J. Chem. Soc. Faraday Trans. I*, 79:975, 1983.
43. B. Eriksson, S. Learder, and M. Ohlin, 15th International Symposium on Controlled Release of Bioactive Materials, paper 219, Basel, Switzerland, August 1988.
44. G. Rehmberg, A. Andreasson, and J. E. Löfroth, 15th International Symposium on Controlled Release of Bioactive Materials, paper 218, Basel, Switzerland, August 1988.
45. S. Wahlgren, A. L. Lindstrom, and S. E. Friberg, *J. Pharm. Sci.* 73:1484, 1984.
46. M. A. El-Nokaly, L. D. Ford, S. E. Friberg, and D. W. Larsen, *J. Colloid Interface Sci.*, 84:228, 1981.
47. M. El-Nokaly, S. E. Friberg, and D. W. Larsen, in *Liquid Crystals and Ordered Fluids*, Vol. 4. (A. C. Griffin, and J. F. Johnson, eds.). Plenum, New York, pp. 441–450, 1984.
48. S. E. Friberg, C. S. Wohn, and F. E. Lockwood, *J. Pharm. Sci.*, 74:771, 1985.
49. J. Ziegenmeyer, and C. Fuhrer, *Acta Pharm. Technol.*, 26:273, 1980. (Translated from German)
50. D. W. Osborne, A. J. I. Ward, and K. J. O'Neill, *Drug. Dev. Ind. Pharm.*, 14:1203, 1988.
51. D. W. Osborne, A. J. I. Ward, and K. J. O'Neill (in preparation).
52. D. W. Osborne, A. J. I. Ward, and K. J. O'Neill (in preparation).
53. B. Lindman, and P. Stilbs, in *Microemulsion Structure and Dynamics.* CRC Press, Boca Raton, pp. 119–152, 1987.
54. T. D. Sokoloski, D. W. Osborne, R. G. Stehle, K. J. Stefanski, N. F. H. Ho, and D. Gleason Paper PD 826, AAPS National Meeting, October 30, 1988, Orlando, Fla.

18

Gel Dosage Forms: Theory, Formulation, and Processing

LORRAINE E. PENA   *The Upjohn Company, Kalamazoo, Michigan*

I. THEORY

Gels are transparent to opaque semisolids containing a high ratio of solvent to gelling agent. When dispersed in an appropriate solvent, gelling agents merge or entangle to form a three-dimensional colloidal network structure. This network limits fluid flow by entrapment and immobilization of the solvent molecules. The network structure is also responsible for a gel's resistance to deformation and, therefore, its viscoelastic properties.

A variety of structures are associated with gel networks. Figure 1 illustrates some of the most common ones. *Random coils* are the least ordered and occur most frequently with synthetic polymers such as resins and cellulose derivatives. The *helix* is a more ordered structure formed from the intertwining of two polymer chains. Xanthan gum and starch are typical examples. *Stacks*, or the egg-box model, as it is sometimes called, results from cross-linking of polymer chains by divalent cations. Calcium alginate is a classic example. The *house of cards* structure is characteristic of gel-forming colloidal particles such as bentonite and Veegum. In the case of Veegum, thestructure results from the alignment of the positively charge edges with the negatively charged flat surfaces of the clay particles (1). Because most gels that are used in the pharmaceutical industry are associated with the random coil network, further discussion will be centered around that structure.

Random coil gelation mechanisms are rooted in the polymer—polymer and polymer—solvent interactions. With a given polymer, the gel network forms through successive increases in concentration.

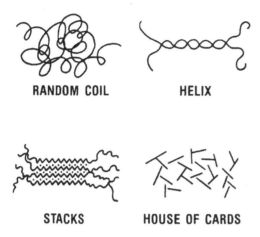

RANDOM COIL   HELIX

STACKS   HOUSE OF CARDS

Figure 18.1 Gel structures.

This results in a reduction of the interparticle distances, which
subsequently leads to chain entanglement and the development of
cross-links. As the number of cross-links increases, the chains
lock, solvent mobility is reduced, and a gel forms. Continued polymer addition strengthens the gel network and results in increased
resiliency and viscoelasticity.

     Although the gel network is basically formed through polymeric
interactions, the nature of the polymer–solvent affinity actually determines the integrity of the gel.   Classical gel theory distinguishes
between three categories of solvents: (1) free solvent that is very
mobile; (2) solvent bound as a solvation layer, usually through hydrogen bonding; and (3) solvent entrapped within the network
structure.   The ratios of the three solvent types in a given gel are
dependent on the polymer concentration and the solvent affinity for
the polymer. Solvent affinity governs extension of the random coil.
The greater the solvent affinity, the more the coil expands and en-
tangles with adjacent coils to form cross-links. In a good solvent,
the polymer chains are interpenetrated by solvent molecules, and
the solvation layer is enhanced. This facilitates random coil expansion and network formation. In a poor solvent, the polymer chains
contract to minimize solvent contact, thereby reducing the effective
number of cross-links and weakening the gel network structure.

II. FORMULATION

Gelation theory can be readily applied when formulating gel products.
However, before discussing this topic, presentation of some desira-

ble gel characteristics is in order. For optimum consumer appeal, the gel should have good optical clarity and sparkle. A high viscosity and a high-yield value are essential, but how high is primarily a matter of intended product application. To preserve product integrity, the gel should maintain its viscosity at all temperatures that may be encountered during shipment and storage. During formulation, there is frequently a trade-off between these optimum characteristics and the chemical requirements of active ingredients.

Carbomer is a commonly used gelling agent that produces gels having a number of these desirable characteristics. The grade, carbomer 934 P NF, is most commonly used in the pharmaceutical industry and has been selected as the exemplary polymer for formulation discussion. Chemically, carbomer 934 P is a cross-linked acrylic acid polymer having a molecular weight of approximately 3×10^6 (2). The gelation mechanism depends on neutralization of the carboxylic acid moiety to form a soluble salt. The polymer is hydrophilic and produces sparkling clear gels when neutralized. Although carbomer tolerates large amounts of alcohol, it does so with decreased viscosity and clarity. Gel viscosity is strongly dependent on pH and the presence of electrolytes. A maximum of approximately 3% electrolyte can be tolerated before precipitation occurs as a rubbery mass. Carbomer gels possess good thermal stability in that gel viscosity and yield value are essentially unaffected by temperature. As a topical product, carbomer gels possess optimum rheological properties. The inherent pseudoplastic flow permits immediate recovery of viscosity when shear is terminated and the high-yield value and quick break make it ideal for dispensing.

The viscosity building effects of carbomer are readily apparent when examining the gelation mechanism in association with the colloidal network structure. As Figure 2 illustrates, before neutralization, carbomer in water exists in its un-ionized form and yields a thin opalescent dispersion of approximately pH 3 (3). At this pH, the polymer is very flexible and behaves like a random coil. Addition of sodium hydroxide or a neutralizing amine to the dispersion shifts the ionic equilibrium in favor of the soluble salt form. This results in ionic repulsion of the carboxylate groups and the polymer becomes stiff and rigid, thereby increasing the viscosity of the water. Overneutralization and excess salts reduce the viscosity of carbomer gels or cause precipitation by the counterion effect.

Figure 3 shows the dramatic effect of pH on the viscosity development of carbomer gels. As pH increases and the carboxylic acid moieties of the polymer are neutralized, viscosity and clarity increase. Acceptable gel clarity and viscosity occur at approximately pH 4.5 to 5.0, but optimum viscosity and clarity are at pH 7. Overneutralization results in a decrease in viscosity that cannot be reversed by addition of acid to lower the pH because an electrolyte is formed.

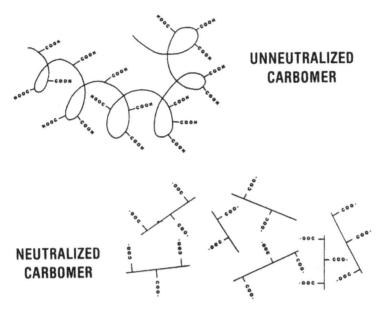

Figure 18.2 Carbomer gel network structures.

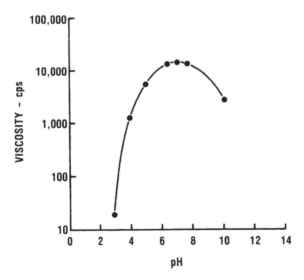

Figure 18.3   pH−viscosity profile of 0.5% (w/w) carbomer 934 P NF
in water.

Table 18.1 Alcohol Effects on Carbomer Gels

Carbomer % (w/w)	Alcohol % (w/w)	pH	Viscosity (cP)[a]
0.75	0.0	5.5	12,800
1.25	50.0[b]	5.5	12,800
0.50	20.0[c]	8.2	12,800
0.85	40.0[c]	8.2	12,800

[a]Shear rate 7.61 sec^{-1}.
[b]Isopropanol 70% (v/v).
[c]Alcohol USP, 95% (v/v) ethanol.

Although carbomer can be used to gel formulations with a large proportion of alcohol, the dehydration effects of the alcohol on the polymer are still substantial. As Table 1 indicates, at pH 5.5, the use of 50% isopropanol in a formulation requires 0.5% more carbomer to produce a viscosity equivalent to that of an aqueous gel. At pH 8.2, doubling the concentration of alcohol USP requires 0.35% more carbomer to produce an equivalent gel viscosity. The viscosity responses of the gels to alcohol may be interpreted by its action as a nonsolvent. Because the solvent affinity is reduced, the polymer contracts, with a consequential increase in the interparticle distance and subsequent decrease in the number of entanglements and cross-links. To decrease the interparticle distances and restore the integrity of the gel network structure, a greater concentration of polymer must be used. The reduction in solvent affinity and change in the polymer conformation results in increased haziness of the gel as the alcohol content increases.

The rheograms of carbomer 934 P gels in hydroalcoholic and aqueous formulations are presented in Figure 4 using 0.5% (w/w) and 1.0% (w/w) carbomer, respectively. The characteristic concavity of the rheogram toward the shear rate axis indicates that the gels exhibit pseudoplastic flow. This pseudoplasticity results from a colloidal network structure that aligns itself in the direction of shear, thereby decreasing the viscosity as the shear rate increases. The rheograms also show that the gels exhibit substantial yield values at 666 and 866 dynes/cm^2 for the hydroalcoholic and aqueous formulations, respectively. The yield value is an indication of the extent of formation of a three-dimensional colloidal network structure. The variation in the yield values of the two gels is a reflection of the higher carbomer concentration in the aqueous formulation.

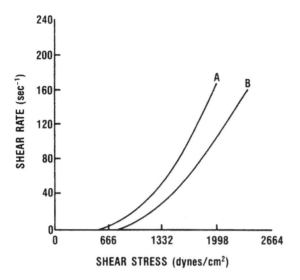

Figure 18.4 Rheograms of carbomer 934 P NF gels. (A) hydroal-coholic gel; (B) aqueous gel.

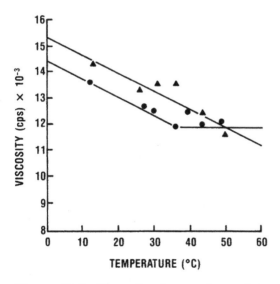

Figure 18.5 Viscosity—temperature plots of carbomer 934 P NF gels: circle, hydroalcoholic gel; triangle, aqueous gel.

The viscosity—temperature relationships of the aqueous and hydroalcoholic gels are shown in Figure 5. The hydroalcoholic gel shows a plateau at 37°C, whereas the aqueous gel continues to show a general decline in viscosity. As the graphs indicate, the magnitude of the viscosity decline is not substantial over the temperature range covering refrigerated to elevated storage conditions.

III. PROCESSING

To manufacture clear, uniform, air-free gels, certain key processing characteristics must be provided. The nature of carbomer requires initial high-shear mixing to form a uniform smooth dispersion, followed by low-shear planetary mixing during the neutralization—gelling process. Air entrainment during the neutralization process can be minimized by subsurface addition of liquids in conjunction with low-shear mixing. In addition, mixing under vacuum, if available, will withdraw entrapped air from the dispersion during manufacture and prevent further air entrainment by incidental surface breaks. Minimization of air entrainment is necessary from the aesthetic standpoint and, most importantly, from the aspect of controlling fill weights during packaging operations.

A large variety of planetary mixers with and without vacuum capability are available for gel manufacture. The Nauta mixer (4) has a conical mix tank and consists of a lumpbreaker, located at the base of the tank, for high-shear mixing and a rotating auger on a pivot arm for low-shear planetary mixing. The lumpbreaker is flush with the side of the tank and consists of four flat blades spinning at either 1800 rpm or 3600 rpm. The auger can be operated to spiral either upward or downward, depending on whether one desires to deaerate or incorporate materials from the surface into the batch. Both the auger and lumpbreaker are controlled separately. Air entrapment is minimized with the Nauta by mixing under vacuum and by subsurface addition of materials through a port near the base of the mixer.

The Agi-mixer (5) has a hemispherical mix tank. High-shear mixing is provided by the homomixer located at the base of the tank. Counterrotating paddles equipped with scraper blades provide the low-shear planetary mixing action. The homomixer can be operated at only two speeds, whereas the paddles are infinitely variable over the range 0 to 60 rpm. As with the Nauta mixer, aeration can be controlled by mixing under vacuum and by subsurface addition of materials through a basal port.

ACKNOWLEDGMENTS

The assistance of C. M. Horton, B. L. Lee, and J. F. Stearns in obtaining some of the carbomer gel data presented here is greatly appreciated.

REFERENCES

1. B. C. Carson, *Cosmet. Toiletries*, *92*(7):81, 1977.
2. *Carbopol Water Soluble Resins*, Bulletin Number GC-67, B. F. Goodrich Company, Cleveland, Ohio.
3. C. A. Dittmar, *Drug Cosmet. Ind.*, *81*:447, 1957.
4. Day Mixing Company, 4932 Beech St., Cincinnati, Ohio 45212-2397.
5. Greerco Corporation, Executive Drive, P.O. Box 187, Hudson, New Hampshire 03051.

19

Using Silicones in Topical Products

MICHAEL S. STARCH   *Dow Corning Corporation, Midland, Michigan*

I. INTRODUCTION

A. What Are Silicones?

The term *silicone* refers to a class of synthetic polymers that are
based on alternating silicon−oxygen, or siloxane (-Si-O-) units.
The silicones that are most commonly used in topical products are
polydiorganosiloxanes in which two organic groups are bonded to
each silicon atom. For commercial polydiorganosiloxanes, the organic
groups are nearly always methyl groups, and such materials are referred to as *polydimethylsiloxanes* (PDMS). Other types of silicones
that are used in topical products can be thought of as derivatives
of PDMS in which some of the methyl groups have been replaced
with other organic groups.

Polymethylsiloxanes are produced in one of two forms: linear or
cyclic. Both are widely used in topical formulations and together
account for most of the volume of silicones used in these applications.
Linear PDMS is a colorless, odorless oil that is available in a wide
range of viscosities. The viscosity of linear PDMS is directly related
to molecular weight, and can range from less that 1 centistoke to
over 1 million centistokes (cs). Cyclic PDMS is an odorless, colorless, low-viscosity, volatile oil. Commercially produced cyclic PDMS
is available in a fairly narrow range of molecular weights, corresponding to the cyclic species with four, five, or six dimethylsilox-
ane units in the ring. As expected, volatility of cyclic PDMS is inversely related to molecular weight. Unlike linear PDMS, cyclic
PDMS is a fairly good solvent, and this is one reason for its popularity among formulators.

The chemical and physical properties of silicones that contain organic substituents other than methyl vary according to the nature of the organic substituent. For example, substitution of some methyl groups on PDMS with polyethylene oxide chains dramatically increases the viscosity of the material because of the polarity of the polyethylene oxide. The chemical properties of the organic substituents are generally not affected by the siloxane backbone. Consequently, polydiorganosiloxanes exhibit the chemical reactivity that would be associated with the organic substituents. The two types of silicones with nonmethyl substituents that are used in topical products are discussed in Sections II.C and II.D.

B.   Benefits of Silicones in Topical Products

Silicones exhibit a unique combination of properties that, to a large extent, arise from the inorganic siloxane backbone. The silicon—oxygen bonds that form this backbone are quite strong, yet they are very flexible. This results in polymers that are particularly stable and exhibit unusual surface activity because of the ability of the siloxane chain to present its pendant organic groups at interfaces (1). In PDMS, the inorganic siloxane backbone is covered by nonpolar methyl groups, leading to extremely low intermolecular forces. As a result, PDMS is a liquid over a wide range of molecular weights. In fact, PDMS will exhibit viscous flow even at molecular weights above 700,000. Low intermolecular forces are also responsible for the low surface tension of PDMS, which is in the range of 20 to 21 dyn/cm. The low surface tension and relatively high molecular weight of PDMS give rise to a number of properties that are beneficial in topical products: excellent spreading, detackification, defoaming, and water repellency. A PDMS that has been substituted with nonmethyl organic groups will provide these benefits to varying degrees, depending on the amount of substitution and the nature of the nonmethyl organic groups.

II.   TYPES OF SILICONES USED IN TOPICAL PRODUCTS AND THEIR PROPERTIES

A.   Linear Polydimethylsiloxanes

Ingredients that are used in topical products are often referred to by nomenclature established by the Cosmetics Toiletries, and Fragrance Association (CTFA). These CFTA names (2) have been established primarily for the purpose of ingredient labeling. The CTFA has chosen the term *dimethicone* for linear PDMS that conform to the general structure given in Figure 1. This structure is representative of the linear PDMS available from all silicone manufacturers.

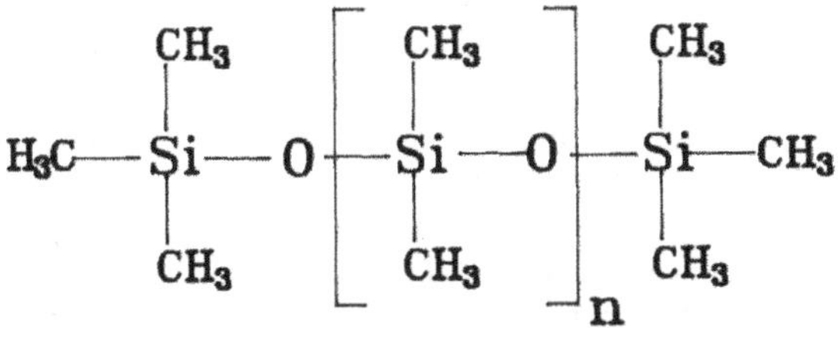

$$\text{H}_3\text{C} - \underset{\underset{\text{CH}_3}{|}}{\overset{\overset{\text{CH}_3}{|}}{\text{Si}}} - \text{O} - \left[\underset{\underset{\text{CH}_3}{|}}{\overset{\overset{\text{CH}_3}{|}}{\text{Si}}} - \text{O} \right]_n \underset{\underset{\text{CH}_3}{|}}{\overset{\overset{\text{CH}_3}{|}}{\text{Si}}} - \text{CH}_3$$

Figure 19.1 Dimethicone.

Dimethicone was the first silicone used in a commercial topical product, and it continues to be widely used. Until the late 1970s, however, dimethicone was generally used only in small amounts (less than 0.5%) for its defoaming properties. At this level, dimethicone is very effective in preventing the objectionable foam generated when cream or lotion is spread onto the skin. This foaming or lathering is a problem, particularly for creams and lotions containing a soap (e.g., triethanolamine stearate). A small amount of dimethicone in such formulations effectively disrupts the foam because the silicone rapidly spreads over the foam, causing it to break (3).

Beginning in the mid-1970s, formulators began to take advantage of the other benefits of dimethicone that become apparent at higher use levels. The general approach was to substitute dimethicone for a portion of organic emollients traditionally used in topical products such as mineral oil, petrolatum, and fatty acid esters. Although this increased the cost of the formulation, it could be justified on the basis of improved aesthetics. Dimethicone provides a less-greasy feel on the skin compared with organics of comparable viscosity, an effect that can be attributed to the higher molecular weight of the silicone and to its surface properties. The use of dimethicone also helps to reduce the stickiness associated with some ingredients because it tends to form a film over the other ingredients when the formulation is applied to the skin. Another consequence of the high molecular weight of dimethicone relative to organic ingredients is improved water repellency and wash-off resistance. This is particularly important for topical products, such as sunscreens, for which it is desirable to retain the active ingredient on the skin during water immersion.

The increasing use levels of dimethicone in topical products has been prompted, in part, by the regulation of over-the-counter (OTC) drug products by the Food and Drug Administration (FDA). In 1978, the FDA issued the first draft of proposed regulations (4) based on recommendations from their advisory panel for OTC topical analgesic, antirheumatic, otic, burn, and sunburn prevention and treatment drug products. These regulations categorize dimethicone as a safe and effective ingredient for such products. Specifically, the panel concluded that dimethicone was a safe and effective ingredient for a topical drug product used to provide temporary relief of minor skin irritations when used at a concentration from 1% to 30%. After reviewing comments from the public, the FDA issued a notice of proposed rule-making in 1983 in which the status of dimethicone as a safe an effective ingredient was retained (5).

The dimethicones most commonly used in topical formulations are in the viscosity range of 100 to 1000 cs. This represents a compromise between ease of formulation and the benefits that are typically sought by the inclusion of silicone in the formulation. High-viscosity dimethicones have increased skin protection benefits and are more substantive than low-viscosity dimethicones, but they are more difficult to incorporate into topical formulations. Dimethicones with viscosities below 100 cs are sometimes used when optimum spreading and lubricity are desired, but cyclic PDMS has largely replaced dimethicone in these instances because of its ease of formulation.

To the formulator who is attempting to incorporate unfamiliar ingredients into a topical formulation, one useful rule is that additives that are soluble in one or more ingredients of a known formulation can generally be included in small amounts without making any other formulation changes. The rule applies to both homogeneous (solution) and heterogeneous (emulsion) products. Vaughan has published a solubility parameter for dimethicone of 5.92 (6). A general rule (also given in the same publication) is that materials with solubility parameters within 2 units of the material of interest will be soluble. This suggests that dimethicone should be soluble in several nonpolar oils commonly used in topical formulations: mineral oil and petrolatum. In practice, however, this is not true for dimethicone, because solubility parameters do not work well for predicting the solubility of high polymers owing to factors arising from the entropy of mixing. Isopropyl myristate is one of the few commonly used organic ingredients that is miscible with dimethicone. The solubility of dimethicone in other organic ingredients is generally poor. The only exceptions are very low molecular weight dimethicones with viscosities of less than 5 cs.

Despite the insolubility of most dimethicones in typical topical formulation ingredients, they are not especially difficult to include in topical formulations. For emulsion products, a number of factors work in the formulators favor. First, the emulsifiers commonly used to stabilize topical formulations can usually accommodate a small amount of dimethicone, even though it has different solubility properties than the other oils used in the formulation. Second, most topical formulations are sufficiently viscous to stabilize the added dimethicone, which might separate from a thinner product. Finally, the low intermolecular forces in dimethicone cause it to be broken down to very small droplets under the mixing conditions normally encountered in the manufacture of a topical emulsion product. Stoke's law predicts that reducing the size of emulsion droplets leads to a more stable emulsion, and this works to the advantage of the formulator when using silicones. For the few topical products that are solutions, it is often possible to use isopropyl myristate or another cosolvent to get dimethicone into the product.

B. Cyclic Polydimethylsiloxanes

The CTFA (2) has established the name *cyclomethicone* for cyclic polydimethylsiloxanes that are used in topical products. The chemical structure for the cyclomethicone tetramer is shown in Figure 2. The cyclomethicone tetramer is the most widely used because it has the highest evaporation rate of the homologous series, but the cyclomethicone pentamer has somewhat better solubility properties. A cyclomethicone trimer does exist, but it is a solid at room temperature and is not used in topical products. The higher homologues, such as the cyclomethicone hexamer, are normally present in commercial cyclomethicone, but in small amounts; hence, their lower volatility generally is not a concern. Typical commercial cyclomethicones are mixtures of the cyclomethicone tetramer and pentamer which are made by distillation of crude polydimethylsiloxanes.

Cyclomethicones provide benefits that would be expected from a low-molecular-weight PDMS: low toxicity, low odor, excellent spread-

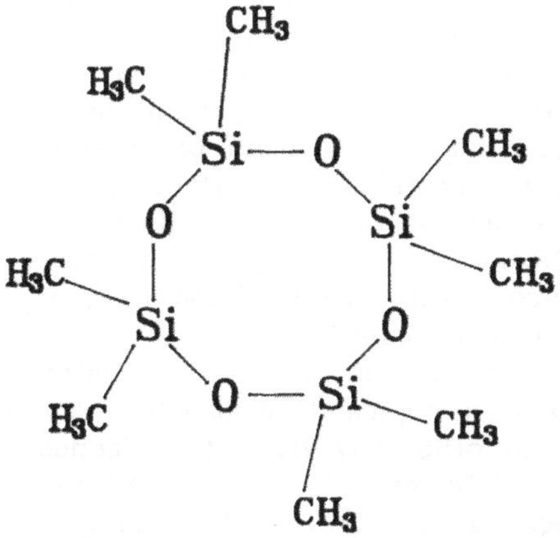

Figure 19.2 Cyclomethicone (tetramer).

ing, and lubricity. These properties, together with their volatility and mild solvent properties, have made cyclomethicone the highest-volume silicone product used in the cosmetics and toiletries industry. They were first used in the late 1970s as a vehicle for antiperspirant salts in underarm topical products. Since that ime, cyclomethicones have been used in virtually every category of topical product. One reason for the popularity of cyclomethicones as a formulation vehicle is their low heat of vaporization. The heat of vaporization for the cyclomethicone tetramer is only 32 cal/g, which contrasts sharply with that of water (520 cal/g) and ethanol (210 cal/g). This means that topical formulations based on cyclomethicone have a pleasant "dry" feel when they are applied to the skin.

Unlike the dimethicones, cyclomethicones are soluble in a variety of ingredients commonly used in topical products. Table 1 lists the solubility of cyclomethicone in a number of these ingredients. For ingredients that are solids at room temperature, the solubility given in Table 1 is for a mixture that has been heated to approximately 80°C. Solubility of cyclomethicone is generally limited to predominantly nonpolar organic materials, although it is soluble in some polar organic solvents such as ethanol and isopropanol. Because cyclomethicone is more soluble in other ingredients than dimethicone, it is easier to formulate into an existing topical product. This is particularly true for formulations that are solutions. When cyclomethicone functions as the primary formulation vehicle and is the major component in a topical formulation, it is often necessary to employ some special ingredients to accommodate the active ingredients.

C. Phenyltrimethylsiloxane

Phenyl trimethicone is the name that the CTFA (2) has assigned to silicones that conform to the structure shown in Figure 3. Commercially available phenyl trimethicone is a mixture of species corresponding to the structures indicated by Figure 3 for which the value of *n* ranges from one to three. The increased number of organic substituents on the silicon in phenyl trimethicone greatly increases its solubility in organic ingredients relative to dimethyl silicones. Phenyl trimethicone is the only silicone that is miscible in all proportions with 95% ethanol. It is also miscible with all commonly used nonpolar organic ingredients in topical formulations.

Because of its solubility in other ingredients, phenyl trimethicone is easy to incorporate into existing topical formulations. Typically it provides the same benefits as other low-viscosity silicones. Phenyl trimethicone has the additional benefit of a somewhat higher refractive index than dimethyl silicones (1.46 vs. 1.39). Hence, it is used in topical formulations for which gloss is desired, such as in a hair dressing.

Table 19.1 Solubility of Topical Formulation
Ingredients in Cyclomethicone

Ingredient	Solubility
Mineral oil	Soluble
Petrolatum	Soluble (80°C)
Paraffin wax	Soluble (80°C)
Stearyl alcohol	Soluble (80°C)
Stearic acid	Soluble (80°C)
Isopropyl myristate	Soluble
Isopropyl palmitate	Soluble
Glycerol monostearate	Soluble (80°C)
PEG 8 stearate	Insoluble (80°C)
PEG 40 stearate	Insoluble (80°C)
PEG 400	Insoluble
Glycerin	Insoluble

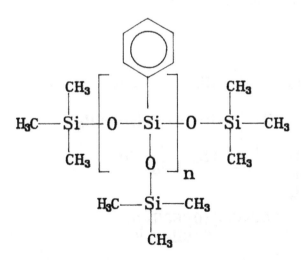

Figure 19.3 Phenyl trimethicone.

D. Polydimethylsiloxane-Polyalkylene Oxide Copolymers

When some of the methyl groups on PDMS are substituted for side
chains of polyalkylene oxide, a very surface-active class of silicones
is produced. For commercial silicones of this type, the polyalkylene
oxide are polymers of either ethylene oxide (EO), propylene oxide
(PO), or copolymers of ethylene and propylene oxide (EO/PO). The
CTFA (2) has established the term *dimethicone copolyol* for these
materials. A general structure for the most commonly encountered
dimethicone copolyols is shown in Figure 4. The large difference in
polarity between the PDMS and polyalkylene oxide segments is responsible for the surface activity of the dimethicone copolyols. Unlike conventional nonionic surface-active agents, which generally
consist of a single hydrophobic (hydrocarbon) segment connected to
the polyalkylene oxide segment, dimethicone copolyols often have
multiple polyalkylene oxide segments attached along the siloxane
chain. This gives rise to a large variety of possible dimethicone
copolyol structures, with concomitant variations in physical properties and surface activity. Dimethicone copolyols range from liquids
to waxy solids, depending primarily on molecular weight and degree
of substitution with polyalkylene oxide.

Dimethicone copolyols that contain large amounts of polyethylene
oxide are water-soluble, and these are most often the ones used in
topical formulations. Water-soluble dimethicone copolyols are easy
to add to aqueous topical formulations, and they do not exhibit defoaming behavior. They can also be used in certain situations as
emulsifiers for other silicones. The disadvantage of using water-
soluble dimethicone copolyols is that they generally provide few of
the benefits associated with dimethyl silicones.

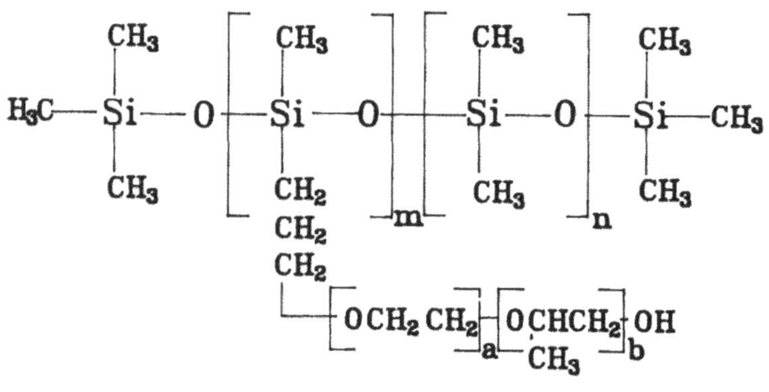

Figure 19.4 Dimethicone copolyol.

III. FORMULATING SILICONES INTO CONVENTIONAL TOPICAL FORMULATIONS

A. How to Use the Example Formulations

In the following sections, examples are provided that illustrate how silicones discussed in the preceding sections can be used in topical formulations. These examples are relatively simple formulations that have proved to be stable and easy to prepare using mixing equipment commonly available in a formulation laboratory. They demonstrate how silicones can be used with organic ingredients and are intended to be used as a starting point for formulators who wish to develop a silicone-containing topical formulation. Alternatively, the example formulations can suggest which silicone may be appropriate for a known formulation that is similar to one of the examples.

The CTFA nomenclature has been used for all the ingredients in the example formulations. The CTFA Ingredient Dictionary (2) provides a chemical description and supplier/tradename reference for all the materials listed in the example formulations. A tradename and supplier has been indicated for materials that are available in several grades or molecular weights. Tradenames are also indicated when the ingredient is a blend of several components. Here, substitution for the indicated material by an similar material from another vendor may be satisfactory; however, this may affect formulation viscosity and stability. If no tradename is indicated, then any suitable grade of material (e.g., USP, NF) conforming to the CTFA description can be used.

In many cases, the examples are similar to commercial formulations that contain silicones; however, they are not finished formulations that would be suitable for commercial use, because several important ingredient types have been omitted in the interest of simplicity. The most significant omission is preservatives, which are added to virtually all commercial topical formulations to prevent the growth of microorganisms in the product. The selection of a proper preservative or preservatives depends on a number of factors, including the intended shelf life, storage conditions, and the types of organisms that might contaminate the product. Perfumes and colors have also been omitted because proper selection requires knowledge of the intended use for the formulation. Such considerations are beyond the scope of this chapter.

B. Compatibility of Silicones with Other Formulation Components

Silicones that are used in topical formulations are chemically nonreactive under conditions encountered during the preparation and use of these formulations. Consequently, there are no ingredients that are incompatible with silicones because of formation of undesired re-

action products, or degradation of an ingredient by silicone. The
only troublesome interaction of silicones with other ingredients is
their well-known defoaming properties, which can be very undesirable in a product that is intended to foam! If foaming is an important formulation attribute, then dimethicone copolyol should be used
because this class of silicones exhibits defoaming behavior only at
elevated temperatures.

C. The Use of Silicones in Topical Creams and Lotions

The term *lotion* is generally used to describe topical formulations
that are pourable emulsions. *Creams* are similar to lotions, but
higher in viscosity and not pourable. Often a lotion can be modified to produce a cream simply by increasing the proportion of oils
used in the formulation. In describing the preparation of topical
creams and lotions, it is convenient to group the ingredients into
two categories: water-insoluble ingredients, which are referred to
collectively as the *oil phase*, and water-soluble ingredients, which
are collectively referred to as the *water phase*. Topical creams and
lotions contain emulsifiers that allow stable mixtures of the oil phase
and water phase to be prepared. Although emulsifiers, by their
nature, are partially soluble in both phases, they are usually added
as part of the oil phase.

Table 2 lists the ingredients for a lotion formulation based on
the widely used stearate—cetyl alcohol emulsifier combination. The
primary emulsifier for this formulation is triethanolamine stearate,
which is formed from triethanolamine and stearic acid when the formulation components are mixed together. Cetyl alcohol is included
as a secondary emulsifier to improve the stability of the lotion and
provide a smooth texture. The other ingredients: dimethicone, mineral oil, and petrolatum could be considered the "active ingredients,"
as these are the most commonly used emollients in commercial skin
care lotions. The same OTC panel that concluded that dimethicone
is safe and effective for use in skin protectants (see Sect. II.A)
also approved mineral oil and petrolatum for this purpose.

The basic lotion formulation is made by weighing the oil-phase
ingredients and water-phase ingredients in separate containers.
Each phase should be heated to approximately 70°C, and mixed until
uniform. The hot oil phase is then added to the hot water phase
while stirring with a propeller-type mixer. The mixer speed and
addition rate should be such that the oil phase is rapidly dispersed
in the water phase during the addition. However, the mixer speed
should not be so high that a vortex is created that will draw air
bubbles into the mixture. Air bubbles will interfere with the emul-
sification of the oil phase. Mixing must be continued as the lotion

Table 19.2 Basic Lotion Formulation

Ingredient	Wt%	Tradename/supplier
Oil phase		
dimethicone (500 cs)	3.0	
mineral oil	1.5	Klearol / Witco Chemical
petrolatum	1.0	Sono-Jell No.9 / Witco Chemical
stearic acid (50% min.)	3.0	
cetyl alcohol	1.0	
Water phase		
water	89.3	
triethanolamine (99%)	1.2	

is allowed to cool until it reaches 40°C. Heat-sensitive ingredients (e.g., fragrance) can be mixed in at this point befoŕ the formulation is packaged.

The viscosity of the formulation in Table 2 can be adjusted, to a certain extent, by changing the water/oil-phase component ratios. To increase the formulation viscosity, the amount of water is reduced. This provides a way to make a cream formulation variant that is based on the same ingredients. Another topical cream formulation containing silicones is given in Table 3. The Arlacel 186 is the primary emulsifier for this formulation. Additional stabilization and thickening are provided by the ozokerite and beeswax. The formulation is a suitable starting point for a "cleansing cream" that would be used to remove oily soils (e.g., cosmetics) from the skin. A novel feature of this cream formulation is that it is a water-in-oil emulsion and, therefore, should be appropriate for delivering oil-soluble therapeutic agents. The preparation is essentially the same as for the previous lotion formulation, except that the water phase is added to the oil phase while mixing.

The formulation listed in Table 4 is for a topical antiperspirant that is designed to deliver the active ingredient, aluminum chlorohydrate, to the skin, where it will suppress the activity of the sweat-producing glands. The formulation is a low-viscosity emulsion suitable for packaging in rolling ball (roll-on) applicator. The inclusion of cyclomethicone eliminates the stickiness that is characteristic of aqueous solutions of aluminum chlorohydrate and other antiperspi-

Table 19.3 Water-in-Oil Cream

Ingredient	Wt%	Tradename / supplier
Oil phase		
cyclomethicone	4.0	345 Fluid / Dow Corning
dimethicone (1000 cs)	1.0	
synthetic beeswax	2.0	Syncrowax BB4 / Croda
ozokerite wax	2.0	
glyceryl oleate (and)		
propylene glycol	2.0	Arlacel 186 / ICI Americas
mineral oil	15.0	
Water phase		
water	56.0	
glycerin	18.0	

Table 19.4 Aqueous Antiperspirant "Roll-on"

Ingredient	Wt%	Tradename / supplier
Magnesium aluminum silicate	1.0	Veegum / R. T. Vanderbilt Co.
Water	33.2	
Aluminum chlorohydrate (powder)	50.0	
Glyceryl stearate (and)		
PEG 100 stearate	8.0	Arlacel 165 / ICI Americas
Glycerin	0.8	
Cyclomethicone	7.0	

rant salts. Arlacel 165, a nonionic emulsifier, is used to stabilize
the cyclomethicone, and an inorganic thickener (Veegum) is included
to improve emulsion stability. This type of formulation would be
suitable for delivering many other water-soluble active ingredients.

The antiperspirant formulation is prepared by first dispersing
the Veegum in hot (75°C) water using a Cowles mixing blade or some
other type of high-shear stirrer. The Veegum and water should be
mixed at 75°C for about 1 hr to ensure that the Veegum becomes
fully hydrated. Next, the aluminum chlorohydrate is dissolved in
the hot mixture, followed by the glycerin and Arlacel 165. The
mixture is allowed to cool, with continuous mixing, until the temperature drops to 30°C. Finally, the cyclomethicone is emulsified by
slowly pouring it into the warm mixture with vigorous stirring.

D. The Use of Silicones in Anhydrous Topical Formulations

Anhydrous topical formulations are, in many respects, less complex
than the emulsion formulations discussed in the previous section.
They are often homogeneous blends that are prepared by simply
blending the ingredients. Organic oils serve as the vehicle to deliver therapeutic or other active ingredients to the skin. The key
to successful formulation of this type of formulation is to find the
oil, or combination of oils that will solubilize the desired active ingredient. Tables 5 and 6 list two sample formulations: a sunscreen
oil and an oil base. In the sunscreen, the octyldimethyl-PABA is
an absorber of ultraviolet light which provides protection from the
sun. It is soluble in both the cyclomethicone and the isopropyl
myristate, so the amount of protection can be easily adjusted by
changing the concentration in the formulation. The formulation in
Table 6 is a base formulation that was developed as the starting
point for a scented bath oil. Phenyl trimethicone and the lactate
ester blend (Ceraphyl 41) are used to solubilize the fragrance in
the cyclomethicone. This base mixture will accommodate up to 15%
fragrance or other relatively nonpolar ingredients.

The last example of a homogeneous oil-based formulation is given
in Table 7. In this formulation, the active ingredient, benzocaine,
is solubilized in the cyclomethicone vehicle with two fatty alcohol poly-
propylene glycol (PPG) ethers. This formulation is an alternative
to the more traditional approach in which the benzocaine is delivered
by a hydroalcoholic vehicle. The advantage of an alcohol-free vehicle
is that the stinging and drying effects are eliminated.

The formulation given in Table 8 is a special case in the category of anhydrous topical formulations. Here the active ingredient
is an antiperspirant salt that is partially suspended in the cyclo-

Table 19.5   Clear Sunscreen Oil

Ingredient	Wt%	Tradename / supplier
Cyclomethicone	16.0	344 Fluid / Dow Corning
Isopropyl myristate	13.0	
Mineral oil	68.0	Carnation / Witco Chemical
Octyldimethyl-PABA	3.0	

Table 19.6   Clear Oil Base

Ingredient	Wt%	Tradename / supplier
Cyclomethicone	17.6	344 Fluid / Dow Corning
Phenyl trimethicone	6.0	
Mineral oil	58.8	Carnation / Witco Chemical
C12-15 alcohols lactate	17.6	Ceraphyl 41 / Van Dyk & Company

Table 19.7   Clear Benzocaine Lotion

Ingredient	Wt%	Tradename / supplier
Cyclomethicone	36.4	344 Fluid / Dow Corning
PPG-15 stearyl ether	43.9	Arlamol E / ICI Americas
PPG-10 cetyl ether	13.6	Procetyl 10 / Croda
Benzocaine	6.1	

Table 19.8 Anhydrous Antiperspirant "Roll-on"

Ingredient	Wt%	Tradename / supplier
Dimethicone (10 cs)	5.0	
Cyclomethicone (and)		
Quaternium 18 hectorite		Bentone Gel VS-5 /
(and) ethanol	3.0	NL Chemicals
Cyclomethicone	70.0	
Ethanol (SDA 40)	2.0	
Aluminum chlorohydrate (powder)	20.0	

methicone base. The salt will settle to the bottom, but is easily re-dispersed by shaking. This type of formulation has improved aes-thetics relative to antiperspirant shown in Table 4 because the elim-ination of water gives the formulation a pleasant dry feel when it is applied to the skin. Other active ingredients that are available in the form of finely divided powders could be substituted for the alu-minum chlorohydrate to produce a variety of different formulations.

E. The Use of Silicones in Clear Aqueous Formulations

Unlike the water-based formulations that were discussed in Section III.B, the formulations in this section are intended to deliver active ingredients that are completely soluble in water. Such formulations are less common than emulsion forms, because often the active in-gredients are not water-soluble. The formulation listed in Table 9 can be thought of as a vehicle to deliver glycerin to the skin; how-ever, other ingredients that are glycerin-soluble can be included. The formulation is a gel that is thickened with carbomer 934, a form of polyacrylic acid. It is prepared by first dispersing the carbomer powder in water. The temperature is not an important factor for this step, but good mixing is essential because the carbomer tends to form lumps when it hydrates. Once the carbomer is dissolved, it will be somewhat hazy, but the solution viscosity will be low. Thickening will occur when the carbomer is neutralized with the triethanolamine. The triethanolamine should be dissolved in a small portion of the water to facilitate mixing it with the carbomer solu-tion. The other ingredients are then stirred into the gel. Di-methicone copolyol is included in the formulation to reduce the sticky feel associated with the glycerin.

Table 19.9 Clear Aqueous Gel

Ingredient	Wt%	Tradename / supplier
Water	59.8	
Carbomer 934	0.5	Carbopol 934 / B.F. Goodrich
Triethanolamine (99%)	1.2	
Glycerin	34.2	
Propylene glycol	2.0	
Dimethicone copolyol	2.3	193 Surfactant / Dow Corning

The active ingredients for the formulation listed in Table 10 are surfactants designed for cleaning. The formulation is similar to a shampoo but is suitable for cleaning skin as well because the surfactant combination used has very low skin irritation. Dimethicone copolyol is the only type of silicone that is suitable for use in this type of formulation because it will not interfere with foaming. The viscosity of the formulation is controlled by the amount of sodium chloride used. If a higher viscosity is desired, then the sodium chloride level should be increased. The formulation is prepared by first dissolving the sodium chloride in the water, and then mixing in the other ingredients in the order listed. The two surfactants, sodium laureth sulfate and cocamidopropyl betaine, are usually sold as water solutions. For this formulation, the specific concentrations of surfactants in water are given. Surfactants from other manufacturers may differ in concentration, so adjustments should be made to compensate.

F. The Use of Silicones in Solid Delivery Forms (Sticks)

In certain categories of topical formulations, solid, or stick formulations are well suited for delivering the active ingredient. This is especially true for antiperspirants and deodorants. The formulations given in Tables 11 and 12 are based on a deodorant and an antiperspirant, respectively. The clear stick is suitable for delivering active ingredients that are soluble in one or more of the liquid ingredients. Sodium stearate is the gellant for the clear stick. A typical deodorant stick is simply a solution of biocide (e.g., triclosan) in a mixture of water and ethanol gelled with sodium stearate. To accommodate the cyclomethicone while maintaining clarity, a combination of propylene glycol, isostearyl alcohol, and PPG 10

Table 19.10 Foaming Cleanser

Ingredient	Wt%	Tradename / supplier
Sodium chloride	2.0	
Water	31.0	
Cocamide DEA	2.0	
Dimethicone copolyol	4.5	193 Surfactant / Dow Corning
Cocamidopropyl betaine (31%)	14.0	Lonzaine C / Lonza Inc.
Sodium laureth sulfate (28%)	46.5	

cetyl ether is used. Because the formulation must be heated to dissolve the sodium stearate, it should be prepared in some type of covered container equipped with a condenser to prevent significant losses of volatile components. The formulation is prepared by mixing together all the ingredients except the sodium stearate and heating them to 65°C. The sodium stearate is then dissolved by stirring it in the hot mixture. The formulation should be poured into molds while it is hot because it gels as it cools.

The other stick formulation, listed in Table 12, is for an antiperspirant. This formulation is similar to that given in Table 8, except that the antiperspirant salt is suspended in a mixture of molten ingredients that form a stick when cooled. Because this formulation must be heated, the same precautions recommended for the previous formulation to minimize loss of volatile material should be made. To prepare this formulation, the stearic acid and cetyl alco-

Table 19.11 Clear Stick Base

Ingredient	Wt%	Tradename / supplier
Propylene glycol	14.9	
Ethanol (SDA 40)	7.5	
PPG 10 cetyl ether	10.0	Procetyl 10 / Croda
Isostearyl alcohol	19.8	Adol 66 / Sherex
Cyclomethicone	39.8	345 Fluid / Dow Corning
Sodium stearate	8.0	

hol are heated to about 80°C, and the aluminum chlorohydrate is stirred in. After the aluminum chlorohydrate is dispersed, the cyclomethicone is added, and the mixture is stirred until it is uniform. The formulation should be allowed to cool, with mixing, until just above the solidification temperature (50–60°C) and then poured into molds.

IV. NONTRADITIONAL TOPICAL FORMULA- TIONS BASED ON SILICONES

The topical formulations containing silicones discussed in the previous sections are, for the most part, variations of commercial formulations that have changed little over the last 10 to 20 years. This is especially true for the emulsion formulations discussed in Section III.C. Recently, the development of new silicone surfactants (dimethicone copolyols) has provided the formulator with emulsifiers that allow novel silicone-based formulations to be made. One such emulsifier is a dimethicone copolyol that can be used to make a variety of water-in-oil emulsions based on cyclomethicone (7). These formulations are quite different from water-in-oil emulsions, such as the one given in Table 3, that must be thickened to provide a stable formulation. Formulations of the type listed in Tables 13 and 14, on the other hand, are stable over a wide viscosity range, giving the formulator more flexibility in producing a formulation with the desired viscosity. These formulations can accommodate ingredients that are soluble in either water or cyclomethicone. Because the external phase of the emulsion is cyclomethicone, all have the pleasant dry feel associated with the silicone.

The formulation given in Table 13 is an example of a skin lotion that delivers mineral oil and glycerin as the active ingredients. Other emollients such as petrolatum can be added to the oil phase, or substituted for the mineral oil. A small amount of electrolyte is

Table 19.12 Antiperspirant Stick

Ingredient	Wt%	Tradename / supplier
Stearic acid	15.0	
Cetyl alcohol	15.0	
Aluminum chlorohydrate (powder)	20.0	
Cyclomethicone	50.0	345 Fluid / Dow Corning

Table 19.13 Water-in-Oil Skin Lotion

Ingredient	Wt%	Tradename / supplier
Oil phase		
cyclomethicone (and)		3225C Formulation Aid /
dimethicone copolyol	7.2	Dow Corning
cyclomethicone	9.6	
mineral oil	7.4	
pareth-15-3	0.4	Tergitol 15-S-3 /
		Union Carbide
Water phase		
glycerin	20.2	
water	53.2	
sodium chloride	2.0	

Table 19.14 Water-in-Oil Antiperspirant Lotion

Ingredient	Wt%	Tradename / supplier
Oil phase		
cyclomethicone (and)		3225C Formulation Aid /
dimethicone copolyol	6.0	Dow Corning
cyclomethicone	27.0	
polysorbate-20	1.0	Tween 20 / ICI Americas
Water phase		
aluminum chlorohydrate (powder)	20.0	
water	46.0	

needed for emulsion stability, and sodium chloride has been used in this example. The nonionic surfactant pareth-15-3 is included to improve emulsion stability. The viscosity of the formulation can be controlled by adjusting the oil-phase/water-phase ratio. Preparation of the formulation is similar to other topical emulsion formulations, except that no heating is needed. The oil phase and water phase are mixed in separate containers and the emulsion is made by slowly adding the water phase to the oil phase with rapid mixing. A high-shear mixer, such as an Eppenbach or Silverson, is desirable for making this type of emulsion, but a stirrer equipped with a Cowles blade can be used. For large quantities of emulsion, the best procedure is to prepare the batch with a conventional stirrer and then pass it through a colloid mill to ensure that the entire batch is adequately mixed.

Table 14 lists a starting formulation for an antiperspirant emulsion. In contrast with the formulation given in Table 8, the antiperspirant salt will not settle out because it is dissolved in the water phase. No additional electrolyte is required in the formulation because of the high level of antiperspirant salt used. The formulation is prepared using the same method given for the previous example.

REFERENCES

1. M. J. Owen, *Chemtech*, *11*:288, 1981.
2. *CTFA Cosmetic Ingredient Dictionary*, 3rd ed. The Cosmetic, Toiletry, and Fragrance Association, Washington, D.C., 1982.
3. *Encyclopedia of Polymer Science and Engineering*, Vol. 2, 2nd ed. John Wiley & Sons, New York, p. 66, 1985.
4. *Federal Register*, *43*:3468, 1978.
5. *Federal Register*, *48*:6820, 1983.
6. C. D. Vaughan, *J. Soc. Cosmet. Chem.*, *36*:329, 1985.
7. A. J. DiSapio, M. S. Starch, *Cosmet. Toiletries*, *96*(8):55, 1981.

APPENDIX

Suppliers

Witco Chemical Corporation
Sonneborn Division
520 Madison Avenue
New York, New York 10022
(212) 605-3908

Dow Corning Corporation
220 West Salzburg Road
Midland, Michigan 48640
(517) 496-4000

Croda Incorporated
183 Madison Avenue
New York, New York 10016
(212) 683-3089

ICI Americas
Wilmington, Delaware 19897
(800) 441-7757

R. T. Vanderbilt Company Incorporated
Main & William Streets
Belleville, New Jersey 07109
(201) 759-3225

NL Industries Incorporated
NL Chemcals Division
Wycoff Mills Road
Hightstown, New Jersey 08520
(609) 443-2000

B. F. Goodrich Chemical Company
6100 Oak Tree Boulevard
Cleveland, Ohio 44131
(216) 524-0200

Union Carbide Corporation
Old Ridgebury Road
Danbury, Connecticut 06817
(203) 794-2000

Sherex Chemical Company
P.O. Box 646
Dublin, Ohio 43017
(614) 764-6500

Lonza Incorporated
22-10 Route 208
Fairlawn, New Jersey 07410
(201) 791-7500

Index